Benchmark Papers in Behavior Series

Editor: Martin W. Schein—West Virginia University
Carol Sue Carter-Porges—University of Illinois at Urbana-Champaign

EVOLUTION OF PLAY BEHAVIOR / *Dietland Müller-Schwarze*
PARENTAL BEHAVIOR IN BIRDS / *Rae Silver*
CRITICAL PERIODS / *John Paul Scott*
THERMOREGULATION / *Evelyn Satinoff*
LEARNING IN ANIMALS / *Robert W. Hendersen*
MAMMALIAN SEXUAL BEHAVIOR: Foundations for Contemporary Research / *Donald A. Dewsbury*
BEHAVIOR-GENETIC ANALYSIS / *Jerry Hirsch and Terry R. McGuire*
FOUNDATIONS OF COMPARATIVE PSYCHOLOGY / *Donald A. Dewsbury*
MALNUTRITION AND HUMAN BEHAVIOR: Experimental, Clinical, and Community Studies / *Josef Brozek*
FOUNDATIONS OF COMPARATIVE ETHOLOGY / *Gordon M.Burghardt*

Related Titles

AGING IN NONHUMAN PRIMATES / *Douglas M. Bowden*
COMPARATIVE PSYCHOLOGY IN THE TWENTIETH CENTURY / *Donald A. Dewsbury*
CAPTIVITY AND BEHAVIOR IN THE PRIMATES / *Joseph Erwin*
THE MACAQUES / *Donald G. Lindburg*
GORILLA BEHAVIOR / *Terry Maple*
ORANGUTAN BEHAVIOR / *Terry Maple*
BEHAVIORAL SEX DIFFERENCES IN NON-HUMAN PRIMATES / *Gary Mitchell*
HUMAN SEX DIFFERENCES / *Gary Mitchell*
PRIMATE PATERNALISM / *David M. Taub*

FOUNDATIONS OF COMPARATIVE ETHOLOGY

Edited by

GORDON M. BURGHARDT
University of Tennessee at Knoxville

A Hutchinson Ross Publication

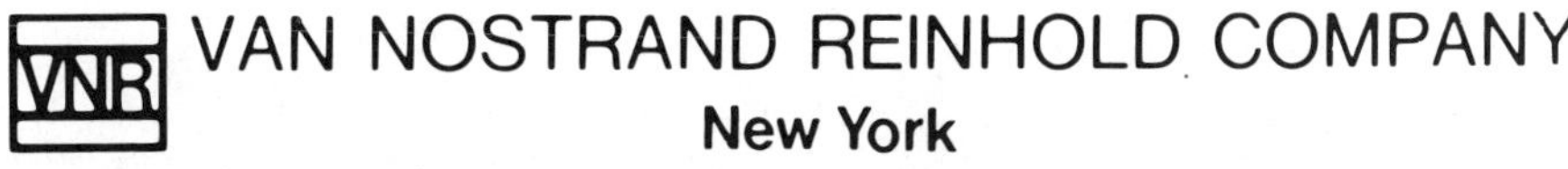

WILLIAM MADISON RANDALL LIBRARY UNC AT WILMINGTON

Copyright © 1985 by **Van Nostrand Reinhold Company Inc.**
Benchmark Papers in Behavior, Volume 19
Library of Congress Catalog Card Number: 85-3211
ISBN: 0-442-21061-2

All rights reserved. No part of this work covered by the copyrights hereon may be reproduced or used in any form or by any means—graphic, electronic, or mechanical, including photocopying, recording, taping, or information storage and retrieval systems—without permission of the publisher.

Manufactured in the United States of America.

Published by Van Nostrand Reinhold Company Inc.
115 Fifth Avenue
New York, New York 10003

Van Nostrand Reinhold Company Limited
Molly Millars Lane
Wokingham, Berkshire RG11 2PY, England

Van Nostrand Reinhold
480 Latrobe Street
Melbourne, Victoria 3000, Australia

Macmillan of Canada
Division of Gage Publishing Limited
164 Commander Boulevard
Agincourt, Ontario MIS 3C7, Canada

15 14 13 12 11 10 9 8 7 6 5 4 3 2 1

Library of Congress Cataloging in Publication Data
Main entry under title:
Foundations of comparative ethology.
(Benchmark papers in behavior; v. 19)
"A Hutchinson Ross benchmark book."
Bibliography: p.
Includes indexes.
1. Animal behavior—Addresses, essays, lectures.
I. Burghardt, Gordon M., 1941- II. Series.
QL751.6.F67 1985 591'.051 85-3211
ISBN 0-442-21061-2

VAN NOSTRAND REINHOLD COMPANY

350 Main Street, Box 699, Stroudsburg, Pa. 18360 • Telephone (717) 421-4060

Errata for FOUNDATIONS OF COMPARATIVE ETHOLOGY, edited by Gordon M. Burghardt

Page 124, line 7 from the bottom should read: ". . . psychology society, as well as the writer of the influential book Whitman attacked. Mills was perhaps more careful than Whitman . . . ."

Page 239, line 21 should read: "checked and verified in every detail by Parker with experiments."

Page 339, line 8 from the bottom should read: ". . . at times linked to urination . . . ."

In both the author citation and subject indexes, references to page numbers 344 and higher are off by one number; that is, 344 should read 343, 345 should read 344, and so on.

QL751
.6
.F67
1985

For Eckhard
A rock in the tide of contrary opinions

CONTENTS

SERIES EDITORS' FOREWORD

It was not too many years ago that virtually all research publications dealing with animal behavior could be housed within the covers of a very few hard-bound volumes that were easily accessible to the few workers in the field. Times have changed! The present-day students of behavior have all they can do to keep abreast of developments within their own area of special interest, let alone in the field as a whole; and of course we have long since given up attempts to maintain more than a superficial awareness of what is happening "in biology," in psychology," "in sociology," or in any of the broad fields touching upon or encompassing the behavioral sciences.

It was even fewer years ago that those who taught animal behavior courses could easily choose a suitable textbook from among the very few that were available; all "covered" the field, according to the bias of the author. Students working on a special project used the text and the journal as reference sources, and for the most part successfully covered their assigned topics. Times have changed! The present-day teacher of animal behavior is confronted with a bewildering array of books to choose among, some purported to be all-encompassing, others confessing to strictly delimited coverage, and still others being simply collections of recent and profound writings.

In response to the problem of the steadily increasing and overwhelming volume of information in the area, the Benchmark Papers in Behavior was launched as a series of single-topic volumes designed to be some things to some people. Each volume contains a collection of what an expert considers to be *the* significant research papers in a given topic area. Each volume, then, serves several purposes. To teachers, a Benchmark volume serves as a supplement to other written materials assigned to students; it permits in-depth consideration of a particular topic while at the same time confronting students (often for the first time) with original research papers of outstanding quality. To researchers, a Benchmark volume serves to save countless hours digging through the various journals to find *the* basic articles in their area of interest; often the journals are not easily available. To students, a Benchmark volume provides a readily accessible set of original papers on the topic, a set that forms the core of the more extensive bibliography that they are likely to compile; it also permits them to see at first hand what an "expert" thinks is important in the area, and to react accordingly. Finally, to librarians, a Benchmark volume represents a collection of important papers from many diverse sources, thus making readily available materials that might otherwise not be economically possible to obtain or physically possible to keep in stock.

The selection of volume topics for this series is no small task. Each of us could generate a long list of possible topics and then search for potential volume editors. Alternatively, we could draw up a long list of recognized and prominent scholars and try to persuade them to do a volume on the topic of their choice. For the most part, we have followed a mix of both approaches, matching a distinguished researcher with a desired topic, and the results have been impressive.

Given a strong background and training in classical ethology, and given further an avid interest in the history of ideas, Gordon Burghardt is clearly a most appropriate choice to serve as editor of the present volume. The fact that he is a distinguished and well-respected research scientist and scholar—and currently serves as a coeditor of the prestigious *Zeitschrift für Tierpsychologie/Journal of Comparative Ethology*—makes this volume all the more desirable, as well as valuable and useful. We are most pleased that he agreed to do the job.

MARTIN W. SCHEIN
SUE CARTER-PORGES

FOREWORD

It is particularly satisfactory to an old dyed-in-the-wool ethologist to find unearthed what may seriously be considered as the foundation of ethology. As an ethologist I want to keep clear the conceptions of comparative ethology and comparative psychology in general. Although I have contributed to his book[1] I cannot agree with Dr. Dewsbury's history book[2] that seems to contend that comparative psychologists have really been doing ethological work all along. Ethology is a comparative science in the same sense as comparative morphology is. Comparative ethology, therefore, could not even begin to exist before something was discovered that could be compared in the same sense as bones, teeth or feathers.

It is particularly satisfactory that classical observers like D. A. Spalding and B. Altum are valued. Altum was scientifically and personally an enemy of Alfred E. Brehm who, by his great book "Brehm's Tierleben," became much more famous in Germany and Europe than Altum ever did. The tragedy of the enmity between these two men really is pathetic. Brehm, convinced of the close affinity of men and animals, unhesitatingly committed gross anthropomorphisms by attributing to birds the highest moral qualities. Altum, a Catholic priest, was a caustic critic of Brehm's anthropomorphical explanations, so he sought strenuously for physiological explanations of even the most complicated activities of birds—and found them. Contrary to Brehm, his erroneous premise—that of the uniqueness of Man—led him to many correct deductions concerning the social behavior of birds.

The fundamental discoveries that set off ethology as a special branch of science were anticipated by Spalding, but more clearly formulated by Charles Otis Whitman and Oskar Heinroth. In fact, they discovered that behavior could be compared: A certain motor pattern, for instance, the grunt-whistle, was an activity found in the true ducks but missing in the blue-winged group like the gargany, the cinnamon teal, and others. It should be remembered that these all-important facts were revealed to people

[1]Dewsbury, D. A., ed., 1985, *Leaders in the Study of Animal Behavior: Autobiographical Perspectives,* Bucknell University Press, Lewisburg, Pennsylvania.

[2]Dewsbury, D. A., 1984, *Comparative Psychology in the Twentieth Century,* Hutchinson Ross, Stroudsburg, Pennsylvania.

investigating the taxonomy of small groups and whose insatiable quest for more and more comparable characters drove them to include behavioral characters in their research.

These taxonomical successes, however, do not explain the success that ethology has attained in analyzing animal and human behavior. These physiological advances seem to me at least as important as the taxonomical successes of ethology.

If I should be requested to fix the point at which comparative ethology took on the character of a physiological science, I should name as a break-through on the one hand Erich von Holst's investigations of the stimulus production in the nervous system, particularly of earthworms, and, on the other hand, the first analysis of threshold lowering and intensity difference, which Alfred Seitz performed at my institute. In fact, the physiological nature of the comparable motor patterns, which von Holst and Seitz discovered, interested neither C. O. Whitman nor O. Heinroth. It was only by concentrating on a certain type of behavioral process that they drew attention to the fact that all the motor patterns they were interested in were "action and reaction in one." Tribute is due to A. F. J. Portielje for stating plainly that this kind of process actually is the element of behavior. It is, of course, an original discovery that the grunt-whistle of the mallard or the threat gesture of a stallion are taxonomically exploitable characters just as teeth, feathers, and bones are. Their extensive use in taxonomy begun by O. Heinroth and continued by Ernst Mayr, Jean Delacour, Frank McKinney, and many others has produced important new taxonomical knowledge. However, the physiological fact that all of these highly specialized motor patterns are action-and-reaction-in-one seems to me the most important revelation contributed by ethology to the understanding of animal behavior.

Coming back to the sources of ethology and delving into the past to ascertain, without any bias, who was aware of what and at what time is indeed an undertaking meriting great admiration.

KONRAD Z. LORENZ
Altenberg, Austria
June 3, 1985

PREFACE

The documents reprinted here, many in English translation for the first time, encompass most of the themes that flowed together in the "classical synthesis" of European ethology. This distinctive theoretical and methodological approach to behavior is most closely associated with Konrad Lorenz and Niko Tinbergen. However, many other scientists contributed greatly to their views and the later development, extension, and modification of the field. Without being extremely superficial it would be impossible to provide sample writings by all those who contributed to comparative ethology up to 1950, the period of the first two seminal books in the field (Danielli and Brown, 1950; Tinbergen, 1951).

This book contains writings by scientists whose work influenced Lorenz in his important theoretical papers prior to 1940. It begins with an influential pre-*Origin of Species* treatment of instinct; the penultimate reading is a physiological contribution by von Holst (Paper 22). The time span is exactly 100 years. Through these readings one can appreciate how much the various strands of Lorenz's thought were dependent upon others, something Lorenz always acknowledged, and derive a better understanding of his initial positions and the context from which he expounded his often controversial views. Also contained in this book are important early writings on instinct and learning that place the strictly ethological writings in a firmer context.

Whether or not one is familiar with classical ethology, reading early writings of Lorenz and Tinbergen in conjunction with this book will help the reader in adequately appreciating the importance each paper played in the classical synthesis (see also the Introduction, p. 1). To aid in this endeavor, a previously untranslated paper by Lorenz ends the book, a paper that outlines virtually the entire early scheme and the sources influencing it.

Today, fifty years after the most recent source paper reprinted here, a retrospective evaluation is in order. Although I have provided some commentary in the text, I have avoided extensive historical analysis or criticism drawn from today's knowledge and attitudes. But a comprehensive treatment similar to those recently appearing on comparative psychology (Boakes, 1984; Dewsbury, 1984) is nevertheless necessary. I must point out, however, that the early writings of Lorenz were often misunderstood by critics and supporters alike. Today, after years of polemical controversies over use of the word 'innate,' the role of ontogeny, the reality of behavioral homologies, the evidence from physiology, and in spite of the many new directions and technical advances characteristic of the field today, the value and flexibility of the classical views are having a continued impact on modern ethology.

Many of the early ethologists were most interested in birds; thus, birds provide most of the examples in the readings. But while birds possessed many characteristics favorable to the observation and discovery of basic phenomena, it is the processes and principles of animal behavior as formulated by the first ethologists that interest us, as well as their current legacy in modern ethology and its various contemporary subfields such as sociobiology, behavioral ecology, neuroethology, and comparative psychology.

Other books in this series cover admirably the early post-Darwinian history of animal-learning studies, comparative psychology, and selected topics such as territory, social dominance, imprinting, play, social organization, and parental care. Many papers influential in the intellectual milieu of classical ethology will be found in those volumes. The recent collection of autobiographical sketches (Dewsbury, 1985) appeared too late to be used, but it is a fascinating source for historical notes on the last 50 years.

Considerable selectivity had to be exercised in choosing specific works and excerpted passages. Basically, I used pre-1936 works cited and used in the seminal ethological documents (1935–1952) by Lorenz and Tinbergen. I wish that more little-known foreign authors could have been translated.

I have benefited over the years from association with many established ethologists who have directly or indirectly shaped my interests or directed me to specific papers. Eckhard H. Hess deserves first mention here since his courses at the University of Chicago from 1962–1966 stimulated my interest in the history of ethology as well as in evolution, instinctive behavior, and naturalistically-based research, areas then still largely off limits in American psychology. His review of basic ethology and its history (Hess, 1962) was of major importance in introducing a generation of psychologists to ethology. After I moved to Tennessee, William S. Verplanck (also an early advocate of ethology in America) served as an important sounding board and critic, and freely provided his personal experiences with behaviorism and ethology.

Naturally my fascination with the writings of Lorenz and Tinbergen was essential to the development of this book. Lorenz provided hospitality and insight during a stay at Seewiesen in 1963 just after I had finished my undergraduate training. Paul Leyhausen deserves a special note of thanks for his suggestions early in the planning of this book. He, along with Katharina Heinroth, Amélie Koehler, and Konrad Lorenz sent treasured supportive letters. Frank McKinney provided unpublished translations of early Heinroth papers, although the present translation (Paper 14) is new.

I am grateful to the students in seminars who, over the years, have shown a continuous and growing interest in the history of their chosen field and the controversies that have both enriched and plagued it. It was their positive response that encouraged me to undertake this task, for which Charles Hutchinson and the Benchmark series provided the opportunity. Bruce Batts, Frank McKinney, Martin Schein, William Verplanck, Sandra Twardosz, and the translators generously commented on the editorial matter. Bruce Batts also provided references and much stimulating discussion, especially on nineteenth-century psychology. Wolfgang Schleidt provided the footnote to Paper 16. Continued support from the National Science Foundation is gratefully acknowledged.

Finally, I am most indebted to Doris Gove and Jeff Mellor, who cheerfully undertook the task of translating the original German writings. Amélie Koehler, Paul and Barbara Leyhausen, and Frank Mckinney made useful suggestions on the translations; they are not responsible for the sometimes difficult, even controversial, choices of words and phrases that had to be made.

The naturalistic study of animals, focusing on activities essential to their survival and informed by an evolutionary attitude, is the hallmark of ethology. Although ethologists are fallible, the understanding of many species, their future and our future, may depend upon how closely we adhere to the ethological mission: refining, extending, applying . . . .

GORDON M. BURGHARDT

REFERENCES

Boakes, R., 1984, *From Darwin to Behaviourism,* Cambridge Univ., Cambridge, 279p.

Danielli, J. F., and R. Brown, eds., 1950, *Physiological Mechanisms in Animal Behavior,* Symp. Soc. Exper. Biol. No. 4, Cambridge Univ., Cambridge, 482p.

Dewsbury, D. A., 1984, *Comparative Psychology in the Twentieth Century,* Hutchinson Ross, Stroudsburg, Pa., 413p.

Dewsbury, D. A., 1985, *Leaders in the Study of Animal Behavior: Autobiographical Perspectives,* Bucknell University Press, Lewisburg, Pennsylvania, 512p.

Hess, E. H., 1962, Ethology: An Approach Toward the Complete Analysis of Behavior, in *New Directions in Psychology,* R. Brown, E. Galanter, E. H. Hess, and G. Mandler, eds., Holt, Rinehart, and Winston, New York, pp. 157–226.

Tinbergen, N., 1951, *The Study of Instinct,* Clarendon Press, Oxford, 228p.

CONTENTS BY AUTHOR

FOUNDATIONS OF COMPARATIVE ETHOLOGY

INTRODUCTION

The lives of humans and other animals have always been intimately connected. Animals have been our prey, predators, parasites, and competitors. They have provided food, clothing, transportation, muscle power, religious symbols, and moral messages. Early humans did develop ways of dealing with animals based on observation, induction, and deduction. Successful hunting as well as the domestication of animals necessitated extensive behavioral knowledge. Although the functional unity of economic, social, religious, and recreational life precluded the development of an objective science of animal behavior, animistic beliefs and close association with the larger mammals could have fostered the view that humans and animals were not all that different. Intellectuals of Eastern and Western civilizations often did use animals as a mirror to human conduct. But it was the controversy centered on evolution and Darwinian natural selection that forced people to look at animal behavior in a new light.

Overviews of early thinking on animal behavior are available in diverse sources (e.g., Warden, 1928; Hess, 1962; Gray, 1968; Diamond, 1971; Heinroth and Burghardt, 1977; McFarland, 1981; Sparks, 1982) and will not be repeated here. Historical and biographical information on the various approaches to animal behavior since the time of Darwin are found in the above sources as well as in Boring (1950), Kortlandt (1955), Burghardt (1973), Klopfer and Hailman (1967), Stamm and Zeier (1978), Hearst (1979), Thorpe (1979), Boakes (1984), Dewsbury (1984*a*), and Burghardt (in press); references to original writings and other historical treatments are cited therein. The first two sections below are partially based on a translation of Burghardt (1978).

In this introduction I will (1) present views on instinct and animal behavior in England prior to Darwin, (2) briefly discuss the period between Darwin and the emergence of ethology as a self-conscious discipline in the mid-1930s, and (3) outline the major characteristics of ethology circa 1940. The commentaries and readings in this book basically cover the second of these views; the other two provide context.

INSTINCT BEFORE DARWIN

Western thinkers, from the time of the Greeks through to the present, have generally viewed animals as operating by instinct and humans as operating by reason. The exact terms fluctuated and the views were influenced by many philosophical and theological issues. Descartes' version of mind-body dualism in the seventeenth century is the most famous expression of the dichotomy that has provided a number of attempts at consistent philosophical, theological, and scientific schemes.

How animal behavior was viewed by the literate public in England during the century before Darwin is exemplified in a book by Charles Owen (1742,9). In his smorgasbord blending of Greek, medieval, and contemporary science as well as biblical references and Christian theology he wrote, after reviewing numerous examples of marvelous animal behavior:

> Thus we see no Creatures so mean in our View, but a Ray of divine Wisdom shines in their Foresight and Contrivance: When we consider how wonderfully these inferior Creatures are conducted in their Operations, how punctually they obey the laws of their Creator, how solicitously every one propagates his Kind, and makes proper Provisions for his Family; it looks as if it were done by some Principle that's more perfect than the common Reason of Man. Nevertheless 'tis past doubt, that Brutes of the highest Order, and most refin'd, are but Brutes, *i.e.* irrational, and it's well for us they are so.
>
> This is call'd *instinct,* a natural Disposition, or Sagacity wherewith Animals are endued; by virtue whereof they are enabled to provide for themselves, know what is good for them, and are determin'd to propagate and preserve their Species. *Instinct* bears some Analogy to Reason or Understanding, and supplies the Defect of it in Brutes.

This excerpt touches on many issues that were to concern students of animal behavior in years to come. These include:

1. Animals seem to know how to behave in order to insure individual survival, reproduction, and preservation of the species.
2. The apparent wisdom and rationality in animals is just that, an appearance.
3. Humans operate by an imperfect but pervasive and qualitatively distinct principle called reason.
4. Animals are inferior beings, although instinct is a divine mechanism allowing animals to survive.
5. Humans have no need of instinct; the bare suggestion would question our superior moral position in the universe.
6. To allow reason and understanding to be present in animals is

dangerous to our conception of ourselves, especially as unique moral beings created in God's own image.

Almost 100 years later Kirby (1835) presented a view of animal behavior embedded in natural theology (Paper 1) that was considerably more sophisticated. Animals had some intelligence and, most importantly, possessed emotions. (This proposition was carried to extreme anecdotal lengths by Thompson [1851], compared to which Romanes [1882], the archetypal anecdotalist according to modern texts, is restrained.) Kirby held that instincts were not perfect, nor were humans without them. In addition instincts stemmed from the operation of physical causes. But the theological straitjacket still constrained thinking about animal behavior, as it did with many scientists even well after 1859.

Darwin (Paper 2) played an absolutely critical role in objectifying the study of instincts and emphasizing the actions of the animal rather than the impulses or forces behind them or their divine origin. He knew it was crucial to demonstrate that the wonderfully complex behavior of animals could be viewed as products of natural selection rather than as evidence of God's handiwork. Thus he pointed out that not only was all behavior variable, adaptive, and fallible, but that heredity influenced behavior and its variations as well as corporeal structure. He also stressed the need for careful observation, the pertinent experiment, and a comparative approach. Later Darwin extended the evolutionary perspective to humans and emphasized the similarity of animal and human behavior. Thus the question of 'instinct versus intelligence' took on great scientific, not to mention philosophical, religious, and political, import. Disputes on these issues rage yet today.

ISSUES IN COMPARATIVE PSYCHOLOGY AND THE EMERGENCE OF ETHOLOGY

Studies of animal instinct and reason, in the post-Darwinian era, took many directions. Some of these are represented in the present readings. The resulting fragmentation led to some relatively consistent approaches, but none of them adequately combined a naturalistic approach, empirical observation, experimentation, appropriate behavioral units and measures, motivation analysis, a wide comparative swath, and an evolutionary framework until the classical synthesis of ethology in the 1930s. Some works may have approached it: to wit, the studies on arthropods of Lubbock, Wheeler, Fabre, and the Peckhams, the 'lower' invertebrate studies of Jennings, the theoretical framework of C. Lloyd Morgan (Papers 7A-7E), the early work on vertebrates of Mills, Yerkes, Thorndike (Paper 12), and Watson. But an effective combination of the seven critical elements did not take place.

There were several problems. The lack of accurate objective descriptive and experimental data and methods fostered uncritical anecdotalism and anthropomorphism. This was partially rectified by Lloyd Morgan in his famous Canon on parsimony (Morgan, 1894; Burghardt, 1985, in press; see also Whitman, Paper 9).

Mendelian genetics was not rediscovered until after 1900, and when it was, more controversies erupted on genetic mechanisms, presenting further difficulties to those trying to characterize how behavior evolved. A related problem was the endurance of Lamarckian views, not just on heredity in general but especially on the possible evolution of instinct from habit. Darwin, Lewes, Romanes, Spalding, Wundt (founder of the first psychology laboratory), and Spencer all subscribed to Lamarckian views to some degree. Morgan, Poulton, Weismann, Groos, Thorndike (Paper 11), and Whitman (Paper 10), among others, argued against the inheritance of acquired habits and for the critical role of natural selection; although their goal was largely accomplished by 1900, echoes still resound down to the present. The psychologist McDougall was an advocate of Lamarckian heredity, even performing suggestive (but flawed) experiments in the 1930s.

Probably the most significant problem in the late nineteenth century was the lack of clarity in the concepts of both intelligence and instinct. The analysis of the reflex was just moving to the laboratory; the study of tropisms and conditioned reflexes were still in the future. Some theorists were interpreting all behavior and mental events by a comprehensive associationism elegant in theory and uncontaminated by data (e.g., Spencer, Paper 6). Trial-and-error, or instrumental, learning was just about to be formalized by Thorndike (e.g., Paper 12). Instinct was not broken down into components; motivational and behavioral aspects were entangled, often viewed as incompatible. Theorists wedded to subjectivism were uncomfortable about replacing a concern with animal consciousness and its significance with the banalities of mere behavior. It is not surprising that it would take years for the threads to sort themselves and a viable weaving (synthesis) to occur.

The case can be made that the first major unifying conceptual watershed after Darwin affecting all areas of behavioral study was ethology. Traditional comparative psychology, in spite of isolated methodological advances and theoretical insights, failed to provide a general model with adequate substance for a comparative science. Holmes (1911) wrote a sensible, balanced, scholarly text that gave virtually equal coverage to zoology and psychology with generous treatments of European work. Thorndike and Whitman were cited almost an equal number of times. Yet this book had apparently little effect; neither Boakes (1984) nor Dewsbury (1984*a*) cite it.

Perhaps comparative psychology was not ready to be a synthetic field. Behaviorism, arising around 1913 as an essential extension of the parsimony of Lloyd Morgan's Canon, provided a more attractive, seemingly more scientific, vision; but ultimately a simplistic vision compromised by an uncertain, often antagonistic, stance toward field studies and comparative analysis, as well as toward evolution in general and heredity in particular. The true comparative psychologists that remained were too few in number and professionally advanced. Most of their students and the new generation of experimental psychologists were lured by the promise of a shortcut to a general animal psychology via the study of the rat. Lashley's presidential address to the American Psychological Association (Lashley, 1938; Dewsbury, 1984*b;* see also Burghardt, 1973) recognized the ensuing limitations of behaviorism and made many points with which ethologists would have agreed (Thorpe, 1979). A detailed treatment by Oppenheim (1982) documents how development, embryology, and epigenesis were misinterpreted during this period and fostered the discrediting of hereditary factors in behavior by branding all such proposals covert preformationism.

Modern ethology is now the most diverse approach to the study of animal behavior, although initially based on a small framework of core ideas (Lorenz, 1981). The ethological synthesis occurred about the same time as the new evolutionary syntheses of J. Huxley, Dobshansky, and Mayr. Today many use the terms ethology and animal behavior interchangeably in referring to the academic discipline. Fifty years ago this would have been unheard of, for just fifty years ago there appeared the seminal paper by Konrad Lorenz (1935) that sparked a revolution in the study of animal behavior and set the first program for ethology. Nonetheless, the term ethology, the theoretical bases, and the methodology advocated were not really new, not even in 1935. What was new, was the synthesis.

ABOUT THIS BOOK

The papers Lorenz published up to 1941 had a great impact (Thorpe, 1979), which would have been accelerated in America and England if political differences culminating in World War II and its aftermath had not intervened. These papers, most of which are available in translation in Schiller (1957) and in the more complete form in Lorenz (1970), show the reliance Lorenz placed on the work of others in achieving his synthesis.

This volume makes available in one convenient source the writings of those who had an important influence on Lorenz; it includes translations from German of works hitherto unavailable in English. A secondary goal has been to demonstrate the links between early American protoethology (e.g., Whitman and Craig) and experimental animal psychology (e.g., Thorndike). A lack of awareness of these links has hindered the appreciation of ethological ideas by comparative psychologists and of the contributions of animal psychology by ethologists. These connections have recently been emphasized in a monograph by Dewsbury (1984*a*). A third goal has been to present writings on instinct roughly contemporary with those of Charles Darwin to more clearly and accurately punctuate the insights stemming from the theory of natural selection.

This book does not represent a broad overview of concepts of "instinct" down through the ages. While it could be said that ethology made the study of "instinct" again respectable, I think it more accurate to say that ethology, in addressing the phenomena earlier cataloged as instincts, made them scientifically tenable by developing a coherent and rather comprehensive comparative framework within which to describe and analyze instinctive *behaviors* and the perceptual (environmental) and motivational (internal) processes involved. All this was informed by the naturalist's devotion to biological diversity and to detailed observation, coupled with an abiding commitment to evolutionary analysis.

This collection, then, is a look backward—but not from the 1980s. Historians rightly criticize such ahistorical sins, although most students and professionals find it most interesting to read about those predecessors who anticipated or influenced the field as it is today. Nor is this an attempt to demonstrate the *Zeitgeist* of the late nineteenth and early twentieth centuries, within which these papers emerged. Rather, it is a look backward, from the 1930s and 1940s when Lorenz, Tinbergen, and their many colleagues and students were developing their ideas and strengthening their position. They were concerned with protecting their flanks from mechanistic, reductionistic, and naively environmentalistic approaches, as well as vitalism and the anthropocentric and subjective European psychology that had little use for experimental and objective approaches to either the human or animal mind.

The classical or core (Lorenz, 1981) ethological synthesis was the product of many European researchers from the 1930s through the early 1950s. Lorenz was one of many theoreticians of note (see, for example, Kortlandt, 1955). Tinbergen (1951) wrote the first synthetic review of the field. While the selection of Lorenz is to some extent arbitrary, I think that there is little doubt he was the original catalyst, with Tinbergen providing an essential experimental focus, developing more systematically the basic concepts, and influencing by his research examples a large number of ethological studies (e.g., Baerends, Beer, and Manning, 1975), including my own. The selection of Lorenz as the focal point does not imply that all his analyses and conclusions were valid or correct in the 1930s, much less today.

If one took the early writings of Tinbergen one would come up with a somewhat different set of influential papers. For example, his early insect orientation and landmark use studies in wasps were influenced by von Frisch, who appears to have had little direct effect on Lorenz. The two edited volumes by Klopfer and Hailman (1972*a*, 1972*b*) are useful in documenting the post 1940 period. But a comprehensive volume is needed on early exprimental approaches, field studies, theoretical developments, and critiques of ethology from 1936–1956 including such notables as Allee, Baerends, Bierens de Haan, Carpenter, Eibl-Eibesfeldt, Hess, Hinde, von Iersel, O. Koehler, Kramer, Leyhausen, Marler, Mittlestaedt, Noble, Schleidt, Schneirla, Seitz, and Thorpe.

Originally, the essential ideas of the ethological synthesis of Lorenz and Tinbergen were not to be presented here in original papers. Most of these works have been reissued in English; the basic outline of the approach is presented, more or less accurately, in many current texts as well as in a recent authoritative treatment by Lorenz (1981). Although some modifications of the classical views are incorporated in the latter, it is remarkably true to the spirit of the original writings and is only modestly informed by recent work (Burghardt, 1982). The "Introductory History" section is useful background for this book.

After the selections were chosen and the book was nearing publication, Paul Leyhausen brought to my attention a paper by Lorenz (1937) that turned out never to have been translated. A symposium address that summarizes Lorenz's views as of 1936, it is the first paper in which Lorenz incorporated the findings of von Holst (e.g. Paper 22); thus, it can be argued that this paper is the first in which Lorenz included all the major theoretical elements. It has been translated and added as an appropriate final reading (Paper 23). This paper is complex and short on the observational detail present in most other of Lorenz's early papers. Thus a brief résumé of the classical synthesis in more current terms might not only help place the included papers in a better perspective but also aid in the appreciation of the Lorenz paper.

Many of Lorenz's concepts are most familiar to us in the terminology used by Tinbergen (1951), a terminology which, while eventually largely adopted by German speaking ethologists including Lorenz himself, somewhat altered the meaning of the original (Burghardt, 1973) and seems to have fostered some of the controversy over ethological ideas (e.g., Lehrman, 1953; partially reprinted in Dewsbury, 1984*b*).

SYNOPSIS OF CLASSICAL ETHOLOGICAL POSITIONS

The focus of animal behavior research should be on understanding normal behavior. Animals should be observed either under natural conditions or simulated ones in which typical behavior can be observed. Theorists of all persuasions have had a far too limited base of carefully observed and described facts about animal behavior.

Experimentation must be based on a knowledge of a species' normal behavior patterns (ethogram) and of their adaptations to the environment. Only with this information can insights into the dynamics of behavior be gained through the study of animals reared or maintained in environments poor in stimuli of various sorts (conspecifics, nesting material, predators, prey, and the like), and where specific experiences are restricted.

Animal behavior is a product of a species' evolutionary history and ecological contingencies analogous to the species' anatomy and physiology. Thus comparative ethology is a field paralleling the various branches of comparative zoology. A direct corollary is that comparative studies should initially incorporate the observation of closely related species with a goal of identifying homologous behavioral adaptations.

Heredity and environment both affect the phenotypic expression of behavior. Hereditary (innate, maturational) and environmental (learning, experience) effects are theoretically separable, and empirically so with some favorable preparations.

"Instinct" is a confusing term, having teleological, subjective, and anthropomorphic connotations. However, instinctive *behavior* is a valid and most important component of animal behavior that requires analysis. Reductionistic approaches based on reflexes, conditioning, sterile laboratory work, or the neglect of genetics result in misleading theories and misrepresented facts. Theories crediting animals with "goals," "insight," or nonanalyzable instincts are also premature. Thus a major aim should be to steer a middle course between reductionistic and vitalistic approaches.

The particulate unit of instinctive behavior is the Fixed Action Pattern (FAP). The FAP is generally unlearned, stereotyped, performed similarly by all conspecifics, and released by specific features of natural objects. The FAP can be independent of sensory feedback once it is set into motion (e.g., egg-rolling in geese, swimming in fish), but taxic (feedback) adjustments can occur throughout performance. While sharing some features with reflexes, more central control is indicated by such phenomena as stimulus threshold lowering and variable intensity of performance (e.g., intention movements).

All animals live in a world surrounded by many environmental features containing potential stimuli. Most are unrecognized due to the lack of appropriate sensory equipment. In addition, often only features of the environment crucial to the organism are perceived and discriminated. These features are sign or key stimuli that through evolution have become paired with the performance of FAPs. Recognition of these stimuli can be innate or acquired during critical periods in development. The Innate Releasing Mechanism (IRM) is the hypothetical central mechanism linking perception of sign stimuli and performance of FAPs.

Imprinting is an important process whereby releasers, sign stimuli for social responses, especially species recognition for later sexual choice, are acquired. Such a process in some birds seems to be rapid, time dependent, and relatively permanent. It is one example of instinct-learning intercalation, processes that give the impression that "instincts" are modifiable, whereas it is not the FAP but other aspects of "instinct," including sign stimuli, that are altered.

Sequences of instinctive behavior (in the broad sense) can often be partitioned into appetitive and consummatory phases. The former result from physiologically based drives or mental states that set the animal "searching" for the stimuli appropriate to release the consummatory act. It is the encounter with releasing stimuli and the performance of the consummatory acts, not the biological consequences of food, water, or mates, that are an animal's goal. Appetitive behavior generally involves random locomotion, orientation, and other often highly variable activities in which learning and plasticity are evident, and from which advanced cognitive abilities such as insight may have evolved. Consummatory acts are always FAPs released by sign stimuli via operation of the Innate Releasing Mechanism (IRM). However, FAPs (as in locomotion) can be employed in the appetitive phases as well.

Each FAP, or tightly linked series, has its own source of endogenous motivation called (Re)action Specific Energy. This can be viewed as a reservoir containing gas under pressure (or later as a liquid in a vat) in which performance of an FAP is controlled by both the sign stimulus triggering the FAP via the IRM and the level of pressure "dammed up" inside. As the pressure accumulates, the intensity of the behavior increases, the threshold for effectiveness of the sign stimulus decreases, and the FAP may even occur without any discernible stimulus (vacuum activity). The joint role of endogenous and environmental factors in determining behavioral performance was eventually called the Method of Dual Quantification.

Social displays are often the result of ritualization, in which movements that originally evolved for use in one context (such as drinking or nest building) are utilized in another, albeit often in a stereotyped or exaggerated form. Intention movements and conflicts between differing motivational states are often the source of such emancipation. 'Displacement' occurs when behaviors "spark over" from one motivational context to another.

Set in this form the scheme circa 1940 begs many questions, is incomplete, and omits many later emendations such as Tinbergen's hierarchical model (1951) and the four ethological queries (1963) as well as Lorenz's later more sophisticated ideas on sources of information, habituation, and experiential effects not based on traditional learning models (summarized in Lorenz, 1981). It serves, however, to enable the reader to assess the contributions made by the authors in this book to the ethological conception of behavior. The details of this conception are still being debated vigorously (e.g., Hoyle, 1984 and comments therein), but its general thrust has influenced many areas of current behaviorial research, and still repays careful consideration.

My introductions to the readings point out some links among them and their anticipation of later and current research areas and controversies; there are many other links that could have been discussed. I hope that the richness and diversity of these and other early writings on animal behavior will, along with their limitations, stimulate more appreciation of the roots of comparative animal behavior and their often hidden, but pervasive, influence on current research in psychology and ethology.

REFERENCES

Baerends, G., C. Beer, and A. Manning, eds., 1975, *Function and Evolution in Behaviour; Essays in Honor of Professor Niko Tinbergen,* Clarendon Press,Oxford, 393p.

Boakes, R., 1984, *From Darwin to Behaviorism,* Cambridge Univ., Cambridge, 279p.

Boring, E. G., 1950, *A History of Experimental Psychology,* Appleton-Century-Crofts, New York, 777p.

Burghardt, G. M., 1973, Instinct and Innate Behavior: Toward an Ethological Psychology, in *The Study of Behavior,* J. A. Nevin, ed., Scott Foresman, Glenview, Ill., pp. 321–400.

Burghardt, G. M., 1978, Die Geschichte der Tierpsychologie, mit besonderer Berücksichtigung von England und Amerika, in *Lorenz und die Folgen, Band VI, Die Psychologie des 20. Jahrhunderts,* R. A. Stamm and H. Zeier, eds., Kindler- Verlag, Zurich, pp. 20–28.

Burghardt, G. M., 1982, Ethology: A Reiteration, *Science* **216:**170–171.

Burghardt, G. M., 1985, Animal Awareness: Current Perceptions and Historical Perspective, *Amer. Psychol.* (in press).

Dewsbury, D. A., 1984*a, Comparative Psychology in the Twentieth Century,* Hutchinson Ross, Stroudsburg, Pa., 413p.

Dewsbury, D. A., ed., 1984*b, Foundations of Comparative Psychology,* Benchmark Papers in Behavior Series, Van Nostrand Reinhold, New York, 365p.

Diamond, S., 1971, Gestation of the Instinct Concept, *J. Hist. Behav. Sci.* **7:**323–336.

Gray, P. H., 1968, The Early Animal Behaviorists: Prolegomenon to Ethology, *Isis* **59:**372–383.

Heinroth, K., and G. M. Burghardt, 1977, History of Ethology, in *Grzimek's Encyclopedia of Ethology,* K. Immelmann, ed., Van Nostrand Reinhold, New York, pp. 1–22.

Hearst, E., ed., 1979, *The First Century of Experimental Psychology,* Erlbaum, Hillsdale, New Jersey, 693p.

Hess, E. H., 1962, Ethology, an Approach Toward the Complete Analysis of Behavior, in *New Directions in Psychology,* R. Brown, E. Galanter, E. H. Hess, and G. Mandler, eds., Holt, Rinehart, and Winston, New York, pp. 157–226.

Holmes, S. J., 1911, *The Evolution of Animal Intelligence,* Holt, New York, 296p.

Hoyle, G., 1984, The Scope of Neuroethology, *Behav. Brain Sci.* **7:**367–412.

Kirby, W., 1835, *On the Power Wisdom and Goodness of God as manifested in the Creation of Animals and in their History Habits and Instincts,* two volumes, William Pickering, London, 406p., 542p.

Klopfer, P. H., and J. P. Hailman, eds., 1967, *An Introduction to Animal Behavior: Ethology's First Century,* Prentice-Hall, Englewood Cliffs, New Jersey, 297p.

Klopfer, P. H., and J. P. Hailman, eds., 1972*a, Control and Development of Behavior: An Historical Sample from the Pens of Ethologists,* Addison-Wesley, Reading, Mass., 281p.

Klopfer, P. H., and J. P. Hailman, eds., 1972*b, Function and Evolution of Behavior: An Historical Sample from the Pens of Ethologists,* Addison-Wesley, Reading, Mass., 404p.

Kortlandt, A., 1955, Aspects and Prospects of the Concept of Instinct (Vicissitudes of the Hierarchy Theory), *Arch. Nèerland. Zool.* **11:**155–284.

Lashley, K., 1938, Experimental Analysis of Instinctive Behavior, *Psychol. Rev.* **45:**445–471.

Lorenz, K., 1935, Der Kumpan in der Umwelt des Vogels. Der Artgenosse als auslösendes Moment sozialer Verhaltungsweisen, *J. Ornith.* **83:**137–214, 289–413.

Lorenz, K., 1937, Über den Begriff der Instinkthandlung, *Folia Biotheoretica* **2:**17–50.

Lorenz, K., 1970, *Studies in Animal and Human Behavior,* two vols., Harvard, Cambridge, Mass., 403p. 366p.
Lorenz, K., 1981, *The Foundations of Ethology,* Springer-Verlag, New York, 380p.
McFarland, D., ed., 1981, *The Oxford Companion to Animal Behaviour,* Oxford Univ., Oxford, 657p.
Morgan, C. L., 1894, *An Introduction to Comparative Psychology,* Walter Scott, London 386p.
Oppenheim, R. W., 1982, Preformation and Epigenesis in the Origins of the Nervous System and Behavior: Issues, Concepts, and Their History, *Perspectives Ethol.* **5:**1-100.
Owen, C., 1742, *An Essay towards a Natural History of Serpents,* privately published, London, 252p.
Romanes, G. J., 1882, *Animal Intelligence,* Kegan, Paul, Tench, Trübner & Co., London, 520p.
Schiller, C. H., ed., 1957, *Instinctive Behavior,* International Universities Press, New York, 328p.
Sparks, J., 1982, *The Discovery of Animal Behaviour,* Little Brown, Boston, 288p.
Stamm, R. A., and H. Zeier, eds., 1978, *Lorenz und die Folgen, Band VI, Die Psychologie des 20. Jahrhunderts,* Kindler-Verlag, Zurich, 1218p.
Thompson, E. P., 1851, *The Passions of Animals,* Chapman and Hall, London, 414p.
Thorpe, W. H., 1979, *The Origins and Rise of Ethology,* Heinemann, London, 174p.
Tinbergen, N., 1951, *The Study of Instinct,* Clarendon Press, Oxford, 228p.
Tinbergen, N., 1963, On Aims and Methods of Ethology, *Z. Tierpsychol.* **20:**410–429.
Warden, C. J., 1928, The Development of Modern Comparative Psychology, *Quart. Rev. Biol.* **3:**486–522.

Part I

DARWIN AND THE ANALYSIS OF INSTINCT

Editor's Comments on Papers 1 Through 5

The era during which Darwin reached scientific maturity was marked by an outpouring of new information about biology and the natural world in general. Systematics, paleontology, physiology, and morphology were developing rapidly as respected fields. Baron Cuvier, Richard Owen, and Louis Agassiz were popular public figures. Waterton's *Wanderings in South America* (Waterton, 1909), first appearing in 1825, was enjoyed by a fascinated public. Careful observation became more common in the study of "habits" and "instinct." But interpretation was difficult because it was impossible to describe animal behavior without thinking about human counterparts or similarities, at least long enough to deny anything other than a superficial resemblance. The natural theology of Charles Owen (see Introduction) had to give way to a more critical scientific approach, even though scientists themselves were imbued with a religious faith that seemed to require all animals *and their behavior* to be created and controlled by a Divine Being. But debates began on whether the Deity influenced behavior directly through supernatural agency or secondarily through natural physical processes.

An early provocative treatise on instinctive behavior is that of Kirby and Spence (1817), who produced a multivolume work on entomology that was the standard for many years; Darwin used it heavily. Instincts were discussed at length, but the interpretation was metaphysical. Kirby later wrote a prestigious Bridgewater Treatise on behavior. While the aim of the series was to support creationism and natural theology, Kirby (Paper 1) set out to refine the concept of instinct. Thus he contrasts direct divine control with indirect intervention, physical agencies being the intermediary, and supports the latter. In the course of the argument he disagrees with his previous co-author Spence (also pointing out that Spence alone wrote the chapters he was now questioning) and notes examples of intelligence and learning in animals as well as reason, loyalty, and emotion. He argues that instinct is not perfect, that misfirings occur, and even cites crossfostering imprinting experiments between chickens and ducks. Although he argues that intellect is not the principle behind instinct, he also holds that "all animals gifted with the ordinary organs of sensation, more or less employ their intellect in the whole routine of their instinctive operations" (p. 240). He notes, as both Darwin and Lorenz did later, that the animals with the lowest intellect also had the fewest instincts.

Kirby did not exclude humans from having instincts. Indeed, he was modern enough to use observations of the human infant to make his point: "The newborn babe has no other teacher to tell it that its mother's breast will supply it with its proper nutriment; it cries for it; it spontaneously applies its mouth to it; and presses it under the bidding of appetite resulting from its organization" (p. 256). Nevertheless, the theological straitjacket compromised the consistency of Kirby's thought. Thus the intelligence, memory, and loyalty shown by dogs, while not a mystical instinct, was due to Divine Providence as ". . . the Deity, it may be presumed, with a secret hand, guides some to fulfill his will, instructing them, as it were, because their unaided instinct would not alone avail, in the decree they are to execute" (p. 227). Humans do not fall on a continuum with animals, being created primarily for a spiritual world, although they are masters of this one. Appalled by Linnaeus' placing of humans with the other primates, Kirby declares that the reason primates were created ". . . seems to be to hold the mirror to man, that he may see how ugly and disgusting an object becomes when he gives himself up to vice and the slave of his passions" (p. 517).

Darwin's chapter on instinct in the first edition of *The Origin of Species* (Paper 2) shows the command that he had of animal behavior as well as his judicious use of the largely anecdotal evidence then available. As he makes clear, behavior had to be explicable by gradual evolutionary processes if his theory was to be convincing. He tackled disconcerting phenomena such as nonreproductive insect castes and slavemaking in ants with surprisingly modern arguments. The contrast with Kirby is remarkable; the last paragraph was surely written with the Kirbys of his day in mind.

Yet writers such as Kirby helped prepare the way for an evolutionary approach to animal behavior, even if they could not accept it themselves. Richards (1981) documents the seriousness with which Darwin took the arguments of the natural theologians. Their writings eventually helped him realize that the origin of instinct from inherited habits could not work for much behavior. Insect castes and the evolution of behavior in sterile workers was the most challenging obstacle Darwin had to overcome. Richards argues that not until Darwin came up with his community (kin) selection argument did he feel confident enough to publish his theory of natural selection. This, rather than the common view that Darwin was afraid of being labeled a materialist, was the reason for Darwin's caution. A final note: Although Lorenz and Tinbergen frequently invoked Darwin's name and natural selection, specific citations of his writings were rare in their early papers.

Bernard Altum was a talented German ornithologist who observed birds carefully and made many important discoveries not given due credit today. (He is not cited in Klopfer and Hailman, 1967; Thorpe, 1979; or other comparable reviews). Paper 3, from his 1868 book, discusses intermale aggression and the function of territory well before H. E. Howard (1920, see also Stokes, 1974) who is usually credited. Altum is a fine representative of German natural historians; his years of careful observation shine through. The discussion is modern in many respects, especially in his concern with behavioral ecology, sex ratios, polygyny, and dispersal. Today we might follow most of his views up to the final argument on dispersed unmated males. Lorenz (1937) refers to him but does not cite a publication.

In addition to careful observation, experiments seeking to disentangle instinct and experience were carried out in the early years following publication of *The Origin of Species*. D. A. Spalding's studies on imprinting and the perceptual and motor performance aspects of instinctive movements have been widely recognized (e.g., Thorpe, 1979). Papers 4 and 5 are two of his concise and today little known articles. Spalding's work was much commented upon in the late nineteenth century by both critics and supporters, although his creative use of the deprivation experiment, which was later to cause so much controversy in ethology (see Lorenz, 1965), was not sufficiently recognized. Haldane (1954) in "rediscovering" Spalding's work claimed that only his early death prevented him from becoming the "father of ethology" (also acknowledged by Lorenz, pers. comm.).

Actually, Spalding's work was well-cited into the 1920s, as in Carmichael (1927). The latter paper was the sequel to an earlier (1926) paper on the development of normal swimming movements of tadpoles reared in drugs inhibiting all muscular movement (reprinted in

Dewsbury, 1984), a paper that Lorenz cited frequently. Both Lorenz (e.g., 1937) and Tinbergen (1951) cite sources in which Spalding was discussed at length, but they apparently neither recognized his significance nor consulted his papers. Indeed, Lorenz summarized work by his student Grohmann on flying in pigeons raised with limited opportunity to exercise their wings, an experiment anticipated by Spalding. Spalding's claim that the study of neonate birds and mammals proves that complex perceptual and motor abilities of an adaptive nature are present at birth was not totally accepted (e.g., Carmichael, 1927), but as suggested by the papers in Part II of this book, verbal arguments were soon to again largely replace careful experimentation. And when psychologists also again started using the deprivation experiment in studies of mammalian perception, Spalding's arguments were forgotten. Hebb's influential theory of perception (Hebb, 1949) relied on the same misinterpretation of cataract removal effects in humans that Spalding criticized in Carpenter's work (Paper 5).

Spalding, like Spencer, held that instincts originated in "inherited associations;" this is emphasized at the end of both of his short papers (Papers 4 and 5). But between the appearance of these two papers, Thomas Huxley's conscious automata theory was set forth (Huxley, 1874/1901) and became incorporated into Spalding's view of instinct (Paper 5). A few more comments on Spalding, courtesy of the unpublished historical research of Bruce Batts, are appropriate. In spite of his nativistic conclusions, Spalding, who once described himself as a disciple of John Stuart Mill, was motivated to undertake his deprivation studies because of Mill's call for such observations as a means of settling debates over which mental phenomena were 'original' and which were acquired. Mill also advocated systematic observations on both children and "young of other animals" (see Mill, 1859/1978, p. 350). Spalding's concern with testing Berkeley's theory of distance perception reflects the controversy, of which Mill was at the center, over this theory among nineteenth-century associationist philosophers and psychologists. Mill advocated the *tabula rasa* or blank slate theory of human behavior; one wonders what he thought of the findings of this disciple!

REFERENCES

Carmichael, L., 1926, The Development of Behavior in Vertebrates Experimentally Removed from the Influence of External Stimulation, *Psychol. Rev.* **33:**51-58.

Carmichael, L., 1927, A Further Study of the Development of Behavior in Vertebrates Experimentally Removed from the Influence of External Stimulation, *Psychol. Rev.* **34:**34-47.

Dewsbury, D. A., 1984, *Foundations of Comparative Psychology,* Benchmark Papers in Behavior Series, Van Nostrand Reinhold, New York, 365p.
Haldane, J. B., 1954, Introducing Douglas Spalding, *Brit. J. Anim. Behav.* **2:**1-2.
Hebb, D. O., 1949, *The Organization of Behavior,* Wiley, New York, 335p.
Huxley, T. H., 1874/1901, On the Hypothesis that Animals Are Automata, and Its History, in *Collected Essays,* vol. 1, *Methods and Results* (T. H. Huxley, ed.), Macmillan, London, pp. 199-250.
Howard, H. E., 1920, *Territory in Bird Life,* John Murray, London.
Kirby, W., 1835, *On the Power Wisdom and Goodness of God as manifested in the Creation of Animals and in their History, Habits and Instincts,* vol. 2, William Pickering, London, 542p.
Kirby, W., and W. Spence, 1817, *An Introduction to Entomology,* vol. 3, Longman, London, 529p.
Klopfer, P. H., and J. P. Hailman, 1967, *An Introduction to Animal Behavior. Ethology's First Century,* Prentice Hall, New York, 297p.
Lorenz, K., 1937, The Nature of Instinct, translated in *Instinctive Behavior,* C. H. Schiller, ed., International Universities Press, New York, pp. 129-175.
Lorenz, K., 1965, *Evolution and Modification of Behavior,* University of Chicago, Chicago, 121p.
Mill, J. S., 1859/1978, Bain's Psychology, in *Collected Works of John Stuart Mill,* vol. 10, Toronto Univ., Toronto.
Richards, R. J., 1981, Instinct and Intelligence in British Natural Theology: Some Contributions to Darwin's Theory of the Evolution of Behavior, *J. Hist. Biol.* **14:**193-230.
Stokes, A. W., 1974, *Territory,* Benchmark Papers in Animal Behavior, Dowden, Hutchinson & Ross, Stroudsburg, Pa., 398p.
Thorpe, W. H., 1979, *The Origins and Rise of Ethology,* Heinemann, London, 174p.
Tinbergen, N., 1951, *The Study of Instinct,* Clarendon Press, Oxford, 228p.
Waterton, C., 1825/1909, *Wanderings in South America,* Macmillan, London, 520p.

1

Reprinted from pages 220–221, 229–231, 239–240, 257–259, and 273–275 of *On the Power Wisdom and Goodness of God as Manifested in the Creation of Animals and in Their History Habits and Instincts.* Bridgewater Treatise VII, vol. 2, William Pickering, London, 1835, 570p.

ON INSTINCT

W. Kirby

THERE is no department of Zoological Science that furnishes stronger proofs of the being and attributes of the Deity, than that which relates to the *Instincts* of animals, and the more so, because where reason and intellect are most powerful and sufficient as guides, as in man, and most of the higher grades of animals, there usually instinct is weakest and least wonderful, while, as we descend in the scale, we come to tribes that exhibit, in an almost miraculous manner, the workings of a Divine Power, and perform operations that the intellect and skill of man would in vain attempt to rival or to imitate. Yet there is no question, concerning which the Natural Historian and Physiologist seems more at a loss than when he is asked—what is INSTINCT? So much has been ably written upon the subject, so many hypotheses have been broached, that it seems wonderful so thick a cloud should still rest upon it. It must not be expected, where so many eminent men have more or less failed, that one of less powers should be enabled to throw much new light upon this palpable obscure, or dissipate all the darkness that envelopes the *secondary* or intermediate *cause* of Instinct. Could even the bee or the ant tell us what it is that goads them to their several labours, and instructs them how to perform them, perhaps we might still have much to learn before we should have any right to cry with the Syracusan Mathematician, Ἑυρηκα, I have unveiled the mystery. Still, however unequal to the task, I cannot duly discharge the duty incumbent upon me, who may be said to be *officially* engaged to prove the great truths of Natural Religion from the *Instincts* of the animal creation, to leave the subject of Instinct, considered in the abstract, exactly as I found it; a field, in which whoever perambulates, may wander " in endless mazes lost."

[*Editor's Note:* Material has been omitted at this point.]

. . . no class of facts so loudly proclaim their Great Author as those which are the result of the nice balancing of conflicting energies and operations observable in the different departments of the animal kingdom.

We may observe, however, that when our Saviour says to his disciples concerning sparrows—*One of them shall not fall to the ground without your Father. But the very hairs of your head are all numbered;*[1]—the observation implies that nothing escapes the notice, or is too mean, or insignificant, to be below the attention and care of Him who is all eye, all ear, all intellect; who directeth all things to answer his purposes, *according to the good pleasure of his will*,[2] which is the universal good of his creatures.

Having premised these general observations, I shall now proceed to inquire into the proximate cause of instinct; admitting, as proved, that every kind of instinct has its origin in the will of the Deity, and that the animal exhibiting it, was expressly organized by Him for it at its creation.

The proximate cause of instinct must be either metaphysical or physical, or a compound of both characters.

1. If *metaphysical*, it must either be the *im-*

[1] Matth. x. 29, 30. [2] *Ephes.* i. 5.

mediate action of the Deity, or the action of some *intermediate* intelligence employed by him, or the *intellect* of the animal exhibiting it.

2. If *physical*, it must be the action or stimulus of some physical power or agent employed by the Deity, and under his guidance, so as to work His will upon the organization of the animal, which must be so constructed as to respond to that action in a certain way; or by the exhibition of certain phenomena peculiar to the individual genus or species.

3. If *compound* or *mixed*, it will be subject occasionally to variations from the general law, when the intelligent agent sees fit.

1. With respect to the *first* Hypothesis, one of the principal promulgators and patrons of which is Addison,[1] it nearly amounts to this, as that amiable writer confesses, that "God is the soul of brutes." It is contrary, however, to the general plan of Divine Providence, which usually produces effects indirectly, and by the intervention and action of means or secondary causes, to suppose that it acts *immediately* upon insects and other animals, and is so intimately connected with them as to direct their instinctive operations; such an action, it should seem, would be infallible, and never at fault, whereas

[1] See *Spectator*, ii. p. 121.

observation has proved that animals are sometimes mistaken, where their instinct should direct them. For, if God were their *immediate* instructor, would it be possible for the flesh-fly, as I have seen that she does, to mistake the blossom of the carrion-plant[1] for a piece of flesh, and lay her eggs in it; or for a hen to sit upon a piece of chalk, as they are stated to do,[2] instead of an egg? Still all instincts are from God, He decreed them, and organized animals to act according to that decree, and employed means to impel them to do so.

[*Editor's Note:* Material has been omitted at this point.]

[1] *Stapelia hirsuta.* [2] Spectator, ii. n. 120.

Indeed, if intellect was the sole fountain of those operations usually denominated instinctive, animals, though they sought the same end, would vary more or less in the path they severally took to arrive at it; they would require some instruction and practice before they could be perfect in their operations; the new born bee would not immediately be able to rear a cell, nor know where to go for the materials, till some one of riper experience had directed her. But experience and observation have nothing to do with her proceedings. She feels an indomitable appetite which compels her to take her flight from the hive when the state of the atmosphere is favourable to her purpose. Her organs of sight—which though not gifted with any power of motion, are so situated as to enable her to see whatever passes above, below, and on each side of her—enable her to avoid any obstacles, and to thread her devious way through the numerous and intertwining branches of shrubs and flowers; some other sense directs her to those which contain the precious articles she is in quest of. But though her senses guide her in her flight, and indicate to her where she may most profitably exercise her talent, they must then yield her to the impulse and direction of her instincts, which this happy and industrious little creature plies with indefatigable diligence and energy, till having completed her lading of nectar and ambrosia, she returns to the common habitation of her people, with whom she unites in labours before described,[1] for the general benefit of the community to which she belongs.

More reasons might be adduced to prove that intellect is not the great principle of instinct, but enough seems to have been said to establish that point. It should be borne in mind, however, that though intellect is not the great principle, yet it must be admitted that all animals gifted with the ordinary organs of sensation, more or less employ their intellect in the whole routine of their instinctive operations, as I shall show under another head.

2. But if no metaphysical power can be satisfactorily demonstrated to be the immediate cause of instinct, then it seems to follow that it must be either a physical one, or one partly physical and partly metaphysical.

In the former case, it must be the action of some physical power or agent, employed by the Deity, and under his guidance so as to work his will, upon the organization of the animal; which must be so constructed as to respond to that action in a certain way, or by the exhibition of certain phenomena peculiar to the individual genus or species.

1 See above, p. 187, and *Introd. to Ent.* ii. 173.

[*Editor's Note:* Material has been omitted at this point.]

We may divide instincts into *three* general heads:—

a. Those relating to the multiplication of the species, especially the care of animals for their young both before and after birth.

β. Those relating to their food.

γ. Those relating to their Hybernation.

a. The pairing of animals usually begins to take place in the spring, when the winter is passed, the earth is covered with verdure and adorned by the various flowers that now expand their blossoms, in proportion as the great centre of light and heat more and more manifests his power over the earth; the birds sing their love-songs; the nightingale is now—" Most musical, most melancholy;"—the cuckoo repeats his monotonous note; and every other animal seems to partake of the universal joy. All this appears the result of a *physical* rather than a *metaphysical* excitement.

As to their care of their future progeny, a great variety of circumstances take place. Viviparous animals have generally to give suck to their young for a time; oviparous ones either to construct a nest to receive their eggs, and, after hatching, to provide them with appropriate food during a certain period, or to deposit their eggs where their young progeny, as soon as hatched, may infallibly find it. But first, I must say something of that *Storge*, or instinctive affection, which is almost universally exhibited by females for their progeny both before and after parturition; a feeling of affection not generally common to the males, or rather only in a few instances, as where the male bird assists the female in incubation. Yet this instinctive fondness, as soon as it ceases to be necessary, vanishes; except, as was before observed,[1] in the human species; a fact that seems to prove that it is not the result of the association of ideas, but of an impress of the Creator interwoven with the frame. But that this impress is by means of a physical interagent, seems to follow from this circumstance—that the *hen* shows the same instinctive attachment to the young *ducklings* that have been hatched under her, that she would do to chickens, the produce of her own eggs; and if the new-born offspring of any mammiferous animal is abstracted from her, and another substituted, even of a *different* kind,

[1] See above, p. 238.

the same affectionate tenderness is manifested towards it, as its own real offspring would have experienced. Now was it a metaphysical, and not a physical, impulse, surely this would not be the case. This is only one of many instances, which prove that instinct is not infallible: and, in truth, with regard to the higher animals, many associations may take place between the child and parent that help to endear the former to the latter. In the first place, the very circumstance of its being the fruit of her own bowels, and fed with milk from her own breast must bind it to her by the tenderest of ties; especially as, at the same time, it relieves her from what is troublesome. There is something also in infant helplessness, and infant gambols, calculated to win upon the doting mother. The subsequent alienation and estrangement of the female from her young, which takes place in all animals except man, appears, in the first instance, to be produced by their becoming troublesome and annoying to her; which, in some degree, may account for her desire to cast them off. Examining the subject, therefore, on all sides, in the highest grades of animals, and those in whom maternal affection appears most intense, intellect and associations may be a good deal mixed with instinct in producing it.

[*Editor's Note:* Material has been omitted at this point.]

Thus we see the Almighty and All-wise manifests his *goodness*, as well as his wisdom and power, in providing for the wants of all the creatures that he has made; fitting each with peculiar organs adapted to its assigned kind of food, both for procuring it, preparing it, digesting it, assimilating it, and for rejecting the residuum of all these operations. A physical action upon each of these organs and systems, fitted by him to receive and respond to it, is all that the case seems to require in the majority of instances: in those, however, that depend upon artifice and stratagem for their food, the exciting cause is less obvious. These, indeed, belong to the higher instincts considered under the *first* head.

γ. That class of Instincts which relates to the *hybernation* of animals having been considered in another place,[1] I shall only observe here, that the action of a physical cause is in no department of the history of animals more evidently made out.

My learned friend and coadjutor, Mr. Spence, has, in the *Introduction to Entomology*, produced several facts, as not easily reconcilable to the hypothesis with respect to the cause of

[1] See above, p. 248.

Instinct which I am now considering; and probably a great many more might be brought forward; but my object here is merely to consider the general principle; it would, indeed, be needless and endless to discuss particular cases, and fully to account for all aberrations, which, in the present state of our knowledge, it would not be possible to do.

But there is one circumstance of a less confined nature, and upon which a good deal of the question hinges, to which it will be proper to advert. I mean the change that has been observed in the nervous system of some insects in their passage from one state to another. It is contended that this change has nothing to do with any alterations that then take place in their instincts, but only with those in their organs of sense or motion.[1] In confirmation of this opinion it is further affirmed, that in three whole Orders,[2] the structure of the nervous chord is not altered, and yet they acquire new instincts.

But though no change has been *noticed* to take place in the number of ganglions of these Orders, there must necessarily be a developement in those that render nerves to the wings and reproductive organs; so that, though some ganglions may not become confluent, as in the *Lepidoptera*, yet the range of their nerves

1 *Introd. to Ent.* iv. 27, 28.

2 Viz. *Orthoptera*, *Hemiptera*, and *Neuroptera*.

is increased. In this respect, they are in much the same situation with the higher animals, though their nervous system, as to its organization, undergoes no material change, yet from the period of their birth, it is gradually more and more developed till they arrive at the age of puberty, when new appetites are experienced and new powers acquired, not by *metaphysical*, but by *physical*, action upon their several systems. In the three Orders referred to by Mr. Spence, there is not that difference between the different states of the insects that compose the majority of them, that there is between those whose pupes are not locomotive. The larves of the locust, for instance, are stated to emigrate, as well as the perfect insect, and live upon the same food ; the only difference is in the locomotive and reproductive powers of the latter, both of which, as I have just said, must be connected with some change in their nervous system, operated gradually by a physical agent.

From what has been stated, with respect to these several classes of instincts, it appears, that, as far as can be judged from circumstances, they have their beginning in consequence of the action of an intermediate physical cause upon the organization of the animal, which certainly renders it extremely probable that such is the general proximate cause of the phenomena

. . .

[*Editor's Note:* In the original, material follows this excerpt.]

2

Reprinted from pages 207-244 of *On the Origin of Species by Means of Natural Selection or the Preservation of Favoured Races in the Struggle for Life*, Murray, London, 1859, 502p.

INSTINCT

Charles Darwin

Instincts comparable with habits, but different in their origin—Instincts graduated — Aphides and ants — Instincts variable—Domestic instincts, their origin—Natural instincts of the cuckoo, ostrich, and parasitic bees — Slave-making ants — Hive-bee, its cell-making instinct—Difficulties on the theory of the Natural Selection of instincts—Neuter or sterile insects—Summary.

THE subject of instinct might have been worked into the previous chapters; but I have thought that it would be more convenient to treat the subject separately, especially as so wonderful an instinct as that of the hive-bee making its cells will probably have occurred to many readers, as a difficulty sufficient to overthrow my whole theory. I must premise, that I have nothing to do with the origin of the primary mental powers, any more than I have with that of life itself. We are concerned only with the diversities of instinct and of the other mental qualities of animals within the same class.

I will not attempt any definition of instinct. It would be easy to show that several distinct mental actions are commonly embraced by this term; but every one understands what is meant, when it is said that instinct impels the cuckoo to migrate and to lay her eggs in other birds' nests. An action, which we ourselves should require experience to enable us to perform, when performed by an animal, more especially by a very young one, without any experience, and when performed by many individuals in the same way, without their knowing for what purpose it is performed, is usually said to be instinctive.

But I could show that none of these characters of instinct are universal. A little dose, as Pierre Huber expresses it, of judgment or reason, often comes into play, even in animals very low in the scale of nature.

Frederick Cuvier and several of the older metaphysicians have compared instinct with habit. This comparison gives, I think, a remarkably accurate notion of the frame of mind under which an instinctive action is performed, but not of its origin. How unconsciously many habitual actions are performed, indeed not rarely in direct opposition to our conscious will! yet they may be modified by the will or reason. Habits easily become associated with other habits, and with certain periods of time and states of the body. When once acquired, they often remain constant throughout life. Several other points of resemblance between instincts and habits could be pointed out. As in repeating a well-known song, so in instincts, one action follows another by a sort of rhythm; if a person be interrupted in a song, or in repeating anything by rote, he is generally forced to go back to recover the habitual train of thought: so P. Huber found it was with a caterpillar, which makes a very complicated hammock; for if he took a caterpillar which had completed its hammock up to, say, the sixth stage of construction, and put it into a hammock completed up only to the third stage, the caterpillar simply re-performed the fourth, fifth, and sixth stages of construction. If, however, a caterpillar were taken out of a hammock made up, for instance, to the third stage, and were put into one finished up to the sixth stage, so that much of its work was already done for it, far from feeling the benefit of this, it was much embarrassed, and, in order to complete its hammock, seemed forced to start from the third stage, where it had left off, and thus tried to complete the already finished work.

If we suppose any habitual action to become inherited—and I think it can be shown that this does sometimes happen—then the resemblance between what originally was a habit and an instinct becomes so close as not to be distinguished. If Mozart, instead of playing the pianoforte at three years old with wonderfully little practice, had played a tune with no practice at all, he might truly be said to have done so instinctively. But it would be the most serious error to suppose that the greater number of instincts have been acquired by habit in one generation, and then transmitted by inheritance to succeeding generations. It can be clearly shown that the most wonderful instincts with which we are acquainted, namely, those of the hive-bee and of many ants, could not possibly have been thus acquired.

It will be universally admitted that instincts are as important as corporeal structure for the welfare of each species, under its present conditions of life. Under changed conditions of life, it is at least possible that slight modifications of instinct might be profitable to a species; and if it can be shown that instincts do vary ever so little, then I can see no difficulty in natural selection preserving and continually accumulating variations of instinct to any extent that may be profitable. It is thus, as I believe, that all the most complex and wonderful instincts have originated. As modifications of corporeal structure arise from, and are increased by, use or habit, and are diminished or lost by disuse, so I do not doubt it has been with instincts. But I believe that the effects of habit are of quite subordinate importance to the effects of the natural selection of what may be called accidental variations of instincts;—that is of variations produced by the same unknown causes which produce slight deviations of bodily structure.

No complex instinct can possibly be produced through

natural selection, except by the slow and gradual accumulation of numerous, slight, yet profitable, variations. Hence, as in the case of corporeal structures, we ought to find in nature, not the actual transitional gradations by which each complex instinct has been acquired—for these could be found only in the lineal ancestors of each species—but we ought to find in the collateral lines of descent some evidence of such gradations; or we ought at least to be able to show that gradations of some kind are possible; and this we certainly can do. I have been surprised to find, making allowance for the instincts of animals having been but little observed except in Europe and North America, and for no instinct being known amongst extinct species, how very generally gradations, leading to the most complex instincts, can be discovered. The canon of "Natura non facit saltum" applies with almost equal force to instincts as to bodily organs. Changes of instinct may sometimes be facilitated by the same species having different instincts at different periods of life, or at different seasons of the year, or when placed under different circumstances, &c.; in which case either one or the other instinct might be preserved by natural selection. And such instances of diversity of instinct in the same species can be shown to occur in nature.

Again as in the case of corporeal structure, and conformably with my theory, the instinct of each species is good for itself, but has never, as far as we can judge, been produced for the exclusive good of others. One of the strongest instances of an animal apparently performing an action for the sole good of another, with which I am acquainted, is that of aphides voluntarily yielding their sweet excretion to ants: that they do so voluntarily, the following facts show. I removed all the ants from a group of about a dozen aphides on a dock-plant, and prevented their attendance during several hours. After this interval, I felt sure that the aphides would want to excrete. I watched them for some time through a lens, but not one excreted; I then tickled and stroked them with a hair in the same manner, as well as I could, as the ants do with their antennæ; but not one excreted. Afterwards I allowed an ant to visit them, and it immediately seemed, by its eager way of running about, to be well aware what a rich flock it had discovered; it then began to play with its antennæ on the abdomen first of one aphis and then of another; and each aphis, as soon as it felt the antennæ, immediately lifted up its abdomen and excreted a limpid drop of sweet juice, which was eagerly devoured by the ant. Even the quite young aphides behaved in this manner, showing that the action was instinctive, and not the result of experience. But as the excretion is extremely viscid, it is probably a convenience to the aphides to have it removed; and therefore probably the aphides do not instinctively excrete for the sole good of the ants. Although I do not believe that any animal in the world performs an action for the exclusive good of another of a distinct species, yet each species tries to take advantage of the instincts of others, as each takes advantage of the weaker bodily structure of others. So again, in some few cases, certain instincts cannot be considered as absolutely perfect; but as details on this and other such points are not indispensable, they may be here passed over.

As some degree of variation in instincts under a state of nature, and the inheritance of such variations, are indispensable for the action of natural selection, as many instances as possible ought to have been here given; but want of space prevents me. I can only assert, that instincts certainly do vary—for instance, the migratory instinct, both in extent and direction, and in its total loss. So it is with the nests of birds, which vary partly in dependence on the situations chosen, and on the nature and temperature of the country inhabited, but often from causes wholly unknown to us: Audubon has given several remarkable cases of differences in nests of the same species in the northern and southern United States. Fear of any particular enemy is certainly an instinctive quality, as may be seen in nestling birds, though it is strengthened by experience, and by the sight of fear of the same enemy in other animals. But fear of man is slowly acquired, as I have elsewhere shown, by various animals inhabiting desert islands; and we may see an instance of this, even in England, in the greater wildness of all our large birds than of our small birds; for the large birds have been most persecuted by man. We may safely attribute the greater wildness of our large birds to this cause; for in uninhabited islands large birds are not more fearful than small; and the magpie, so wary in England, is tame in Norway, as is the hooded crow in Egypt.

That the general disposition of individuals of the same species, born in a state of nature, is extremely diversified, can be shown by a multitude of facts. Several cases also, could be given, of occasional and strange habits in certain species, which might, if advantageous to the species, give rise, through natural selection, to quite new instincts. But I am well aware that these general statements, without facts given in detail, can produce but a feeble effect on the reader's mind. I can only repeat my assurance, that I do not speak without good evidence.

The possibility, or even probability, of inherited variations of instinct in a state of nature will be strengthened by briefly considering a few cases under domestication. We shall thus also be enabled to see the respective parts which habit and the selection of so-called accidental variations have played in modifying the mental qualities of our domestic animals. A number of curious and authentic instances could be given of the inheritance of all shades of disposition and tastes, and likewise of the oddest tricks, associated with certain frames of mind or periods of time. But let us look to the familiar case of the several breeds of dogs: it cannot be doubted that young pointers (I have myself seen a striking instance) will sometimes point and even back other dogs the very first time that they are taken out; retrieving is certainly in some degree inherited by retrievers; and a tendency to run round, instead of at, a flock of sheep, by shepherd-dogs. I cannot see that these actions, performed without experience by the young, and in nearly the same manner by each individual, performed with eager delight by each breed, and without the end being known,—for the young pointer can no more know that he points to aid his master, than the white butterfly knows why she lays her eggs on the leaf of the cabbage,—I cannot see that these actions differ essentially from true instincts. If we were to see one kind of wolf, when young and without any training, as soon as it scented its prey, stand motionless like a statue, and then slowly crawl forward with a peculiar gait; and another kind of wolf rushing round, instead of at, a herd of deer, and driving them to a distant point, we should assuredly call these actions instinctive. Domestic instincts, as they may be called, are certainly far less fixed or invariable than natural instincts; but they have been acted on by far less rigorous selection, and have been transmitted for an incomparably shorter period, under less fixed conditions of life.

How strongly these domestic instincts, habits, and dispositions are inherited, and how curiously they become mingled, is well shown when different breeds of dogs are crossed. Thus it is known that a cross with a bull-dog has affected for many generations the courage and obstinacy of greyhounds; and a cross with a greyhound has given to a whole family of shepherd-dogs a tendency to hunt hares. These domestic instincts, when thus tested by crossing, resemble natural instincts, which in a like manner become curiously blended together, and for a long period exhibit traces of the instincts of either parent: for example, Le Roy describes a dog, whose great-grandfather was a wolf, and this dog showed a trace of its wild parentage only in one way, by not coming in a straight line to his master when called.

Domestic instincts are sometimes spoken of as actions which have become inherited solely from long-continued and compulsory habit, but this, I think, is not true. No one would ever have thought of teaching, or probably could have taught, the tumbler-pigeon to tumble,—an action which, as I have witnessed, is performed by young birds, that have never seen a pigeon tumble. We may believe that some one pigeon showed a slight tendency to this strange habit, and that the long-continued selection of the best individuals in successive generations made tumblers what they now are; and near Glasgow there are house-tumblers, as I hear from Mr. Brent, which cannot fly eighteen inches high without going head over heels. It may be doubted whether any one would have thought of training a dog to point, had not some one dog naturally shown a tendency in this line; and this is known occasionally to happen, as I once saw in a pure terrier. When the first tendency was once displayed, methodical selection and the inherited effects of compulsory training in each successive generation would soon complete the work; and unconscious selection is still at work, as each man tries to procure, without intending to improve the breed, dogs which will stand and hunt best. On the other hand, habit alone in some cases has sufficed; no animal is more difficult to tame than the young of the wild rabbit; scarcely any animal is tamer than the young of the tame rabbit; but I do not suppose that domestic rabbits have ever been selected for tameness; and I presume that we must attribute the whole of the inherited change from extreme wildness to extreme tameness, simply to habit and long-continued close confinement.

Natural instincts are lost under domestication: a remarkable instance of this is seen in those breeds of fowls which very rarely or never become "broody," that is, never wish to sit on their eggs. Familiarity alone prevents our seeing how universally and largely the minds of our domestic animals have been modified by domestication. It is scarcely possible to doubt that the love of man has become instinctive in the dog. All wolves, foxes, jackals, and species of the cat genus, when kept tame, are most eager to attack poultry, sheep, and pigs; and this tendency has been found incurable in dogs which have been brought home as puppies from countries, such as Tierra del Fuego and Australia, where the savages do not keep these domestic animals. How rarely, on the other hand, do our civilised dogs, even when quite young, require to be taught not to attack poultry, sheep, and pigs! No doubt they occasionally do make an attack, and are then beaten; and if not cured, they are destroyed; so that habit, with some degree of selection, has probably concurred in civilising by inheritance our dogs. On the other hand, young chickens have lost, wholly by habit, that fear of the dog and cat which no doubt was originally instinctive in them, in the same way as it is so plainly instinctive in young pheasants, though reared under a hen. It is not that chickens have lost all fear, but fear only of dogs and cats, for if the hen gives the danger-chuckle, they will run (more especially young turkeys) from under her, and conceal themselves in the surrounding grass or thickets; and this is evidently done for the instinctive purpose of allowing, as we see in wild ground-birds, their mother to fly away. But this instinct retained by our chickens has become useless under domestication, for the mother-hen has almost lost by disuse the power of flight.

Hence, we may conclude, that domestic instincts have been acquired and natural instincts have been lost partly by habit, and partly by man selecting and accumulating during successive generations, peculiar mental habits and actions, which at first appeared from what we must in our ignorance call an accident. In some cases compulsory habit alone has sufficed to produce such inherited mental changes; in other cases compulsory habit has done nothing, and all has been the result of selection, pursued both methodically and unconsciously; but in most cases, probably, habit and selection have acted together.

We shall, perhaps, best understand how instincts in a state of nature have become modified by selection, by considering a few cases. I will select only three, out of the several which I shall have to discuss in my future work,—namely, the instinct which leads the cuckoo to lay her eggs in other birds' nests; the slave-making instinct of certain ants; and the comb-making power of the hive-bee: these two latter instincts have generally, and most justly, been ranked by naturalists as the most wonderful of all known instincts.

It is now commonly admitted that the more immediate and final cause of the cuckoo's instinct is, that she lays her eggs, not daily, but at intervals of two or three days; so that, if she were to make her own nest and sit on her own eggs, those first laid would have to be left for some time unincubated, or there would be eggs and young birds of different ages in the same nest. If this were the case, the process of laying and hatching might be inconveniently long, more especially as she has to migrate at a very early period; and the first hatched young would probably have to be fed by the male alone. But the American cuckoo is in this predicament; for she makes her own nest and has eggs and young successively hatched, all at the same time. It has been asserted that the American cuckoo occasionally lays her eggs in other birds' nests; but I hear on the high authority of Dr. Brewer, that this is a mistake. Nevertheless, I could give several instances of various birds which have been known occasionally to lay their eggs in other birds' nests. Now let us suppose that the ancient progenitor of our European cuckoo had the habits of the American cuckoo; but that occasionally she laid an egg in another bird's nest. If the old bird profited by this occasional habit, or if the young were made more vigorous by advantage having been taken of the mistaken maternal instinct of another bird, than by their own mother's care, encumbered as she can hardly fail to be by having eggs and young of different ages at the same time; then the old birds or the fostered young would gain an advantage. And analogy would lead me to believe, that the young thus reared would be apt to follow by inheritance the occasional and aberrant habit of their mother, and in their turn would be apt to lay their eggs in other birds' nests, and thus be successful in rearing their young. By a continued process of this nature, I believe that the strange instinct of our cuckoo could be, and has been, generated. I may add that, according to Dr. Gray and to some other observers, the European cuckoo has not utterly lost all maternal love and care for her own offspring.

The occasional habit of birds laying their eggs in other birds' nests, either of the same or of a distinct species, is not very uncommon with the Gallinaceæ; and this perhaps explains the origin of a singular instinct in the allied group of ostriches. For several hen ostriches, at least in the case of the American species, unite and lay first a few eggs in one nest and then in another; and these are hatched by the males. This instinct may probably be accounted for by the fact of the hens laying a large number of eggs; but, as in the case of the cuckoo, at intervals of two or three days. This instinct, however, of the American ostrich has not as yet been perfected; for a surprising number of eggs lie strewed over the plains, so that in one day's hunting I picked up no less than twenty lost and wasted eggs.

Many bees are parasitic, and always lay their eggs in the nests of bees of other kinds. This case is more remarkable than that of the cuckoo; for these bees have not only their instincts but their structure modified in accordance with their parasitic habits; for they do not possess the pollen-collecting apparatus which would be necessary if they had to store food for their own young. Some species, likewise, of Sphegidæ (wasp-like insects) are parasitic on other species; and M. Fabre has lately shown good reason for believing that although the Tachytes nigra generally makes its own burrow and stores it with paralysed prey for its own larvæ to feed on, yet that when this insect finds a burrow already made and stored by another sphex, it takes advantage of the prize, and becomes for the occasion parasitic. In this case, as with the supposed case of the cuckoo, I can see no difficulty in natural selection making an occasional habit permanent, if of advantage to the species, and if the insect whose nest and stored food are thus feloniously appropriated, be not thus exterminated.

Slave-making instinct.—This remarkable instinct was first discovered in the Formica (Polyerges) rufescens by Pierre Huber, a better observer even than his celebrated father. This ant is absolutely dependent on its slaves; without their aid, the species would certainly become extinct in a single year. The males and fertile females do no work. The workers or sterile females, though most energetic and courageous in capturing slaves, do no other work. They are incapable of making their own nests, or of feeding their own larvæ. When the old nest is found inconvenient, and they have to migrate, it is the slaves which determine the migration, and actually carry their masters in their jaws. So utterly helpless are the masters, that when Huber shut up thirty of them without a slave, but with plenty of the food which they like best, and with their larvæ and pupæ to stimulate them to work, they did nothing; they could not even feed themselves, and many perished of hunger. Huber then introduced a single slave (F. fusca), and she instantly set to work, fed and saved the survivors; made some cells and tended the larvæ, and put all to rights. What can be more extraordinary than these well-ascertained facts? If we had not known of any other slave-making ant, it would have been hopeless to have speculated how so wonderful an instinct could have been perfected.

Formica sanguinea was likewise first discovered by P. Huber to be a slave-making ant. This species is found in the southern parts of England, and its habits have been attended to by Mr. F. Smith, of the British

Museum, to whom I am much indebted for information on this and other subjects. Although fully trusting to the statements of Huber and Mr. Smith, I tried to approach the subject in a sceptical frame of mind, as any one may well be excused for doubting the truth of so extraordinary and odious an instinct as that of making slaves. Hence I will give the observations which I have myself made, in some little detail. I opened fourteen nests of F. sanguinea, and found a few slaves in all. Males and fertile females of the slave-species are found only in their own proper communities, and have never been observed in the nests of F. sanguinea. The slaves are black and not above half the size of their red masters, so that the contrast in their appearance is very great. When the nest is slightly disturbed, the slaves occasionally come out, and like their masters are much agitated and defend the nest: when the nest is much disturbed and the larvæ and pupæ are exposed, the slaves work energetically with their masters in carrying them away to a place of safety. Hence, it is clear, that the slaves feel quite at home. During the months of June and July, on three successive years, I have watched for many hours several nests in Surrey and Sussex, and never saw a slave either leave or enter a nest. As, during these months, the slaves are very few in number, I thought that they might behave differently when more numerous; but Mr. Smith informs me that he has watched the nests at various hours during May, June and August, both in Surrey and Hampshire, and has never seen the slaves, though present in large numbers in August, either leave or enter the nest. Hence he considers them as strictly household slaves. The masters, on the other hand, may be constantly seen bringing in materials for the nest, and food of all kinds. During the present year, however, in the month of July, I came across a community with an unusually large stock of slaves, and I observed a few slaves mingled with their masters leaving the nest, and marching along the same road to a tall Scotch-fir-tree, twenty-five yards distant, which they ascended together, probably in search of aphides or cocci. According to Huber, who had ample opportunities for observation, in Switzerland the slaves habitually work with their masters in making the nest, and they alone open and close the doors in the morning and evening; and, as Huber expressly states, their principal office is to search for aphides. This difference in the usual habits of the masters and slaves in the two countries, probably depends merely on the slaves being captured in greater numbers in Switzerland than in England.

One day I fortunately chanced to witness a migration from one nest to another, and it was a most interesting spectacle to behold the masters carefully carrying, as Huber has described, their slaves in their jaws. Another day my attention was struck by about a score of the slave-makers haunting the same spot, and evidently not in search of food; they approached and were vigorously repulsed by an independent community of the slave species (F. fusca); sometimes as many as three of these ants clinging to the legs of the slave-making F. sanguinea. The latter ruthlessly killed their small opponents, and carried their dead bodies as food to their nest, twenty-nine yards distant; but they were prevented from getting any pupæ to rear as slaves. I then dug up a small parcel of the pupæ of F. fusca from another nest, and put them down on a bare spot near the place of combat; they were eagerly seized, and carried off by the tyrants, who perhaps fancied that, after all, they had been victorious in their late combat.

At the same time I laid on the same place a small parcel of the pupæ of another species, F. flava, with a few of these little yellow ants still clinging to the fragments of the nest. This species is sometimes, though rarely, made into slaves, as has been described by Mr. Smith. Although so small a species, it is very courageous, and I have seen it ferociously attack other ants. In one instance I found to my surprise an independent community of F. flava under a stone beneath a nest of the slave-making F. sanguinea; and when I had accidentally disturbed both nests, the little ants attacked their big neighbours with surprising courage. Now I was curious to ascertain whether F. sanguinea could distinguish the pupæ of F. fusca, which they habitually make into slaves, from those of the little and furious F. flava, which they rarely capture, and it was evident that they did at once distinguish them: for we have seen that they eagerly and instantly seized the pupæ of F. fusca, whereas they were much terrified when they came across the pupæ, or even the earth from the nest of F. flava, and quickly ran away; but in about a quarter of an hour, shortly after all the little yellow ants had crawled away, they took heart and carried off the pupæ.

One evening I visited another community of F. sanguinea, and found a number of these ants entering their nest, carrying the dead bodies of F. fusca (showing that it was not a migration) and numerous pupæ. I traced the returning file burthened with booty, for about forty yards, to a very thick clump of heath, whence I saw the last individual of F. sanguinea emerge, carrying a pupa; but I was not able to find the desolated nest in the thick heath. The nest, however, must have been close at hand, for two or three individuals of F. fusca were rushing about in the greatest agitation, and one was perched motionless with its own pupa in its mouth on the top of a spray of heath over its ravaged home.

Such are the facts, though they did not need confirmation by me, in regard to the wonderful instinct of making slaves. Let it be observed what a contrast the instinctive habits of F. sanguinea present with those of the F. rufescens. The latter does not build its own nest, does not determine its own migrations, does not collect food for itself or its young, and cannot even feed itself: it is absolutely dependent on its numerous slaves. Formica sanguinea, on the other hand, possesses much fewer slaves, and in the early part of the summer extremely few. The masters determine when and where a new nest shall be formed, and when they migrate, the masters carry the slaves. Both in Switzerland and England the slaves seem to have the exclusive care of the larvæ, and the masters alone go on slave-making expeditions. In Switzerland the slaves and masters work together, making and bringing materials for the nest: both, but chiefly the slaves, tend, and milk as it may be called, their aphides; and thus both collect food for the community. In England the masters alone usually leave the nest to collect building materials and food for themselves, their slaves and larvæ. So that the masters in this country receive much less service from their slaves than they do in Switzerland.

By what steps the instinct of F. sanguinea originated I will not pretend to conjecture. But as ants, which are not slave-makers, will, as I have seen, carry off pupæ of other species, if scattered near their nests, it is possible that pupæ originally stored as food might become developed; and the ants thus unintentionally reared would then follow their proper instincts, and do what work they could. If their presence proved useful to the species which had seized them—if it were more advantageous to this species to capture workers than to procreate them—the habit of collecting pupæ originally for food might by natural selection be strengthened and rendered permanent for the very different purpose of raising slaves. When the instinct was once acquired, if carried out to a much less extent even than in our British F. sanguinea, which, as we have seen, is less aided by its slaves than the same species in Switzerland, I can see no difficulty in natural selection increasing and modifying the instinct—always supposing each modification to be of use to the species—until an ant was formed as abjectly dependent on its slaves as is the Formica rufescens.

Cell-making instinct of the Hive-Bee.—I will not here enter on minute details on this subject, but will merely give an outline of the conclusions at which I have arrived. He must be a dull man who can examine the exquisite structure of a comb, so beautifully adapted to its end, without enthusiastic admiration. We hear from mathematicians that bees have practically solved a recondite problem, and have made their cells of the proper shape to hold the greatest possible amount of honey, with the least possible consumption of precious wax in their construction. It has been remarked that a skilful workman, with fitting tools and measures, would find it very difficult to make cells of wax of the true form, though this is perfectly effected by a crowd of bees working in a dark hive. Grant whatever instincts you please, and it seems at first quite inconceivable how they can make all the necessary angles and planes, or even perceive when they are correctly made. But the difficulty is not nearly so great as it at first appears: all this beautiful work can be shown, I think, to follow from a few very simple instincts.

I was led to investigate this subject by Mr. Waterhouse, who has shown that the form of the cell stands in close relation to the presence of adjoining cells; and the following view may, perhaps, be considered only as a modification of his theory. Let us look to the great principle of gradation, and see whether Nature does not reveal to us her method of work. At one end of a short series we have humble-bees, which use their old cocoons to hold honey, sometimes adding to them short tubes of wax, and likewise making separate and very irregular rounded cells of wax. At the other end of the series we have the cells of the hive-bee, placed in a double layer: each cell, as is well known, is an hexagonal prism, with the basal edges of its six sides bevelled so as to join on to a pyramid, formed of three rhombs. These rhombs have certain angles, and the three which form the pyramidal base of a single cell on one side of the comb, enter into the composition of the bases of three adjoining cells on the opposite side. In the series between the extreme perfection of the cells of the hive-bee and the simplicity of those of the humble-bee, we have the cells of the Mexican Melipona domestica, carefully described and figured by Pierre Huber. The Melipona itself is intermediate in structure between the hive and humble bee, but more nearly related to the latter: it forms a nearly regular waxen comb of cylindrical cells, in which the young are hatched, and, in addition, some large cells of wax for holding honey. These latter cells are nearly spherical and of nearly equal sizes, and are aggregated into an irregular mass. But the important point to notice, is that these cells are always made at that degree of nearness to each other, that they would have intersected or broken into each other, if the spheres had been completed; but this is never permitted, the bees building perfectly flat walls of wax between the spheres which thus tend to intersect. Hence each cell consists of an outer spherical portion and of two, three, or more perfectly flat surfaces, according as the cell adjoins two, three, or more other cells. When one cell comes into contact with three other cells, which, from the spheres being nearly of the same size, is very frequently and necessarily the case, the three flat surfaces are united into a pyramid; and this pyramid, as Huber has remarked, is manifestly a gross imitation of the three-sided pyramidal basis of the cell of the hive-bee. As in the cells of the hive-bee, so here, the three plane surfaces in any one cell necessarily enter into the construction of three adjoining cells. It is obvious that the Melipona saves wax by this manner of building; for the flat walls between the adjoining cells are not double, but are of the same thickness as the outer spherical portions, and yet each flat portion forms a part of two cells.

Reflecting on this case, it occurred to me that if the Melipona had made its spheres at some given distance from each other, and had made them of equal sizes and had arranged them symmetrically in a double layer, the resulting structure would probably have been as perfect as the comb of the hive-bee. Accordingly I wrote to Professor Miller, of Cambridge, and this geometer has kindly read over the following statement, drawn up from his information, and tells me that it is strictly correct:—

If a number of equal spheres be described with their centres placed in two parallel layers; with the centre of each sphere at the distance of radius $\times \sqrt{2}$, or radius $\times$ 1·41421 (or at some lesser distance), from the centres of the six surrounding spheres in the same layer; and at the same distance from the centres of the adjoining spheres in the other and parallel layer; then, if planes of intersection between the several spheres in both layers be formed, there will result a double layer of hexagonal prisms united together by pyramidal bases formed of three rhombs; and the rhombs and the sides of the hexagonal prisms will have every angle identically the same with the best measurements which have been made of the cells of the hive-bee.

Hence we may safely conclude that if we could slightly modify the instincts already possessed by the Melipona, and in themselves not very wonderful, this bee would make a structure as wonderfully perfect as that of the hive-bee. We must suppose the Melipona to make her cells truly spherical, and of equal sizes; and this would not be very surprising, seeing that she already does so to a certain extent, and seeing what perfectly cylindrical burrows in wood many insects can make, apparently by turning round on a fixed point. We must suppose the Melipona to arrange her cells in level layers, as she already does her cylindrical cells; and we must further suppose, and this is the greatest difficulty, that she can somehow judge accurately at what distance to stand from her fellow-labourers when several are making their spheres; but she is already so far enabled to judge of distance, that she always describes her spheres so as to intersect largely; and then she unites the points of intersection by perfectly flat surfaces. We have further to suppose, but this is no difficulty, that after hexagonal prisms have been formed by the intersection of adjoining spheres in the same layer, she can prolong the hexagon to any length requisite to hold the stock of honey; in the same way as the rude humble-bee adds cylinders of wax to the circular mouths of her old cocoons. By such modifications of instincts in themselves not very wonderful,—hardly more wonderful than those which guide a bird to make its nest,—I believe that the hive-bee has acquired, through natural selection, her inimitable architectural powers.

But this theory can be tested by experiment. Following the example of Mr. Tegetmeier, I separated two combs, and put between them a long, thick, square strip of wax: the bees instantly began to excavate minute circular pits in it; and as they deepened these little pits, they made them wider and wider until they were converted into shallow basins, appearing to the eye perfectly true or parts of a sphere, and of about the diameter of a cell. It was most interesting to me to observe that wherever several bees had begun to excavate these basins near together, they had begun their work at such a distance from each other, that by the time the basins had acquired the above stated width (*i. e.* about the width of an ordinary cell), and were in depth about one sixth of the diameter of the sphere of which they formed a part, the rims of the basins intersected or broke into each other. As soon as this occurred, the bees ceased to excavate, and began to build up flat walls of wax on the lines of intersection between the basins, so that each hexagonal prism was built upon the festooned edge of a smooth basin, instead of on the straight edges of a three-sided pyramid as in the case of ordinary cells.

I then put into the hive, instead of a thick, square piece of wax, a thin and narrow, knife-edged ridge, coloured with vermilion. The bees instantly began on both sides to excavate little basins near to each other, in the same way as before; but the ridge of wax was so thin, that the bottoms of the basins, if they had been excavated to the same depth as in the former experiment, would have broken into each other from the opposite sides. The bees, however, did not suffer this to happen, and they stopped their excavations in due time; so that the basins, as soon as they had been a little deepened, came to have flat bottoms; and these flat bottoms, formed by thin little plates of the vermilion wax having been left ungnawed, were situated, as far as the eye could judge, exactly along the planes of imaginary intersection between the basins on the opposite sides of the ridge of wax. In parts, only little bits, in other parts, large portions of a rhombic plate had been left between the opposed basins, but the work, from the unnatural state of things, had not been neatly performed. The bees must have worked at very nearly the same rate on the opposite sides of the ridge of vermilion wax, as they circularly gnawed away and deepened the basins on both sides, in order to have succeeded in thus leaving flat plates between the basins, by stopping work along the intermediate planes or planes of intersection.

Considering how flexible thin wax is, I do not see that there is any difficulty in the bees, whilst at work on the two sides of a strip of wax, perceiving when they have gnawed the wax away to the proper thinness, and then stopping their work. In ordinary combs it has appeared to me that the bees do not always succeed in working at exactly the same rate from the opposite sides; for I have noticed half-completed rhombs at the base of a just-commenced cell, which were slightly concave on one side, where I suppose that the bees had excavated too quickly, and convex on the opposed side, where the bees had worked less quickly. In one well-marked instance, I put the comb back into the hive, and allowed the bees to go on working for a short time, and again examined the cell, and I found that the rhombic plate had been completed, and had become *perfectly flat*: it was absolutely impossible, from the extreme thinness of the little rhombic plate, that they could have effected this by gnawing away the convex side; and I suspect that the bees in such cases stand in the opposed cells and push and bend the ductile and warm wax (which as I have tried is easily done) into its proper intermediate plane, and thus flatten it.

From the experiment of the ridge of vermilion wax, we can clearly see that if the bees were to build for themselves a thin wall of wax, they could make their cells of the proper shape, by standing at the proper distance from each other, by excavating at the same rate, and by endeavouring to make equal spherical hollows, but never allowing the spheres to break into each other. Now bees, as may be clearly seen by examining the edge of a growing comb, do make a rough, circumferential wall or rim all round the comb; and they gnaw into this from the opposite sides, always working circularly as they deepen each cell. They do not make the whole three-sided pyramidal base of any one cell at the same time, but only the one rhombic plate which stands on the extreme growing margin, or the two plates, as the case may be; and they never complete the upper edges of the rhombic plates, until the hexagonal walls are commenced. Some of these statements differ from those made by the justly celebrated elder Huber, but I am convinced of their accuracy; and if I had space, I could show that they are conformable with my theory.

Huber's statement that the very first cell is excavated out of a little parallel-sided wall of wax, is not, as far as I have seen, strictly correct; the first commencement having always been a little hood of wax; but I will not here enter on these details. We see how important a part excavation plays in the construction of the cells; but it would be a great error to suppose that the bees cannot build up a rough wall of wax in the proper position—that is, along the plane of intersection between two adjoining spheres. I have several specimens showing clearly that they can do this. Even in the rude circumferential rim or wall of wax round a growing comb, flexures may sometimes be observed, corresponding in position to the planes of the rhombic basal plates of future cells. But the rough wall of wax has in every case to be finished off, by being largely gnawed away on both sides. The manner in which the bees build is curious; they always make the first rough wall from ten to twenty times thicker than the excessively thin finished wall of the cell, which will ultimately be left. We shall understand how they work, by supposing masons first to pile up a broad ridge of cement, and then to begin cutting it away equally on both sides near the ground, till a smooth, very thin wall is left in the middle; the masons always piling up the cut-away cement, and adding fresh cement, on the summit of the ridge. We shall thus have a thin wall steadily growing upward; but always crowned by a gigantic coping. From all the cells, both those just commenced and those completed, being thus crowned by a strong coping of wax, the bees can cluster and crawl over the comb without injuring the delicate hexagonal walls, which are only about one four-hundredth of an inch in thickness; the plates of the pyramidal basis being about twice as thick. By this singular manner of building, strength is continually given to the comb, with the utmost ultimate economy of wax.

It seems at first to add to the difficulty of understanding how the cells are made, that a multitude of bees all work together; one bee after working a short time at one cell going to another, so that, as Huber has stated, a score of individuals work even at the commencement of the first cell. I was able practically to show this fact, by covering the edges of the hexagonal walls of a single cell, or the extreme margin of the circumferential rim of a growing comb, with an extremely thin layer of melted vermilion wax; and I invariably found that the colour was most delicately diffused by the bees —as delicately as a painter could have done with his brush—by atoms of the coloured wax having been taken from the spot on which it had been placed, and worked into the growing edges of the cells all round. The work of construction seems to be a sort of balance struck between many bees, all instinctively standing at the same relative distance from each other, all trying to sweep equal spheres, and then building up, or leaving ungnawed, the planes of intersection between these spheres. It was really curious to note in cases of difficulty, as when two pieces of comb met at an angle, how often the bees would entirely pull down and rebuild in different ways the same cell, sometimes recurring to a shape which they had at first rejected.

When bees have a place on which they can stand in their proper positions for working,—for instance, on a slip of wood, placed directly under the middle of a comb growing downwards so that the comb has to be built over one face of the slip—in this case the bees can lay the foundations of one wall of a new hexagon, in its strictly proper place, projecting beyond the other completed cells. It suffices that the bees should be enabled to stand at their proper relative distances from each other and from the walls of the last completed cells, and then, by striking imaginary spheres, they can build up a wall intermediate between two adjoining spheres; but, as far as I have seen, they never gnaw away and finish off the angles of a cell till a large part both of that cell and of the adjoining cells has been built. This capacity in bees of laying down under certain circumstances a rough wall in its proper place between two just-commenced cells, is important, as it bears on a fact, which seems at first quite subversive of the foregoing theory; namely, that the cells on the extreme margin of wasp-combs are sometimes strictly hexagonal; but I have not space here to enter on this subject. Nor does there seem to me any great difficulty in a single insect (as in the case of a queen-wasp) making hexagonal cells, if she work alternately on the inside and outside of two or three cells commenced at the same time, always standing at the proper relative distance from the parts of the cells just begun, sweeping spheres or cylinders, and building up intermediate planes. It is even conceivable that an insect might, by fixing on a point at which to commence a cell, and then moving outside, first to one point, and then to five other points, at the proper relative distances from the central point and from each other, strike the planes of intersection, and so make an isolated hexagon: but I am not aware that any such case has been observed; nor would any good be derived from a single hexagon being built, as in its construction more materials would be required than for a cylinder.

As natural selection acts only by the accumulation of slight modifications of structure or instinct, each profitable to the individual under its conditions of life, it may reasonably be asked, how a long and graduated succession of modified architectural instincts, all tending towards the present perfect plan of construction, could have profited the progenitors of the hive-bee? I think the answer is not difficult: it is known that bees are often hard pressed to get sufficient nectar; and I am informed by Mr. Tegetmeier that it has been experimentally found that no less than from twelve to fifteen pounds of dry sugar are consumed by a hive of bees for the secretion of each pound of wax; so that a prodigious quantity of fluid nectar must be collected and consumed by the bees in a hive for the secretion of the wax necessary for the construction of their combs. Moreover, many bees have to remain idle for many days during the process of secretion. A large store of honey is indispensable to support a large stock of bees during the winter; and the security of the hive is known mainly to depend on a large number of bees being supported. Hence the saving of wax by largely saving honey must be a most important element of success in any family of bees. Of course the success of any species of bee may be dependent on the number of its parasites or other enemies, or on quite distinct causes, and so be altogether independent of the quantity of honey which the bees could collect. But let us suppose that this latter circumstance determined, as it probably often does determine, the numbers of a humble-bee which could exist in a country; and let us further suppose that the community lived throughout the winter, and consequently required a store of honey: there can in this case be no doubt that it would be an advantage to our humble-bee, if a slight modification of her instinct led her to make her waxen cells near together, so as to intersect a little; for a wall in common even to two adjoining cells, would save some little wax. Hence it would continually be more and more advantageous to our humble-bee, if she were to make her cells more and more regular, nearer together, and aggregated into a mass, like the cells of the Melipona; for in this case a large part of the bounding surface of each cell would serve to bound other cells, and much wax would be saved. Again, from the same cause, it would be advantageous to the Melipona, if she were to make her cells closer together, and more regular in every way than at present; for then, as we have seen, the spherical surfaces would wholly disappear, and would all be replaced by plane surfaces; and the Melipona would make a comb as perfect as that of the hive-bee. Beyond this stage of perfection in architecture, natural selection could not lead; for the comb of the hive-bee, as far as we can see, is absolutely perfect in economising wax.

Thus, as I believe, the most wonderful of all known instincts, that of the hive-bee, can be explained by natural selection having taken advantage of numerous, successive, slight modifications of simpler instincts; natural selection having by slow degrees, more and more perfectly, led the bees to sweep equal spheres at a given distance from each other in a double layer, and to build up and excavate the wax along the planes of intersection. The bees, of course, no more knowing that they swept their spheres at one particular distance from each other, than they know what are the several angles of the hexagonal prisms and of the basal rhombic plates. The motive power of the process of natural selection having been economy of wax; that individual swarm which wasted least honey in the secretion of wax, having succeeded best, and having transmitted by inheritance its newly acquired economical instinct to new swarms, which in their turn will have had the best chance of succeeding in the struggle for existence.

No doubt many instincts of very difficult explanation could be opposed to the theory of natural selection, —cases, in which we cannot see how an instinct could possibly have originated; cases, in which no intermediate gradations are known to exist; cases of instinct of apparently such trifling importance, that they could hardly have been acted on by natural selection; cases of instincts almost identically the same in animals so remote in the scale of nature, that we cannot account for their similarity by inheritance from a common parent, and must therefore believe that they have been acquired by independent acts of natural selection. I will not here enter on these several cases, but will confine myself to one special difficulty, which at first appeared to me insuperable, and actually fatal to my whole theory. I allude to the neuters or sterile females in insect-communities: for these neuters often differ widely in instinct and in structure from both the males and fertile females, and yet, from being sterile, they cannot propagate their kind.

The subject well deserves to be discussed at great length, but I will here take only a single case, that of working or sterile ants. How the workers have been rendered sterile is a difficulty; but not much greater than that of any other striking modification of structure; for it can be shown that some insects and other articulate animals in a state of nature occasionally become sterile; and if such insects had been social, and it had been profitable to the community that a number should have been annually born capable of work, but incapable of procreation, I can see no very great difficulty in this being effected by natural selection. But I must pass over this preliminary difficulty. The great difficulty lies in the working ants differing widely from both the males and the fertile females in structure, as in the shape of the thorax and in being destitute of wings and sometimes of eyes, and in instinct. As far as instinct alone is concerned, the prodigious difference in this respect between the workers and the perfect females, would have been far better exemplified by the hive-bee. If a working ant or other neuter insect had been an animal in the ordinary state, I should have unhesitatingly assumed that all its characters had been slowly acquired through natural selection; namely, by an individual having been born with some slight profitable modification of structure, this being inherited by its offspring, which again varied and were again selected, and so onwards. But with the working ant we have an insect differing greatly from its parents, yet absolutely sterile; so that it could never have transmitted successively acquired modifications of structure or instinct to its progeny. It may well be asked how is it possible to reconcile this case with the theory of natural selection?

First, let it be remembered that we have innumerable instances, both in our domestic productions and in those in a state of nature, of all sorts of differences of structure which have become correlated to certain ages, and to either sex. We have differences correlated not only to one sex, but to that short period alone when the reproductive system is active, as in the nuptial plumage of many birds, and in the hooked jaws of the male salmon. We have even slight differences in the horns of different breeds of cattle in relation to an artificially imperfect state of the male sex; for oxen of certain breeds have longer horns than in other breeds, in comparison with the horns of the bulls or cows of these same breeds. Hence I can see no real difficulty in any character having become correlated with the sterile condition of certain members of insect-communities: the difficulty lies in understanding how such correlated modifications of structure could have been slowly accumulated by natural selection.

This difficulty, though appearing insuperable, is lessened, or, as I believe, disappears, when it is remembered that selection may be applied to the family, as well as to the individual, and may thus gain the desired end. Thus, a well-flavoured vegetable is cooked, and the individual is destroyed; but the horticulturist sows seeds of the same stock, and confidently expects to get nearly the same variety; breeders of cattle wish the flesh and fat to be well marbled together; the animal has been slaughtered, but the breeder goes with confidence to the same family. I have such faith in the powers of selection, that I do not doubt that a breed of cattle, always yielding oxen with extraordinarily long horns, could be slowly formed by carefully watching which individual bulls and cows, when matched, produced oxen with the longest horns; and yet no one ox could ever have propagated its kind. Thus I believe it has been with social insects: a slight modification of structure, or instinct, correlated with the sterile condition of certain members of the community, has been advantageous to the community: consequently the fertile males and females of the same community flourished, and transmitted to their fertile offspring a tendency to produce sterile members having the same modification. And I believe that this process has been repeated, until that prodigious amount of difference between the fertile and sterile females of the same species has been produced, which we see in many social insects.

But we have not as yet touched on the climax of the difficulty; namely, the fact that the neuters of several ants differ, not only from the fertile females and males, but from each other, sometimes to an almost incredible degree, and are thus divided into two or even three castes. The castes, moreover, do not generally graduate into each other, but are perfectly well defined; being as distinct from each other, as are any two species of the same genus, or rather as any two genera of the same family. Thus in Eciton, there are working and soldier neuters, with jaws and instincts extraordinarily different: in Cryptocerus, the workers of one caste alone carry a wonderful sort of shield on their heads, the use of which is quite unknown: in the Mexican Myrmecocystus, the workers of one caste never leave the nest; they are fed by the workers of another caste, and they have an enormously developed abdomen which secretes a sort of honey, supplying the place of that excreted by the aphides, or the domestic cattle as they may be called, which our European ants guard or imprison.

It will indeed be thought that I have an overweening confidence in the principle of natural selection, when I do not admit that such wonderful and well-established facts at once annihilate my theory. In the simpler case of neuter insects all of one caste or of the same kind, which have been rendered by natural selection, as I believe to be quite possible, different from the fertile males and females,—in this case, we may safely conclude from the analogy of ordinary variations, that each successive, slight, profitable modification did not probably at first appear in all the individual neuters in the same nest, but in a few alone; and that by the long-continued selection of the fertile parents which produced most neuters with the profitable modification, all the neuters ultimately came to have the desired character. On this view we ought occasionally to find neuter-insects of the same species, in the same nest, presenting gradations of structure; and this we do find, even often, considering how few neuter-insects out of Europe have been carefully examined. Mr. F. Smith has shown how surprisingly the neuters of several British ants differ from each other in size and sometimes in colour; and that the extreme forms can sometimes be perfectly linked together by individuals taken out of the same nest: I have myself compared perfect gradations of this kind. It often happens that the larger or the smaller sized workers are the most numerous; or that both large and small are numerous, with those of an intermediate size scanty in numbers. Formica flava has larger and smaller workers, with some of intermediate size; and, in this species, as Mr. F. Smith has observed, the larger workers have simple eyes (ocelli), which though small can be plainly distinguished, whereas the smaller workers have their ocelli rudimentary. Having carefully dissected several specimens of these workers, I can affirm that the eyes are far more rudimentary in the smaller workers than can be accounted for merely by their proportionally lesser size; and I fully believe, though I dare not assert so positively, that the workers of intermediate size have their ocelli in an exactly intermediate condition. So that we here have two bodies of sterile workers in the same nest, differing not only in size, but in their organs of vision, yet connected by some few members in an intermediate condition. I may digress by adding, that if the smaller workers had been the most useful to the community, and those males and females had been continually selected, which produced more and more of the smaller workers, until all the workers had come to be in this condition; we should then have had a species of ant with neuters very nearly in the same condition with those of Myrmica. For the workers of Myrmica have not even rudiments of ocelli, though the male and female ants of this genus have well-developed ocelli.

I may give one other case: so confidently did I expect to find gradations in important points of structure between the different castes of neuters in the same species, that I gladly availed myself of Mr. F. Smith's offer of numerous specimens from the same nest of the driver ant (Anomma) of West Africa. The reader will perhaps best appreciate the amount of difference in these workers, by my giving not the actual measurements, but a strictly accurate illustration: the difference was the same as if we were to see a set of workmen building a house of whom many were five feet four inches high, and many sixteen feet high; but we must suppose that the larger workmen had heads four instead of three times as big as those of the smaller men, and jaws nearly five times as big. The jaws, moreover, of the working ants of the several sizes differed wonderfully in shape, and in the form and number of the teeth. But the important fact for us is, that though the workers can be grouped into castes of different sizes, yet they graduate insensibly into each other, as does the widely-different structure of their jaws. I speak confidently on this latter point, as Mr. Lubbock made drawings for me with the camera lucida of the jaws which I had dissected from the workers of the several sizes.

With these facts before me, I believe that natural selection, by acting on the fertile parents, could form a species which should regularly produce neuters, either all of large size with one form of jaw, or all of small size with jaws having a widely different structure; or lastly, and this is our climax of difficulty, one set of workers of one size and structure, and simultaneously another set of workers of a different size and structure; —a graduated series having been first formed, as in the case of the driver ant, and then the extreme forms, from being the most useful to the community, having been produced in greater and greater numbers through the natural selection of the parents which generated them; until none with an intermediate structure were produced.

Thus, as I believe, the wonderful fact of two distinctly defined castes of sterile workers existing in the same nest, both widely different from each other and from their parents, has originated. We can see how useful their production may have been to a social community of insects, on the same principle that the division of labour is useful to civilised man. As ants work by inherited instincts and by inherited tools or weapons, and not by acquired knowledge and manufactured instruments, a perfect division of labour could be effected with them only by the workers being sterile; for had they been fertile, they would have intercrossed, and their instincts and structure would have become blended. And nature has, as I believe, effected this admirable division of labour in the communities of ants, by the means of natural selection. But I am bound to confess, that, with all my faith in this principle, I should never have anticipated that natural selection could have been efficient in so high a degree, had not the case of these neuter insects convinced me of the fact. I have, therefore, discussed this case, at some little but wholly insufficient length, in order to show the power of natural selection, and likewise because this is by far the most serious special difficulty, which my theory has encountered. The case, also, is very interesting, as it proves that with animals, as with plants, any amount of modification in structure can be effected by the accumulation of numerous, slight, and as we must call them accidental, variations, which are in any manner profitable, without exercise or habit having come into play. For no amount of exercise, or habit, or volition, in the utterly sterile members of a community could possibly have affected the structure or instincts of the fertile members, which alone leave descendants. I am surprised that no one has advanced this demonstrative case of neuter insects, against the well-known doctrine of Lamarck.

Summary.—I have endeavoured briefly in this chapter to show that the mental qualities of our domestic animals vary, and that the variations are inherited. Still more briefly I have attempted to show that instincts vary slightly in a state of nature. No one will dispute that instincts are of the highest importance to each animal. Therefore I can see no difficulty, under changing conditions of life, in natural selection accumulating slight modifications of instinct to any extent, in any useful direction. In some cases habit or use and disuse have probably come into play. I do not pretend that the facts given in this chapter strengthen in any great degree my theory; but none of the cases of difficulty, to the best of my judgment, annihilate it. On the other hand, the fact that instincts are not always absolutely perfect and are liable to mistakes;—that no instinct has been produced for the exclusive good of other animals, but that each animal takes advantage of the instincts of others;—that the canon in natural history, of "natura non facit saltum" is applicable to instincts as well as to corporeal structure, and is plainly explicable on the foregoing views, but is otherwise inexplicable,—all tend to corroborate the theory of natural selection.

This theory is, also, strengthened by some few other facts in regard to instincts; as by that common case of closely allied, but certainly distinct, species, when inhabiting distant parts of the world and living under considerably different conditions of life, yet often retaining nearly the same instincts. For instance, we can understand on the principle of inheritance, how it is that the thrush of South America lines its nest with mud, in the same peculiar manner as does our British thrush: how it is that the male wrens (Troglodytes) of North America, build "cock-nests," to roost in, like the males of our distinct Kitty-wrens,—a habit wholly unlike that of any other known bird. Finally, it may not be a logical deduction, but to my imagination it is far more satisfactory to look at such instincts as the young cuckoo ejecting its foster-brothers,—ants making slaves, —the larvæ of ichneumonidæ feeding within the live bodies of caterpillars,—not as specially endowed or created instincts, but as small consequences of one general law, leading to the advancement of all organic beings, namely, multiply, vary, let the strongest live and the weakest die.

3
FIGHTING OF MALES

Bernard Altum

This article was translated expressly for this Benchmark volume by Chauncey J. Mellor and Doris Gove, The University of Tennessee, Knoxville, from pp. 92–98 of Der Vogel und sein Leben, *Verlag von Wilhelm Niemann, Münster, 1868, pp. 92–98. Translation © 1985 by Chauncey J. Mellor and Doris Gove.*

In the foregoing section, we learned that the singing of birds, or in general, their mating call and other things, has two purposes. Not only does it reveal to the females, who only react to the voices of conspecific males, the often well hidden location of the male from a considerable distance, but the singing also serves as a mutual signal for the males so that the distances between nests, which are entirely necessary for many bird species, will be established. Consequently, the requisite size of the respective brood areas will also be established, since the males that approach each other too closely battle each other in *embittered struggle* until the one party has withdrawn to an appropriate distance. Likewise, it was also suggested above that the infringement of the bird on surrounding nature, necessary for the control of the rest of the animal and plant world, is in this way purposefully distributed. As for the purposefulness of such distribution of our song birds, I could especially call attention to the effort that most seed eaters exert against the proliferation of harmful insects by the nourishment of the young that must be fed. For it is precisely these birds, such as chaffinches [*Fringilla coelebs*], linnets [*Carduelis cannabina*], buntings [Genus *Emberiza*], which themselves live almost exclusively on seeds, that feed their young almost just as exclusively with tender insects, especially the larvae, such as caterpillars. Such an ordered, relatively uniform distribution of broods in the entire area is extraordinarily wise and calculated. The birds require an astonishing mass of this nourishment precisely at the time when the harmful insects develop in such monstrous quantities. In the late summer and in autumn, by contrast, when the insect world is not so energetically active and attacking, the grown-up young eat, as do the adults, seeds of which such an overabundance has matured that Nature must give earnest consideration to their diminution. Of the local seed eaters, I know only two that make an exception in the insect nourishment of their young, the greenfinch [*Carduelis chloris*] and the bullfinch [*Pyrrhula pyrrhula*]. All others generally feed with insects. Among these, the much pampered sparrows probably make themselves the least useful.

What has been said, however, does not hold for all bird species. Some—as is well known, it is those which fly far for their food (swallows [Family Hirundinidae], swifts [Genus *Apus*], jackdaws [*Corrus monedula*], etc.) as well as the omnivores (sparrows [Genus *Passer*])—do not need to keep to a specifically delineated brood area. They brood socially. And yet, among the male individuals, a more or less very embittered struggle flares up at the beginning of reproductive activity. In addition, we find this feud often raging even in those birds that immediately run about with their newly hatched young (chickens, many swamp and swimming birds). These birds are therefore not at all bound for an extended period of time to the limited breeding place, the fixed territory. This territory would have to nourish them to the extent that it would be saved by them from the devastating attack of the insects. This battle therefore—if our teleological conception contains the whole truth—must be based on another necessity of nature in addition to the already cited reasons. It is well known that the progeny of those animals that reproduce among themselves in a small group over many generations gradually degenerate. Marriages of people too closely related are forbidden not merely by wise laws of the church, but also, as is well known, by laws of the state. The flaws of the adult animals, family flaws as it were, are passed on to the young and are summed up in them. For that reason, our animal breeders are constantly compelled to freshen up their breeds by crosses with others. For this purpose, a sheep breeder obtains a ram from, say, the acclimatization garden in Paris and gladly pays 500 or even 800 francs for a single animal. Governments and agricultural organizations take a very active interest in caring in the customary manner for the healthy, vigorous status of livestock in their sphere of influence. But how are precautions against this gradual weakening taken out in open nature, in which neither church nor state laws nor knowledgeable animal breeders concern themselves with maintaining healthy, vigorous offspring? Even here selection for reproduction must be made. Nothing short of the strongest, healthiest, heartiest males must be selected, and for this purpose, there is only a single means which assuredly leads to this goal, and that is the just mentioned mutual *battle of adult males* at the beginning of the reproductive period. In some, the so-called polygamous birds, for example, some chicken-like species, like capercaillie [*Tetrao urogallus*], black grouse [*Lyrurus tetrix*] and domestic chicken, it is well known that a single cock suffices for many hens, as a male red deer [*Cervus elaphus*] does for many females. Since therefore a single rooster fertilizes many hens, and thus the progeny in a large area is dependent on him, it is obviously extremely important that the healthiest and strongest of all available roosters is chosen for the designated purpose. And this can only occur through mutual, highly embittered battle. Just as the red deer stag calls others of his kind to the battle arena, just as in the end all the younger sickly weaker ones are defeated, and only the strongest, most powerful one is used for propagation, so also does the healthiest best rooster remain the progenitor of a vigorous posterity in his courtship district. In the birds living in individual pairs, the matter basically is exactly the same as in the just-mentioned cases. There are many more males than females in the birds, as far as my observations extend. For our small birds, warblers [Family Sylviidae], flycatchers [Genera *Muscicapa* and *Ficedula*], finches [Family Fringillidae], buntings [Genus *Emberiza*], larks [Family Alaudidae], thrushes [Family Turdidae] and similar birds, this fact is certain. If a bird catcher has captured the male near a nest, another male in turn finds his way there often after an extraordinarily short time. If this second bird can be caught, it is still easily replaced by a third and this maybe by a fourth and fifth. A female never remains without a male, but many males are without females. I do not doubt that a similar numerical ratio for both sexes exists for all bird groups. The reason for such an unequal ratio is this according to my investigations: the numerous young of the first broods are almost all males, the less numerous of the late ones almost all females. If for example, a pair of the above-mentioned species breeds three times in the summer, the first nest holds on the average five young, of which four are males and one is female, the second nest has four young, two males and two females, and the third nest three young with one male and two females. So of the twelve young of a pair, seven are male and five are female. But even when the pair raises only one extra male per year, as happens in many cases, the excess of males in the course of several years rises noticeably. Since the young of the last brood often waste away and do not survive, the imbalance as a rule is still greater. If the pair breeds only once, then from five eggs there result usually three males and two females. Consequently, even for birds living in so-called monogamy, not all males can achieve reproduction. It lies in the interest of the balance of nature in turn, that these be the weaker individuals. Therefore the selection of the most vigorous ones must be undertaken, and again, the only thing serving this goal is mutual battle. Thus we see most clearly that this battle of the males "for the females," which is pursued by the otherwise quite peaceloving birds with great energy and embitterment almost to the death, establishes itself as an extremely wise and indeed necessary measure not merely for the determination of borders of brooding territories, as we already learned, but also for the continued maintenance of a progeny undiminished in vigor. The bird knows nothing of all this. He fights because he must. He is acting in the service of a higher call. How childlike the anthropomorphic, extremely cheap turns of phrase of our sentimental animal psychologists appear against such deeper natural reasons! For them, "jealousy, rivalry, love," understood of course entirely in the purely human sense, are the real actuating forces. Taking the matter in the precise sense, even the expression "battle of the males for the females" is incorrect. The males battle to determine the breeding territory size, which is not recognized by them as a necessary condition of life, and for the selection of the healthiest animals for reproduction and for nothing else. No further words are needed to say that these three things are necessarily products of intent: the forming of pairs, the undisturbed cohabitation, and the cooperation of the adult birds. But let it be repeated, that the birds themselves intend nothing, do not struggle consciously for anything, do not desire the unchallenged possession of the females and do not intentionally attempt to obtain them by struggle and effort. As pure creatures of nature, they act only according to completely necessary and strict laws of life. Actually, they do not themselves act, but are incited to quite specific expressions of life according to higher laws.

At this point, we can raise the question, *why* after all *more males are produced than can be used for purposes of reproduction*, why, on the contrary, both sexes do not appear in approximately equal numbers. We can answer at the outset that the purpose of constantly keeping the individual animals strong is achieved far more certainly by nature if it can select the healthiest and strongest from a greater number than if it had to confine its choice absolutely to the available material. In the latter case no weak, sickly male would exist without detriment to the generation—an absurd demand. Then too we have already mentioned that only the purposeful distribution of the nest sites exerts the necessary check by the birds on their natural surroundings, on the plants and animals, that only in this way is injurious overgrowing and inundation by overgrowth or destruction or being destroyed prevented, that only in this way is the harmonious condition of the whole assured. This distribution is, as we already know, not the same everywhere according to the productivity of the locale. The nests of the same species stand closer together at one site than at another. In a third, there is only rarely one of a certain bird species. The first place is so fruitful in certain insects and other nourishment that a small brooding territory size suffices to nourish the entire family. The second yields less, so that a greater radius must be picked over. The third is even more parsimonious with its gifts. In addition, there are places where a pair of birds with, say, five young can no longer live, if the adults neither want to nor can make long excursions. There a nest can not and must not be located. One example might explain this. Our firecrest [*Regulus ignicapillus*] is, in its reproductive period, exclusively a Norway spruce [*Picea abies*] bird. It is found only where Norway spruce grows. Where there is a rather large stand of this evergreen species, several nests are found at an interval of 100–200 paces. If the Norway spruce plot is small, if this tree species stands only as a smaller group of 10–20 rather large individuals in a park (even 6–8 suffice), then regularly a pair, but only one pair, of these nice little birds settles down there. In such an area, the firecrest nests are located as far apart as the Norway spruce stands, and in our area we often must hike for an hour and more before we reach Norway spruces again and with them the bird under discussion. But a single tree, or two or three of them, are not enough for a breeding pair, and none settles there. Now if the birds are prime factors in the necessary limitation of the insect world, if they must intervene deeply in the organic development of nature, then their policing action is also essential even where an entire family cannot live. Even a single standing Norway spruce cannot be exposed to the power of its enemies without any protection, even here an occasional firecrest must drop in to pick off the vermin. For that reason, it is fully founded in the economic plan of nature that many individual birds are not bound to house, home and cradle; that they are free and *can be freely directed to all endangered places lying outside the occupied breeding area*. And that is not possible when the one sex is not present in the majority. Having been driven out of all breeding territories, the weaker males move about therefore in the surrounding region. They must not approach a family male too closely if they do not wish immediately to experience the physical consequences of his exercise of domestic rights against them. For this reason, they keep their distance, glean off the territories on the periphery as well as the places lying outside the territories; in brief, they glean off that to which the activity of the breeding pairs no longer extends. Can one imagine a wiser, more purposeful organization? But if some accident or another befalls a breeding male, then the weaker ones immediately take its place, a fact which I could prove by a hundred experiences, and consequently form a likewise very purposefully established reserve fund for the preservation of the brood. Of all the purposeful qualities, indeed of this necessary interlocking of all individual natural phenomena and expressions of life, the individual animal itself knows nothing. It lives and acts only in its narrow circle, constituting as it were a small cog in the driving mechanism of the whole. But the entire thing is set up according to a general higher plan.

4

Reprinted from *Nature* **6:**485–486 (1872)

ON INSTINCT

D. A. Spalding

WITH regard to instinct we have yet to ascertain the facts. Do the animals exhibit untaught skill and innate knowledge? May not the supposed examples of instinct be after all but the results of rapid learning and imitation? The controversy on this subject has been chiefly concerning the perceptions of distance and direction by the eye and the ear. Against the instinctive character of these perceptions it is argued that, as distance means movement, locomotion, the very essence of the idea is such as cannot be taken in by the eye or ear; that what the varying sensations of sight and hearing correspond to, must be got at by moving over the ground by experience. The results, however, of experiments on chickens were wholly in favour of the instinctive nature of these perceptions. Chickens kept in a state of blindness by various devices, from one to three days, when placed in the light under a set of carefully prepared conditions, gave conclusive evidence against the theory that the perceptions of distance and direction by the eye are the result of associations formed in the experience of each individual life. Often, at the end of two minutes, they followed with their eyes the movements of crawling insects, turning their heads with all the precision of an old fowl. In from two to fifteen minutes they pecked at some object, showing not merely an instinctive perception of distance, but an original ability to measure distance with something like infallible accuracy. If beyond the reach of their necks, they walked or ran up to the object of their pursuit, and may be said to have invariably struck it, never missing by more than a hair's-breadth; this, too, when the specks at which they struck were no bigger than the smallest visible dot of an *i*. To seize between the points of the mandible at the very instant of striking seemed a more difficult operation. Though at times they seized and swallowed an insect at the first attempt, more frequently they struck five or six times, lifting once or twice before they succeeded in swallowing their first food. To take, by way of illustration, the observations on a single case a little in detail:—A chicken at the end of six minutes, after having its eyes unveiled, followed with its head the movements of a fly twelve inches distant; at ten minutes, the fly coming within reach of its neck, was seized and swallowed at the first stroke; at the end of twenty minutes it had not attempted to walk a step. It was then placed on rough ground within sight and call of a hen, with chickens of its own age. After standing chirping for about a minute, it went straight towards the hen, displaying as keen a perception of the qualities of the outer world as it was ever likely to possess in after life. It never required to knock its head against a stone to discover that there was "no road that way." It leaped over the smaller obstacles that lay in its path, and ran round the larger, reaching the mother in as nearly a straight line as the nature of the ground would permit. Thus it would seem that, prior to experience, the eye—at least the eye of the chicken—perceives the primary qualities of the external world, all arguments of the purely analytical school of psychology to the contrary, notwithstanding.

Not less decisive were experiments on hearing. Chickens hatched and kept in the dark for a day or two, on being placed in the light nine or ten feet from a box in which a brooding hen was concealed, after standing chirping for a minute or two, uniformly set off straight to the box in answer to the call of the hen which they had never seen and never before heard. This they did struggling through grass and over rough ground, when not able to stand steadily on their legs. Again, chickens that from the first had been denied the use of their eyes by having hoods drawn over their heads while yet in the shell, were while thus blind made the subject of experiment. These, when left to themselves, seldom made a forward step, their movements were round and round and backward; but when placed within five or six feet of the hen mother, they, in answer to her call, became much more lively, began to make little forward journeys, and soon followed her by sound alone, though of course blindly. Another experiment consisted in rendering chickens deaf for a time by sealing their ears with several folds of gum paper before they had escaped from the shell. These, on having their ears opened when two or three days old, and being placed within call of the mother concealed in a box or on the other side of a door, after turning round a few times ran straight to the spot whence came the first sound they had ever heard. Clearly, of these chickens it cannot be said that sounds were to them at first but meaningless sensations.

One or two observations favourable to the opinion that animals have an instinctive knowledge of their enemies may be taken for what they are worth. When twelve days old one of my little *protégés* running about beside me, gave the peculiar chirp whereby they announce the approach of danger. On looking up, a sparrow-hawk was seen hovering at a great height over head. Again, a young hawk was made to fly over a hen with her first brood of chickens, then about a week old. In the twinkling of an eye most of the chickens were hid among grass and bushes. And scarcely had the hawk touched the ground, about twelve yards from where the hen had been sitting, when she fell upon it, and would soon have killed it outright. A young turkey gave even more striking evidence. When ten days old it heard the voice of the hawk for the first time, and just beside it. Like an arrow from the bow it darted off in the opposite direction, and crouched in a corner, remained for ten minutes motionless and dumb with fear. Out of a vast number of experiments with chickens and bees, though the results were not uniform, yet in the great majority of instances the chickens gave evidence of instinctive fear of these sting-bearing insects.

But to return to examples of instinctive skill and knowledge, concerning which I think no doubt can remain, a very useful instinct may be observed in the early attention that chickens pay to their toilet. As soon as they can hold up their heads, when only from four to five hours old, they attempt dressing at their wings, that, too, when they have been denied the use of their eyes. Another incontestable case of instinct may be seen in the art of scraping in search of food. Without any opportunities of imitation, chickens begin to scrape when from two to six days old. Most frequently the circumstances are suggestive; at other times, however, the first attempt, which generally consists of a sort of nervous dance, was made on a smooth table. The unacquired dexterity shown in the capture of insects is very remarkable. A duckling one day old, on being placed in the open air for the first time, almost immediately snapped at, and caught, a fly on the wing. Still more interesting is the instructive art of catching flies peculiar to the turkey. When not a day and a half old I observed a young turkey, which I had adopted while yet in the shell, pointing its beak slowly and deliberately at flies and other small insects without actually pecking at them. In doing this its head could be seen to shake like a hand that is attempted to be held steady by a visible effort. This I recorded when I did not understand its meaning. For it was not until afterwards that I observed a turkey, when it sees a fly settled on any object, steals on the unwary insect with slow and measured step, and, when sufficiently near, advances its head very slowly and steadily until within reach of its prey, which is then seized by a sudden dart. In still further confirmation of the opinion that such wonderful examples of dexterity and cunning are instinctive and not acquired, may be adduced the significant fact that the individuals of each species have little capacity to learn anything not found in the habits of their progenitors. A chicken was made, from the first and for several months, the sole companion of a young turkey. Yet it never showed the slightest tendency to adopt the admirable art of catching flies that it saw practised before its eyes every hour of the day.

The only theory in explanation of the phenomena of instinct that has an air of science about it, is the doctrine of Inherited Association. Instinct in the present generation of animals is the product of the accumulated experiences of past generations. Great difficulty, however, is felt by many in conceiving how anything so impalpable as fear at the sight of a bee should be transmitted from parent to offspring. It should be remembered, however, that the permanence of such associations in the history of an individual life depends on the corresponding impress given to the nervous organisation. We cannot, strictly speaking, experience any individual act of consciousness twice over; but as, by pulling the bell-cord to-day we can, in the language of ordinary discourse, produce the same sound we heard yesterday, so, while the established connections among the nerves and nerve-centres hold, we are enabled to live our experiences over again. Now, why should not those modifications of brain-matter, that, enduring from hour to hour and from day to day, render acquisition possible, be, like any other physical peculiarity, transmitted from parent to offspring? That they are so transmitted is all but proved by the facts of instinct, while these, in their turn, receive their only rational explanation in this theory of Inherited Association.

5

Reprinted from *Nature* **12:**507-508 (1875)

INSTINCT AND ACQUISITION

D. A. Spalding

SO great was the influence of that school of psychology which maintained that we and all other animals had to acquire in the course of our individual lives all the knowledge and skill necessary for our preservation, that many of the very greatest authorities in science refused to believe in those instructive performances of young animals about which the less learned multitude have never had any doubt. For example, Helmholtz, than whom there is not, perhaps, any higher scientific authority, says: "The young chicken very soon pecks at grains of corn, but it pecked while it was still in the shell, and when it hears the hen peck, it pecks again, at first seemingly at random. Then, when it has by chance hit upon a grain, it may, no doubt, learn to notice the field of vision which is at the moment presented to it."

At the meeting of this Association in 1872, I gave a pretty full account of the behaviour of the chicken after its escape from the shell. The facts observed were conclusive against the individual-experience psychology. And they have, as far as I am aware, been received by scientific men without question. I would now add that not only does the chick not require to learn to peck at, to seize, and to swallow small specks of food, but that it is not a fact, as asserted, and generally supposed, that it pecks while still in the shell. The actual mode of self-delivery is just the reverse of pecking. Instead of striking forward and downward (a movement impossible on the part of a bird packed in a shell with its head under its wing), it breaks its way out by vigorously jerking its head upward, while it turns round within the shell, which is cut in two—chipped right round in a perfect circle some distance from the great end.

Though the instincts of animals appear and disappear in such seasonable correspondence with their own wants and the wants of their offspring as to be a standing subject of wonder, they have by no means the fixed and unalterable character by which some would distinguish them from the higher faculties of the human race. They vary in the individuals as does their physical structure. Animals can learn what they did not know by instinct and forget the instinctive knowledge which they never learned, while their instincts will often accommodate themselves to considerable changes in the order of external events. Everybody knows it to be a common practice to hatch ducks' eggs under the common hen, though in such cases the hen has to sit a week longer than on her own eggs. I tried an experiment to ascertain how far the time of sitting could be interfered with in the opposite direction. Two hens became broody on the same day, and I set them on dummies. On the third day I put two chicks a day old to one of the hens. She pecked at them once or twice; seemed rather fidgety, then took to them, called them to her and entered on all the cares of a mother. The other hen was similarly tried, but with a very different result. She pecked at the chickens viciously, and both that day and the next stubbornly refused to have anything to do with them.

The pig is an animal that has its wits about it quite as soon after birth as the chicken. I therefore selected it as a subject of observation. The following are some of my observations:—That vigorous young pigs get up and search for the teat at once, or within one minute after their entrance into the world. That if removed several feet from their mother, when aged only a few minutes, they soon find their way back to her, guided apparently by the grunting she makes in answer to their squeaking. In the case I observed the old sow rose in less than an hour and a half after pigging, and went out to eat; the pigs ran about, tried to eat various matters, followed their mother out, and sucked while she stood eating. One pig I put in a bag the moment it was born and kept it in the dark until it was seven hours old, when I placed it outside the sty, a distance of ten feet from where the sow lay concealed inside the house. The pig soon recognised the low grunting of its mother, went along outside the sty struggling to get under or over the lower bar. At the end of five minutes it succeeded in forcing itself through under the bar at one of the few places where that was possible. No sooner in than it went without a pause into the pig-house to its mother, and was at once like the others in its behaviour. Two little pigs I blindfolded at their birth. One of them I placed with its mother at once: it soon found the teat and began to suck. Six hours later I placed the other a little distance from the sow; it reached her in half a minute, after going about rather vaguely; in half a minute more it found the teat. Next day I found that one of the two left with the mother, blindfolded, had got the blinders off; the other was quite blind, walked about freely, knocking against things. In the afternoon I uncovered its eyes, and it went round and round as if it had had sight, and had suddenly lost it. In ten minutes it was scarcely distinguishable from one that had had sight all along. When placed on a chair it knew the height to require considering, went down on its knees and leapt down. When its eyes had been unveiled twenty minutes I placed it and another twenty feet from the sty. The two reached the mother in five minutes and at the same moment.

Different kinds of creatures, then, bring with them a good deal of cleverness, and a very useful acquaintance with the established order of nature. At the same time all of them later in their lives do a great many things of which they are quite incapable at birth. That these are all matters of pure acquisition appears to me an unwarranted assumption. The human infant cannot masticate; it can move its limbs, but cannot walk, or direct its hands so as to grasp an object held up before it. The kitten just born cannot catch mice. The newly hatched swallow or tomtit can neither walk, nor fly, nor feed itself. They are as helpless as the human infant. Is it as the result of painful learning that the child subsequently seizes an apple and eats it? that the cat lies in wait for the mouse? that the bird finds its proper food and wings its way through the air? We think not. With the development of the physical parts, comes, according to our view, the power to use them, in the ways that have preserved the race through past ages. This is in harmony with all we know. Not so the contrary view. So old is the feud between the cat and the dog, that the kitten knows its enemy even before it is able to see him, and when its fear can in no way serve it. One day last month, after fondling my dog, I put my hand into a basket containing four blind kittens, three days old. The smell my hand had carried with it set them puffing and spitting in a most comical fashion.

That the later developments to which I have referred are not acquisitions can be in some instances demonstrated. Birds do not *learn* to fly. Two years ago I shut up five unfledged swallows in a small box not much larger than the nest from which they were taken. The little box, which had a wire front, was hung on the wall near the nest, and the young swallows were fed by their parents through the wires. In this confinement, where they could not even extend their wings, they were kept until after they were fully fledged. Lord and Lady Amberley liberated the birds and communicated their observations to me, I being in another part of the country at the time. On going to set the prisoners free, one was found dead—they were all alive on the previous day. The remaining four were allowed to escape one at a time. Two of these were perceptibly wavering and unsteady in their flight. One of them, after a flight of about ninety yards, disappeared among some trees; the other, which flew more steadily, made a sweeping circuit in the air, after the manner of its kind, and alighted, or attempted to alight, on a branchless stump of a beech; at least it was no more seen. No. 3 (which was seen on the wing for about half a minute) flew near the ground, first round Wellingtonia, over to the other side of the kitchen-garden, past the bee-house, back to the lawn, round again, and into a beech-tree. No. 4 flew well near the ground, over a hedge twelve feet high to the kitchen-garden through an opening into the beeches, and was last seen close to the ground. The swallows never flew against anything, nor was there, in their avoiding objects, any appreciable difference between them and the old birds. No. 3 swept round the Wellingtonia, and No. 4 rose over the hedge just as we see the old swallows doing every hour of the day. I have this summer verified these observations. Of two swallows I had similarly confined, one, on being set free, flew a yard or two too close to the ground, rose in the direction of a beech-tree, which it gracefully avoided; it was seen for a considerable time sweeping round the beeches and performing magnificent evolutions in the air high above them. The other, which was observed to beat the air with its wings more than usual, was soon lost to sight behind some trees. Titmice, tomtits, and wrens I have made the subjects of a similar experiment and with similar results.

Again, every boy who has brought up nestlings with the hand

must have observed that while for a time they but hold up their heads and open their mouths to be fed, they by-and-by begin quite spontaneously to snap at the food. Here the development may be observed as it proceeds. In the case of the swallow I am inclined to think that they catch insects in the air perfectly well immediately on leaving the nest.

With regard, now, to man, is there any reason to suppose that, unlike all other creatures, his mental constitution has to be in the case of each individual built up from the foundation out of the primitive elements of consciousness? Reason seems to me to be all the other way. The infant is helpless at birth for the same reason that the kitten or swallow is helpless—because of its physical immaturity; and I know of nothing to justify the contrary opinion, as held by some of our distinguished psychologists. Why believe that the sparrow can pick up crumbs by instinct, but that man must learn to interpret his visual sensations and to chew his food? Dr. Carpenter, in his "Mental Physiology," has attempted to answer this argument in the only way in which it could be answered. He has produced facts which appear to him to prove "that the acquirement of the power of visually guiding the muscular movements is experimental in the case of the human infant." More than forty years ago Dr. Carpenter took part in an operation performed on a boy three years old for congenital cataract. The operation was successful. In a few days both pupils were almost clear; but though the boy "clearly recognised the *direction* of a candle or other bright object, he was unable as an infant to apprehend its *distance;* so that when told to lay hold of a watch he groped at it just as a young child lying in its cradle." He gradually began to use his eyes; first in places with which he was not familiar, but it was several months before he trusted to them for guidance as other children of his age would do. No one will doubt the accuracy of any of these statements; but I cannot agree with Dr. Carpenter that he had in the case of the boy anything "exactly parallel" to my experiment of hooding chickens at birth and giving them their sight at the end of one or two days. This boy was couched when three years old. Probably sight would have been at first rather puzzling to my chickens, had they not received it until they were six months old. Dr. Carpenter seems to have forgotten for the moment that instincts as well as acquisitions decay through desuetude, and that this is especially true when the faculties in question have never once been started into action and are of the kind which develop through exercise. Another and vital difference between Dr. Carpenter's experiment and mine is this, that when at the end of two days I gave my chickens sight, I did not do so by poking out or lacerating the crystalline lenses of their eyes with a needle.

The presumption, then, that the progress of the infant is but the unfolding of inherited powers remains as strong as ever. With wings there comes to the bird the power to use them; and why should we believe that because the human infant is born without teeth, it should, when they do make their appearance, have to discover their use by a series of happy accidents?

One word as to the origin of instincts. In common with other evolutionists, I have argued that instinct in the present generation may be regarded as the product of the accumulated experiences of past generations. More peculiar to myself, and giving a special meaning to the word experience, is the view that the question of the origin of the most mysterious instincts is not more difficult than, or different from, but is the same with the problem of the origin of the physical structure of the creatures. For, however they may have come by their bodily organisation, it, in my opinion, carries with it a corresponding mental nature.

In opposition to this view it has been urged that we have only to consider almost any well-marked instinct to see that it could never have been a product of evolution. We, it is said most frequently, cannot conceive the experiences that might by inheritance have become the instincts; and we *can* see very clearly that many instincts are so essential to the preservation of the creatures that without them they could never have lived to acquire them. The answer is easy. Granting our utter inability to go back in imagination through the infinite multitude of forms, with their diversified mental characteristics, that stand between the greyhound and the speck of living jelly to which, according to the theory of evolution, it is related by an unbroken line of descent. Granting that we are, if possible, still less able to picture in imagination the process of change from any one form to another. What then? Not surely that the theory of evolution is false! For the same argument will prove that no man present can possibly be the son of his father. Our ignorance is very great, but it is not a very great argument.

The other objection, that the creatures could never have lived to acquire their more important instincts, rests on a careless misunderstanding of the theory of evolution. It assumes in the drollest possible way that evolutionists must believe that in the course of the evolution of the existing races there must have from time to time appeared whole generations of creatures that could not start on life from the want of instincts that they had not got. There can be no need to say more than that these unfortunate creatures are assumed to have been singularly unlike their parents. The answer is, that it is not the doctrine of evolution that the bodies are evolved first by one set of causes and the minds are put in afterwards by another. This notion is but the still lingering shadow of the individual-experience psychology. As evolutionists, whether we take the more common view and regard the actions of animals as prompted by their feelings and guided by their thoughts, or believe, as I do, that animals and men are conscious automata, in either case we are under no necessity of assuming in explanation of the origin of the most mysterious instincts anything beyond the operation of those laws that we see operating around us, but concerning which we have yet to learn more, perhaps, than we have learned.

Part II

POST-DARWINIAN BRITISH INSTINCT THEORY

Editor's Comments on Papers 6, 7, and 8

6A SPENCER
Excerpts from *Instinct*

6B SPENCER
Excerpts from *Reason*

7A MORGAN
Excerpt from *Preliminary Definitions and Illustrations*

7B MORGAN
Excerpt from *Further Observations on Young Birds*

7C MORGAN
Excerpt from *Observations on Young Mammals*

7D MORGAN
Excerpt from *Consciousness and Instinct*

7E MORGAN
Excerpt from *Intelligence and the Acquisition of Habits*

8A McDOUGALL
Excerpts from *The Behavior of the Lower Animals*

8B McDOUGALL
Excerpt from *Behavior of Insects*

8C McDOUGALL
Excerpts from *Behavior of the Vertebrates*

Spalding and Altum were careful observers who noted the details of behavior. In contrast with their attempts to replace informal anecdotes with facts, there were "grand theorists" mainly interested in developing comprehensive synthetic theories of mental evolution, not excluding humans.

As in the area of natural theology, the English particularly explored the possibilities of such schemes; three representatives are included here. All three writers were cited in Lorenz's early papers. Lorenz took pains to separate himself from some aspects of their theories, even as he seemed to accept many of their other features. For example, McDougall and Morgan both contrasted reflexes with instinct (instinctive actions for Lorenz). Basic to Lorenz's thinking was that these highly influential theories were based on too limited a knowledge about what animals *really* do, too loose a formulation of instinct, and an absence of at least representative detailed comparative information on closely related taxa.

Spencer is quoted but not cited by Lorenz, who repeatedly refers to the "Spencer/Lloyd Morgan Theory." This is not particularly fair to Morgan as Morgan elaborated his own views in many books. Spencer, a nineteenth-century polymath, wrote a series of influential tomes on the evolutionary 'Principles' of many fields including biology, psychology, sociology, anthropology, ethics, education, and political science. He anticipated some of Darwin's views before the publication of *The Origin of Species* in 1859 and was an early advocate of evolutionary continuity in the social sciences. Far more Lamarckian than Darwin, Spencer espoused the inheritance of acquired behavior, which was also advocated by Spalding. Unlike the latter he was a thorough-going associationist, maintaining that a process of association was behind the evolution of behavioral continuity among species, and provided the links between all the mental and behavioral capacities within an organism. The opening statement (Paper 6A) that instinct is "compound reflex action" actually places Spencer close to the "chain reflex" view. In his early writings Lorenz subscribed to a modified chain reflex view. Lorenz was later to attack it using many of McDougall's arguments. The Spencer readings here (Papers 6A and 6B) are from the second edition of his *Principles of Psychology* that first appeared in 1870.

Unlike Spencer, Morgan made original observations on the behavior of birds, dogs, and other animals. Yet he soon slipped into the custom of writing articles and books on animal behavior rather than continuing original studies. He was always very interested in both philosophical issues and human psychology, reworking his early observations repeatedly in his later books. In Papers 7A-7E, from his 1896 book *Habit and Instinct,* Morgan reveals a balanced view of the role of environmental and constitutional factors in animal behavior. Morgan, aware of the need for careful descriptive and experimental work, regrettably proved representative of an era that tried to settle disputes by argument and logic rather than by direct study of nature. Lorenz disagreed most with two aspects of Morgan's and Spencer's views:

that instinct was an evolutionary way station between reflex and intelligence, and that instincts were readily influenced by learning or "association."

McDougall is a problematic person in the history of early twentieth-century psychology. Greatly influenced by William James (the book excerpted here was dedicated to him), he was opposed to all reductionistic theories. Instead he advocated a purposeful or 'hormic' psychology and eventually became a vigorous, but ineffective, opponent of behaviorism. His opposition to an exclusive focus on behavior ("motor mechanisms") was challenged by the early ethologists, who also disagreed with his simplistic catalogue of global instincts (e.g., parental, pugnacious) and the association of discrete emotions with each. Yet it is clear that Lorenz was otherwise sympathetic to much of McDougall's message, including its focus on endogenous factors, the view of animals as complex actors and not bundles of conditioned or unconditioned reflexes, and the willingness to allow animals cognitive and emotional states. E. C. Tolman was an American experimental psychologist who incorporated some of McDougall's thinking in his own behaviorism in a more rigorous manner (e.g., Tolman, 1932; also Tolman and Honzik, 1930, reprinted in Henderson, 1982). Tolman was frequently cited in Lorenz's early theoretical papers.

A remarkable aspect of the McDougall excerpts (Papers 8A, 8B, and 8C) from his 1923 *Outline of Psychology* is the well-worked-out motivational energy model. It is often forgotten that Lorenz's first (Re)action Specific Energy model used a gas, rather than a liquid, to represent accumulated endogenous energy. McDougall also addressed the question of conflicts between instincts within an individual. Note that he, as did Spencer and Morgan, used pecking in chicks to make important points.

REFERENCES

Henderson, R. W., 1982, *Learning in Animals,* Benchmark Papers in Behavior Series, Hutchinson Ross, Stroudsburg, Pa., 350p.

Lorenz, K., 1937/1957, The Nature of Instinct, translated in *Instinctive Behavior,* C. H. Schiller, ed., International Universities Press, New York, pp. 129–175.

Tolman, E. C., 1932, *Purposive Behavior in Animals and Men,* Century, New York, 463p.

Tolman, E. C., and C. H. Honzik, 1930, "Insight" in Rats, *Univ. Calif. Publ. Psychol.* **4:**215–232.

6A

Reprinted from pages 432, 434–437, 439, and 443 of *The Principles of Psychology,* vol. 1, 2nd ed., Williams and Norgate, London, 1878, 640p.

INSTINCT

H. Spencer

§ 194. Not using the word as the vulgar do, to designate all other kinds of intelligence than the human, but restricting it to its proper signification, Instinct may be described as—compound reflex action. I say described rather then defined, since no clear line of demarkation can be drawn between it and simple reflex action. As remarked in the last section, the *dirigo-motor* processes which reflex actions show us, pass by degrees from the simple to the complex; and a cursory inspection of the facts shows us that the *recipio-motor* processes do the like. Nevertheless we may conveniently distinguish, as a higher order of these automatic nervous adjustments, those in which complex stimuli produce complex movements.

That the propriety of thus marking off Instinct from primitive reflex action may be clearly seen, let us take examples. A chick, immediately it comes out of the egg, not only balances itself and runs about, but picks up fragments of food; thus showing us that it can adjust its muscular movements in a way appropriate for grasping an object in a position that is accurately perceived. Obviously this action, which is proved by the circumstances to be purely automatic, implies the combination of many stimuli.

[*Editor's Note:* Material has been omitted at this point.]

§ 195. Instinct is obviously further removed from purely physical life than is simple reflex action. While simple reflex action is common to the internal visceral processes and to the processes of external adjustment, Instinct is not. There are no instincts displayed by the kidneys, the lungs, the liver: they occur only among the actions of that nervo-muscular apparatus which is the agent of psychical life.

Again, the co-ordination of many stimuli into one stimulus is, so far as it goes, a reduction of diffused simultaneous changes into concentrated serial changes. Whether the combined nervous acts which take place when the fly-catcher seizes an insect, are regarded as a series passing through its centre of co-ordination in rapid succession, or as consolidated into two successive states of its centre of co-ordination, it is equally clear that the changes going on in its centre of co-ordination have a much more decided linear arrangement than have the changes going on in the scattered ganglia of a centipede.

In its higher forms, Instinct is probably accompanied by a rudimentary consciousness. There cannot be co-ordination of many stimuli without some ganglion through which they are all brought into relation. In the process of bringing them into relation, this ganglion must be subject to the influence of each—must undergo many changes. And the quick succession of changes in a ganglion, implying as it does perpetual experiences of differences and likenesses, constitutes the raw material of consciousness. The implication is that as fast as Instinct is developed, some kind of consciousness becomes nascent.

Further, the instinctive actions are more removed from the actions of simple bodily life in this, that they answer to external phenomena which are more complex and more special. While the purely physical processes going on throughout the organism respond to those most general relations common to the environment as a whole; while the simple reflex actions respond to some of the general relations common to the individual objects it contains; these compound reflex actions which we class as instincts, respond to those more involved relations which characterize certain orders of objects and actions as distinguished from others.

Greater differentiation of the psychical life from the physical life is thus shown in several ways—in the growing distinction between the action of the vegetative and animal systems; in the increasing seriality of the changes in the animal system; in the consequent rise of incipient consciousness; and in the higher speciality of the outer relations to which inner relations are adjusted: which last is indeed the essence of the advance, to which the others are necessary accompaniments.

§ 196. We are now prepared to inquire how, by accumulated experiences, compound reflex actions may be developed out of simple ones.

Let us begin with some low aquatic creature possessing rudimentary eyes. Sensitive as such eyes are only to marked changes in the quantity of light, they can be affected by opaque bodies moving in the surrounding water, only when such bodies approach close to them. But bodies carried by their motion very near to the organism, will, by their further motion, be brought in contact with it. The cases in which an external object passes by almost at a tangent to that part of the organism where the rudimentary eye is placed, so as nearly to touch the surface but not quite, must be exceptional. In its earliest forms sight is, as before said, little more than anticipatory touch (§ 142): visual impressions are habitually followed by tactual ones. But tactual impressions are, in all these creatures, habitually followed by contractions — contractions which, as was pointed out in § 140, are probably the necessary effects of mechanically accelerating the vital changes — contractions which, under like stimuli, occur even in certain plants, and are so shown to be producible by alterations in the processes of purely physical life. Result as they may, however, it is beyond question that from the zoophytes upwards, touch and contraction form an habitual sequence; and hence, in creatures whose incipient vision amounts to little more than anticipatory touch, there constantly occurs the succession — a visual impression, a tactual impression, a contraction. Now the evolution of a nervous system is a necessary concomitant of that specialization which originates the senses. On the one hand, until the general sensitiveness is in some degree localized, the internuncial function of the nervous system cannot exist; and on the other hand, no such localized sensitiveness can exist without something in the shape of nerves. A nascent sense of sight, therefore, implies a nascent nervous communication. And along with a nascent nervous communication we may see the first illustration of the growth of intelligence. If psychical states (using the term in its widest sense) which follow one another time after time in a certain order, become every time more closely connected in this order, so as eventually to become inseparable; then it must happen that if, in the experience of any species, a visual impression, a tactual impression, and a contraction, are continually repeated in this succession, the several nervous states produced will become so consolidated that the first cannot be caused without the others following —the visual impression will be instantly succeeded by a nervous excitation like that which a tactual impression produces, and this will be instantly succeeded by a contraction. There will thus occur a contraction in anticipation of touch.

What must result from a further development of vision? Evidently the same bodies will be discerned at greater distances, and smaller bodies will be discerned when close to. Both of these must produce obscurations which are faint in comparison with that obscuration produced by a large body about to strike the creature's surface. But now mark the accompanying experience. A faint obscuration will not, like an extreme one, be habitually followed by a strong tactual impression and a subsequent contraction. If caused by a great mass passing at some distance, there will probably be no collision—no tactual impression at all. If caused by a little mass which is very near, the collision that follows will be comparatively slight—so slight as not to excite a violent contraction, but only such tension in the muscular apparatus as is seen in any creature about to seize upon prey. This is by no means an assumption. Among animals in general, ourselves included, a nervous impression which, if slight, simply rouses attention and braces up the muscles, causes convulsive contortions if intense. It is therefore a deduction from a well-established law of the nervo-muscular system, that a creature possessing this somewhat improved vision will, by a partial obscuration of light, have its muscles brought into a state of partial tension—a state fitting them either for the seizure of a small animal should the partial obscuration be caused by the impending collision of one, or for sudden retreat into a shell should the obscuration be increased by the near approach of a larger animal. So that even from this simple advance there arises a somewhat greater speciality and complexity in the inner relations answering to outer relations.

[*Editor's Note:* Material has been omitted at this point.]

Here, then, we see how one of the simpler instincts will, under the requisite conditions, be established by accumulated experiences. Let it be granted that the more frequently psychical states occur in a certain order, the stronger becomes their tendency to cohere in that order, until they at last become inseparable; let it be granted that this tendency is, in however slight a degree, inherited, so that if the experiences remain the same each successive generation bequeaths a somewhat increased tendency; and it follows that, in cases like the one described, there must eventually result an automatic connexion of nervous actions, corresponding to the external relations perpetually experienced. Similarly if, from some change in the environment of any species, its members are frequently brought in contact with a relation having terms a little more involved; if the organization of the species is so far developed as to be impressible by these terms in close succession; then, an inner relation corresponding to this new outer relation will gradually be formed, and will in the end become organic. And so on in subsequent stages of progress.

[*Editor's Note:* Material has been omitted at this point.]

If, as the instincts rise higher and higher, they come to include psychical changes that are less and less coherent with their fundamental ones; there must arrive a time when the co-ordination is no longer perfectly regular. If these compound reflex actions, as they grow more compound, also become less decided; it follows that they will eventually become comparatively undecided. They will begin to lose their distinctly automatic character. That which we call Instinct will merge into something higher.

The facts are thus rendered comprehensible. We see that, if produced by experience, the evolution of Instinct must proceed from the simple to the complex, and that by a progression thus wrought out, it must insensibly pass into a higher order of psychical action; which is just what we find it to do in the higher animals.

6B

Reprinted from pages 456 and 459–460 of *The Principles of Psychology,* vol. 1, 2nd ed., Williams and Norgate, London, 1878, 640p.

REASON

H. Spencer

[*Editor's Note:* In the original, material precedes this excerpt.]

—there is a tendency for the psychical states excited in me by the snarling dog, to be followed by those other psychical states that have before followed them. In other words, there is a nascent excitation of the motor apparatus concerned in picking up and throwing; there is a nascent excitation of all the sensory nerves affected during such acts; and, through these, there is a nascent excitation of the visual nerves, which on previous occasions received impressions of a flying dog. That is, I have the *ideas* of picking up and throwing a stone, and of seeing a dog run away; for these that we call ideas, are nothing else than weak repetitions of the psychical states caused by actual impressions and motions. But what happens further? If there is no antagonist impulse—if no other ideas or partial excitations arise, and if the dog's aggressive demonstrations produce in me feelings of adequate vividness, these partial excitations pass into complete excitations. I go through the previously-imagined actions. The nascent motor changes become real motor changes; and the adjustment of inner relations to outer relations is completed. This, however, is just the process which we saw must arise whenever, from increasing complexity and decreasing frequency, the automatic adjustment of inner to outer relations becomes uncertain or hesitating. Hence it is clear that the actions we call instinctive pass gradually into the actions we call rational.

Further proof is furnished by the converse fact, that the actions we call rational are, by long-continued repetition, rendered automatic or instinctive. By implication, this lapsing of reason into instinct was shown in the last chapter, when exemplifying the lapsing of memory into instinct: the two facts are different aspects of the same fact.

[*Editor's Note:* Material has been omitted at this point.]

We lately saw that while, on the one hand, instinctive actions pass into rational actions when from increasing complexity and infrequency they become imperfectly automatic; on the other hand, rational actions pass, by constant repetition, into automatic or instinctive actions. Similarly, we may here see that while, on the one hand, rational inferences arise when the groups of attributes and relations cognized become such that the impressions of them cannot be simultaneously co-ordinated; on the other hand, rational inferences pass, by constant recurrence, into automatic inferences or organic intuitions. All acquired perceptions exemplify this truth. The numberless cases in which we seem directly to know the distances, forms, solidities, textures, &c., of the things around us, are cases in which psychical states originally answering to phenomena separately perceived, and afterwards connected in thought by inference, have, by repetition, become indissolubly united, so as to constitute a rational knowledge that appears intuitive.

Thus, the experience-hypothesis furnishes an adequate solution. The genesis of instinct, the development of memory and reason out of it, and the consolidation of rational actions and inferences into instinctive ones, are alike explicable on the single principle, that the cohesion between psychical states is proportionate to the frequency with which the relation between the answering external phenomena has been repeated in experience.

§ 206. But does the experience-hypothesis also explain the evolution of the higher forms of rationality out of the lower? It does. Beginning with reasoning from particulars to particulars—familiarly exhibited by children and by domestic animals—the progress to inductive and deductive reasoning is similarly unbroken, as well as similarly determined. And by the accumulation of experiences is also determined the advance from narrow generalizations to generalizations successively wider and wider.

[*Editor's Note:* Material has been omitted at this point.]

7A

Reprinted from pages 24-28 of *Habit and Instinct,* Arnold, London, 1896, 351p.

PRELIMINARY DEFINITIONS AND ILLUSTRATIONS

C. Lloyd Morgan

[*Editor's Note:* In the original, material precedes this excerpt.]

It must not be supposed that the distinction between what is congenital and what is acquired—a distinction which we are endeavouring to draw with the utmost clearness—is invalidated by the fact that there are a great number of activities, such, for example, as the perfected flight of birds, which are of double origin, being in part congenital and in part acquired. Such cases do indeed show that imperfect instincts may be perfected by habit and individual acquisition of skill. But they render the more necessary a careful distinction between the factors which co-operate to produce the ultimate result. And only when the distinction is thus duly emphasized does the question, whether the acquired perfection of one generation tends to lessen the congenital imperfection in the next generation, stand out in its true significance. It is interesting here to compare the flight of birds with the flight of insects. In the latter class we find many cases in which the instinctive factor in flight is relatively much more highly developed towards congenital perfection than it is in the former class, at any rate so far as most birds are concerned. The question therefore arises, Is the greater relative perfection in the instinctive flight of some insects due to the inheritance of acquired skill on the part of their ancestors? Or is it due to the fact that there has been among insects more elimination of those who failed in congenital power of flight, and hence a survival through natural selection of those in which the instinctive flight was better developed? Or is it due to some other cause?

At present we will not attempt to answer such questions as these. We are concerned merely in drawing as clearly as possible a distinction which shall enable us to put the questions in a definite and intelligible form. It is a matter of no small importance accurately to focus the point of such questions.

The distinction between that which is congenital and that which is acquired may be further illustrated from a structural point of view. There is an inherited organic mechanism through the possession of which an animal is fitted to perform certain more or less definite and adaptive activities without learning, with little or no practice (though even in these cases practice helps to make perfect), through no teaching, by no imitation, and without any individual experience on which intelligent choice of the best mode of procedure could be based. This is due to what may be termed *congenital automatism.* On the other hand, there is an organic mechanism which is gradually developed during the individual lifetime, through the due co-ordination and persistent repetition of certain selected activities. These activities, by constant repetition, themselves become automatic as habits. We may term the working of this organic mechanism, which is thus developed during the course of individual life, *acquired automatism.* The biological question is—Does the acquired automatism of one generation contribute, through inheritance, to the congenital automatism of the next?

There is still, however, the difficulty suggested but not removed a few pages back, that after all an animal can only acquire that for acquiring which it inherits a potentiality; and that we must in any case, even when individual acquisition is under consideration, come back to heredity in the last resort. This is indubitably true; and it shows that the more or less definite congenital activities do not by any means exhaust the hereditary possibilities. All that an animal owes to heredity may, indeed, be classified under two heads. Under the first head will fall those relatively definite modes of activity which fit it to deal at once, on their first occurrence, with certain essential or frequently recurring conditions of the environment, and this forms the group here termed "congenital." Under the second head will fall the power of dealing with special circumstances as they arise, and this we may term *innate capacity.* The former may be likened to the inheritance of specific drafts for particular and relatively definite purposes in the conduct of life; the latter may be likened to the inheritance of a legacy which may be drawn upon for any purpose as need arises. If the need become habitual, the animal may, so to speak, instruct his banker to set aside a specific sum to meet this need as often as it arises. But this arrangement is a purely individual matter, and no wise dictated by the terms of the bequest.

In this classification instinctive activities, as we propose to define them, fall under the first head. They display some share of that hereditary definiteness which is characteristic of what we have termed "congenital activities." It must be remembered, however, that, as already mentioned, there is unfortunately no common and accepted agreement so to define the term "instinctive." Professor Wundt, indeed, divides instincts into two classes: (1) those which are congenital, and (2) those which are acquired. So that the distinction we are drawing will by no means be accepted by him and his followers. Where opposing views are in the field, it is necessary to carefully weigh the advantages and disadvantages of each, and to adopt one or the other. I am quite convinced that, from the biological point of view, it is more satisfactory to restrict the term "instinctive" to those activities which are in greater or less degree congenitally definite, and it is in this sense that the phrase "instinctive activities" will be used in this work. No doubt there are in many cases difficulties of interpretation; but these must be met as they arise in the further prosecution of our studies.

It may be convenient to indicate the nature of the suggested classification in tabular form:—

INHERITED.	
Congenitally definite activities, under which those termed "instinctive" are comprised.	Innate capacity, involving (*a*) a power of association, and (*b*) hereditary susceptibilities to pleasure and pain.

ACQUIRED.	
(*a*) Confirmation, or (*b*) Modification, of congenitally definite or instinctive activities so as to render them habitual by repetition. (*c*) Suppression of congenital activities.	Particular application of innate capacity; (*a*) Occasional and under special circumstances; (*b*) Frequently repeated, with the consequent formation of acquired habits.

We may now sum up what has been advanced in the foregoing discussion, and say that from the biological point of view (and it is from this standpoint alone that they have been so far considered) instincts are congenital, adaptive, and co-ordinated activities of relative complexity, and involving the behaviour of the organism as a whole. They are not characteristic of individuals as such, but are similarly performed by all like members of the same more or less restricted group, under circumstances which are either of frequent recurrence or are vitally essential to the continuance of the race. While they are, broadly speaking, constant in character, they are subject to variation analogous to that found in organic structures. They are often periodic in development and serial in character. They are to be distinguished from habits which owe their definiteness to individual acquisition and the repetition of individual performance.

7B

Reprinted from pages 98–100 of *Habit and Instinct,* Arnold, London, 1896, 351p.

FURTHER OBSERVATIONS ON YOUNG BIRDS

C. Lloyd Morgan

[*Editor's Note:* In the original, material precedes this excerpt.]

I have already, perhaps, presumed too largely on the reader's patience. The diary-notes—themselves selected from a considerable body of observations recorded day by day—may well appear in many cases to savour of triviality. It is only, however, by careful and minute observation that we can hope to gauge the length to which heredity runs. In anatomical investigations we must pay patient attention to details of structure; and in investigations into the phenomena of habit and instinct, we must not shrink from the labour and the expenditure of time involved in daily and almost hourly observation, if we would attempt to distinguish between what is inherited in a relatively perfect condition, and what is acquired by experience or through imitation.

Such observations as have been given in this and the preceding chapters require to be extended to other species, and over longer periods of time. If what is here set down should induce others to take up a mode of investigation which will be found full of interest, and in which much still remains to be done, one of the objects in placing on record the details of my own work will be fulfilled. There are many activities and modes of behaviour which are generally assumed to be instinctive and due to heredity, but which may be the result of tradition, handed on by example from parents to offspring. Partridges, for example, when they jug, nestling close together at night, would seem, from the appearance of the droppings, which are generally deposited in a circle of only a few inches in diameter, to arrange themselves in a circle, tails inwards and heads outwards. Is this behaviour instinctive or is it traditional? The young of the colin (*Ortyx virginiana*), we are told,* "when the shades of evening approached, crowded together in a circle on the ground, and prepared themselves for the slumbers of the night by placing their tails all together, with their pretty mottled chins facing to the front in a watchful round-robin." The fact that these birds behave thus, apparently instinctively, would lead one to surmise that the behaviour is instinctive in the partridge. My own birds died at too early an age for this point to be determined.

We may now summarize some of the general conclusions to be drawn from our observations as follows:—

1. That which is congenitally definite as instinctive behaviour is essentially a motor response or train of motor responses. Mr. Herbert Spencer's description of instinct as compound reflex action is thus justified.

2. These often show very accurate and nicely adjusted congenital or hereditary co-ordinations.

3. They are evoked by stimuli, the general type of which is fairly definite, and may, in some cases, be in response to particular objects. Of the latter possibility we have, however, but little satisfactory evidence.

4. There does not seem to be any convincing evidence

* Yarrell, "British Birds," *sub. spec.*

of inherited ideas or knowledge (as the term is popularly used); that is to say, the facts can be equally well explained on the view that what is inherited is of the nature of an organic response.

5. Association of ideas is strong, and is rapidly formed as the result of individual acquisition.

6. Acquired definiteness is built, through association, on the foundation of congenital responses, which are modified, under experience, to meet new circumstances.

7. Acquired definiteness may pass, through frequent repetition, into more or less stereotyped habit.

7C

Reprinted from pages 120–121 of *Habit and Instinct,* Arnold, London, 1896, 351p.

OBSERVATIONS ON YOUNG MAMMALS

C. Lloyd Morgan

[*Editor's Note:* In the original, material precedes this and follows excerpt.]

But though, as Dr. Mills says, in an intelligent animal, there come after a while to be "almost no pure instincts," that which was given as instinct having been utilized, modified, and adapted through experience and acquisition, yet the fundamental distinction between that which is congenital and instinctive, on the one hand, and that which is acquired through individual experience, on the other hand, remains unaltered. Nor does the fact that all acquisition is rendered possible by an innate faculty for acquiring—nay, more, for acquiring in this way or that in accordance with hereditary character—diminish a whit the value of the distinction. The instinctive action is prior to experience; the acquired action is due to experience. And this distinction holds, no matter how hard it may be to decide whether this action or that is in the main instinctive or in the main acquired. The final products of individual development may be, and no doubt generally are, of twofold origin, partly instinctive and partly due to acquisition; but this, I repeat, does not in any way serve to annul the distinction between the two several elements in the final product. If we add water to our whisky, there is no longer pure whisky; the whisky has been more or less modified, but the spirit is still there none the less, and we cannot neglect its presence. So, too, if we add the water of experience to the whisky of instinct, we have a joint product of which so much comes from the bottle and so much comes from the jug. And when Dr. Mills contends that in the kitten "its psychic life is determined by experience," we must take him to mean that the whisky of instinct is, as a matter of fact, always more or less watered down in the course of individual development.

When animals are brought up by their parents, it is often a matter of difficulty to determine how far any specially characteristic behaviour is due to the influence of the parent, whose behaviour is similar, and how far it is due to instinct. But when the young are separated from their parents at an early age, the instinctive basis is often rendered more clear and obvious.

7D

Reprinted from page 136 of *Habit and Instinct*, Arnold, London, 1896, 351p.

CONSCIOUSNESS AND INSTINCT

C. Lloyd Morgan

[*Editor's Note:* In the original, material precedes and follows this excerpt.]

The net result of our discussion of the matter is therefore this, that on the occasion of the first performance of an instinctive activity the co-ordination involved (and it is sometimes quite elaborate) is automatic, and cannot be regarded as under the guidance of consciousness; but that the carrying out of the activity furnishes data to consciousness in the light of which the subsequent performance of a like activity may be perfected, or modified, or checked. From this it follows that only on the occasion of its first performance does such a congenital activity present itself for our study in its instinctive purity. For on subsequent occasions it is more or less modified by the results of the experience acquired by the individual. It then possesses acquired elements in addition to those which are congenital and instinctive; and when such acquired modification is rendered stereotyped and uniform by repetition, it is, so far as thus modified, a habit. In such organisms as birds and young mammals, therefore, instincts are to be regarded as the automatic raw material which will be shaped and moulded under the guidance of consciousness into what may be called *instinct-habits*, if by this compound term we may understand activities founded on a congenital instinctive basis, but modified by acquired experience.

In the case, too, of such an instinctive procedure of the deferred type as that presented by the diving of a young moorhen, though, on the first occasion of its performance the congenital automatism predominates, yet it is difficult to believe, and is in itself improbable, that the individual experience of the young bird does not, even on the first occasion, exercise some influence on the way in which the dive is performed. If we desire to reach a true interpretation of the facts, we must realize the fact that an activity may be of mixed origin.

7E

Reprinted from pages 150–151 of *Habit and Instinct,* Arnold, London, 1896, 351p.

INTELLIGENCE AND THE ACQUISITION OF HABITS

C. Lloyd Morgan

[*Editor's Note:* In the original, material precedes this and follows excerpt.]

But if the cortical augmentation and inhibition are founded in heredity; and if this augmentation and inhibition form the basis upon which all acquisition and all control are based; what becomes of the distinction between instinctive and acquired activities? What, of that between automatic and controlled behaviour?

Let us look again at the facts which we are endeavouring to interpret. A chick sees for the first time in its life a cinnabar caterpillar, instinctively pecks at it under the influence of the visual stimulus; seizes it, and under the influence of the taste-stimulus instinctively shrinks. So far we have instinct and automatism. Presently we throw to it another similar caterpillar. Instinct and automatism alone would lead to a repetition of the previous series of events; seeing, seizing, tasting, shrinking. The oftener the experiment was performed the more smoothly would the organic mechanism work, the more definitely would the same sequence be repeated—seeing, seizing, tasting, shrinking. Is this what we actually observe? Not at all. On the second occasion the chick, under the influence of the previous experience, acts differently. Though he sees, he does not seize, but shrinks without seizing. We believe that there is a revival in memory of the nasty taste. And in this we seem justified, since we may observe that sometimes the chick on such occasions wipes the bill on the ground as he does on experiencing an unpleasant taste, though he have not touched the larva. The chick, then, does not continue to act merely from instinct and like an automaton. His behaviour is modified in the light of previous experience. What, then, has taken place in and through which this modification, born of experience, is introduced? In answering this question we disclose the essential feature of the distinction we have all along been drawing—that between congenital and acquired activities. The answer may be given in two words—*association* and the *suggestion* that arises therefrom. The chick's first experience of the cinnabar caterpillar leads to an association between the appearance of the larva and its taste; or, from the physiological point of view, to a direct connection between the several cortical disturbances. On the second occasion the taste is suggested by the sight of the cinnabar larva; or, physiologically, the disturbance associated with taste is directly called forth by the disturbance associated with sight. It is through association and suggestion that an organism is able to profit by experience, and that its behaviour ceases to be merely instinctive and automatic. And such association would seem to be a purely individual matter—founded, no doubt, on an innate basis, linking activities of the congenital type, but none the less wholly dependent upon the immediate touch of individual experience. Hence the development of consciousness as effective for the guidance of life, takes its origin in the *linkage*, through association, of sentient states. In the absence of such linkage sentience is a mere adjunct accompanying certain organic transactions in the nervous system or elsewhere.

In watching, then, the behaviour of young birds or other animals, we observe a development which is to be interpreted as the result of conscious choice and selection. For the chick to which a handful of mixed caterpillars is thrown, chooses the nice ones, and leaves the nasty untouched. This selection is dependent upon an innate power of association which needs the quickening touch of individual experience to give it actuality and definition, without which it would lie dormant as a mere potentiality.

8A

Copyright © 1923 by Charles Scribner's Sons; copyright renewed 1951 by Anne A. McDougall

Reprinted from pages 51-54, 56-57, and 70-71 of *An Outline of Psychology,* 2nd ed., Methuen, London, 1923, 456p., by permission of Charles Scribner's Sons

THE BEHAVIOR OF THE LOWER ANIMALS

William McDougall

[*Editor's Note:* In the original, material precedes this excerpt.]

In the first chapter I argued that the psychologist should and must choose the fundamental categories appropriate to his science, if he is to make progress toward his proper goal, the better understanding and control of human nature and human behavior. *Purposive action is the most fundamental category of psychology;* just as the motion of a material particle according to the mechanical principles of Newton's laws of motion has long been the fundamental category of physical science. Behavior is always purposive action, or a train or sequence of purposive actions.

Purposive and Reflex Actions Contrasted

Let us look a little more closely at the distinction between *purposive action* and the *reflex action* which the mechanists attempt to put in its place, with the consequences that we have already glanced at; namely, "sensations" or "ideas" confusedly thought of as mysterious entities, dragged passively into and out of that other mysterious entity "consciousness" by the mechanical reflex processes of the brain. I do not mean to assert that reflex processes as conceived by physiology do not occur. Human organisms, as well as the higher animals, do exhibit reactions of the mechanical reflex type. If the reader, comfortably seated, will cross his right knee over the other and sharply tap the tendon below the right knee-cap, his right foot will jerk forward owing to the contraction of the large muscles of the front of the thigh. That is a very simple reflex action.

Many other such tendon-reflexes can be provoked in similar fashion; the observation of them is of great value to the neurologist as throwing light upon the condition of the nervous system. Others can be provoked by gently scratching the skin in various parts; *e. g.*, "the abdominal reflex," a sharp contraction of the muscles of the wall of the abdomen, by scratching the skin of the flank; or the closing of the eyelid at the slightest touch on the skin over the eyeball. Many other reflex processes, resulting in contractions of the muscles of the visceral organs, such as the heart, blood-vessels, respiratory and digestive organs, or in secretion of tears, saliva, gastric and other juices, may be provoked by stimulation of various sensory nerves. Such reflex actions have been elaborately studied in animals whose brains have been destroyed.[1] It has been shown that the simple reflexes are, in many instances, functionally linked in such a way that they naturally succeed one another, producing a more or less complicated train of movement of a serviceable type, the first movement producing a stimulus for the second, the second for the third, and so on. For example, the legs of a dog whose brain has been destroyed may be thrown by stimulation of the soles of the feet into a sequence of movements resembling the movements of walking.[2] Such a sequence of reflex actions is commonly called a chain-reflex; and it is by imagining the arrangements of nerve-paths which subserve such chain reflexes to be multiplied and complicated enormously, but on the same mechanical principles, that the mechanists seek to explain all human behavior. And recently it has been shown that the human spinal cord, when detached from the brain by a wound, is richer in such complicated reflex mechanisms than had previously been supposed.[3]

[1] By no one so thoroughly as by Professor Sir Charles Sherrington, the President of the Royal Society of London, whose great work embodying his studies of reflex action ("The Integrative Action of the Nervous System") should be mastered by every serious student of psychology.

[2] The animal in this condition cannot, however, walk; for walking is a very much more complex process than the mere sequence of leg movements; the balancing of the whole body is involved in it.

[3] Especially by the studies of Doctor Henry Head. Cf. "Studies in Neurology," London, 1920.

But, although such reflex movements are serviceable, in the sense that they are such as the organism makes use of in the course of its normal behavior (for example, the reflex walking movements of the legs of the brainless dog) when we examine them closely, we see that they do not exhibit the characteristic marks of behavior, the objective criteria of purpose.

The reflex lacks (1) the spontaneity of behavior: the legs of the brainless dog only execute the walking reflex when stimulated in a particular manner: the normal dog gets up from sleep in a dark quiet corner and walks, without being excited to this behavior by any assignable stimulus. The reflex lacks (2) the persistency of behavior: that is to say, the reflex movements continue only so long as the stimulus is applied to the appropriate sense-organ. The chain-reflex is no exception to the rule, though it may seem to be so to the casual observer; for each movement produces the stimulus to the next in the sequence; just as, when one shell in a "dump" is exploded, a sequence of explosions may result. (3) The reflex is stereotyped or fixed, the movements evoked by the same stimulus falling on the same sensory nerve are, approximately, the same on all occasions; whereas purposive movements are indefinitely variable. If we saw a dog walking steadily forward over a smooth road, we might be in doubt whether his movements were purely reflex; and, if the movements were long continued, we should be inclined to suspect this; just by reason of the absence of those perpetual variations of movement which are so characteristic of behavior. But, when we see a dog, aroused by his master's whistle, run from door to window and back again, varying his movements in a hundred ways as he does so, we confidently regard this as behavior.

(4) The reflex movements do not present that appearance of seeking a goal which is common to all behavior, and of which the essential feature is the persistence of movements with variation until, and only until, that goal is attained. It is true that the cessation of the stimulus which provokes the reflex might be claimed as the analogue of the goal of purposive behavior; as when the foot is withdrawn from a prick, or the dog's hind-foot scratches its irritated flank. But the natural goal of

behavior (*i. e.*, of a purposive movement) is more than the cessation of a stimulus; its attainment, which brings the movement to a close, involves some positive novelty in the total situation. Thus, if we saw a dog lying in the sunlight and then saw him get up and wander about, we might suppose that the heat of the sun's rays had stimulated him to reflex walking; but, if we saw him walk to a patch of shade and there lie down and resume his slumber, we should confidently infer that this was behavior, a purposive movement attaining its natural goal.

(5) Reflex action does not show that preparation for the coming situation (the situation which will result from the action) which in behavior suggests anticipation of that future situation. Conceiving the dog as a reflex machine, a bundle of reflexes, we might legitimately imagine such a mechanism to be stimulated by a sudden noise, such as the sound of his master's voice, to rouse up from sleep, get upon his feet, and wander about. But nothing that we know of reflex action would justify us in supposing that he could be led by similar reflex processes to make all those preparations for a joyous and riotous welcome with which the normal dog responds to such a sense-impression.

(6) Reflex processes are not improved by repetition, as the movements of behavior are. The same stimulus, applied again and again under the same conditions, repeatedly evokes the same movements or train of movements. It is possible that by repetition a reflex movement may be rendered more fixed or more easily evocable; but even that has not, I think, been shown to be true. Still less has it been shown that any reflex process becomes more nicely adjusted, or more effective on repetition. The scratch-reflex of the brainless dog's hind-foot, which superficially resembles so closely a purposive movement, always shows a lack of nicety of direction toward the spot irritated, as compared with the scratching movements of the normal dog. And there is no evidence to suggest that any amount of repetition of stimulation at any one point of the skin of the brainless dog would result in a nicer direction of the foot toward the spot.

[*Editor's Note:* Material has been omitted at this point.]

A Seventh Mark of Behavior

In contrasting reflex action with purposive action or behavior, we must take notice of yet another distinction of great importance, which perhaps deserves to rank as a seventh objective mark of behavior; namely, a reflex action is always a partial reaction, but a *purposive action is a total reaction of the organism.* Let us examine this distinction more nearly. If your dog is lying idly by your side, perhaps occasionally snapping at a fly, you may, by pulling a hair or otherwise stimulating his flank, repeatedly provoke the scratch-reflex of his hind leg, without interfering with his repose; he may continue to snap, cocking his eye now and then at some interesting object, apparently quite unaware of your stimulus and of the machine-like movements of his leg—with a very comical effect. In a similar way, various reflex reactions may be simultaneously evoked by independent stimuli; and, so long as they do not affect the same organs, they do not interfere with one another; each is a strictly local reaction of a segment (or of some few segments) of the animal.[1] In a similar way in ourselves, many reflexes may be excited simultaneously and independently of one another; the pupil may contract to increase of light, the foot jerk to a tap on the tendon, the respiration or blood-vessels or heart respond to various stimulations.

In purposive action, on the other hand, the whole organism is commonly involved; the processes of all its parts are subordinated and adjusted in such a way as to promote the better pursuit of the natural goal of the action. If, while you amuse yourself by repeatedly exciting the scratch-reflex in your dog, some sound excites him to behavior, then, even though the behavior

[1] A few highly complex reflex movements involve many segments, *e. g.*, the walking movements of the dog's legs; and here, as in all opposed reflexes of the same limb, such as those of extension and flexion, the *law of reciprocal inhibition* holds good; that is to say, the more strongly excited reflex inhibits the less strongly excited and incompatible reflex.

consists in nothing more than assuming an alert attitude with eyes and ears directed toward the disturbing object, your stimulation of his flank becomes ineffective; at the same time the fly is ignored, and even your voice, raised in command, may fail to provoke any sign of obedience. If the sound is followed by the appearance of a stranger (dog or man) your dog springs to his feet with every muscle and organ at work in preparation for attack; and, while this condition continues, all his reflexes are subordinated to this major purposive activity; stimuli which, when he was at rest, would have evoked a wide range of reflexes produce no appreciable effect, and objects which might have provoked him to other trains of behavior are ignored. That is the type of the total reaction. The vital energies of the whole organism are concentrated upon the task in hand.[1]

The Relation of the Human to the Animal Mind

Having now studied the essential peculiarities of purposive action objectively observed, we may turn to review very briefly the actions of animals at various levels in the scale of life. The study of animal behavior teaches four lessons of high importance for psychology: (1). It makes clearer the nature of purposive action and reveals its prevalence throughout the whole of the animal world: (2) it elucidates the very foundations of human nature, by displaying in relative simplicity among the animals the modes of action (namely, instinctive actions) which are fundamental in human behavior, but which in human life are so complicated and obscured by the great development of our intellectual powers that their full importance is only now beginning to be recognized: (3) it shows us how we may conceive the structure of the relatively simple mind of an animal, and so gives us a valuable cue for building up our description of the structure of the human mind: (4) for it reveals some of the stages which the mind must have passed through in the long course of mental evolution from animalcule to man.

[1] For a more detailed discussion of the distinction between reflex and instinctive action, I would refer the senior student to my article, "The Use and Abuse of Instinct in Social Psychology" (*Journal of Abnormal and Social Psychology*, 1922).

[*Editor's Note:* Material has been omitted at this point.]

The behavior of insects is particularly interesting also, because insects are rightly held to illustrate in the richest and purest manner the operation of "Instinct." The problem of "Instinct" is of fundamental importance. Instinctive action, rather than reflex action, is in my view, and in that of many other psychologists, the key to the understanding of human behavior. It is the teaching of this book that human behavior is built up on a basis of innate tendencies which are in all essentials very similar to the instinctive tendencies of animals.

Others, notably Professor Bergson, take a different view. They assert that Instinct and Intelligence are two diverse developments of Mind that have little in common. They suggest that in the course of evolution Mind arrived at a parting of the ways, that the path of evolution was split into two divergent paths, Instinct and Intelligence; and that, while the insects followed the former and developed Instinct in a very high degree and Intelligence hardly at all, the vertebrates and mammals followed the other path and developed Intelligence till it culminated in the intellect of man; Instinct in this line remaining comparatively inactive, until it was rediscovered by Professor Bergson and recognized by him as the essential function involved in philosophic intuition.

The mechanists form a third party who see in Instinct nothing more than complex reflex action, often of the type of the chain-reflex, and sometimes modified or controlled by tropisms.

We have then to keep in mind these three rival views during our study of insect behavior; and we must attempt to find grounds of decision in favor of one or other of them.

The definition and demarcation of Instinct and Intelligence have been much debated of recent years; but no general agreement as to the precise use of the words has yet been reached, nor can be reached until agreement as to the nature of the facts denoted shall have been achieved. But the great majority of all parties would agree that we may properly call "instinctive" those actions of animals which seem to be purposive (*i. e.*, exhibit the marks of behavior) and which are performed by any animal independently of previous experience of similar situations. The surest evidence of such independence of prior experience is the performance of an action immediately after the animal emerges from its egg, or after some similar radical change of environment and mode of life; though in many other cases it is possible to be sure that, in spite of the lapse of time since such a change of life, the animal has not had contact with any object or situation similar to that which evokes the instinctive response. In defining instinctive action, some authorities would add that an instinctive action or tendency is one which is common to all members of the species. This is, no doubt, very generally true, but seems to be an unnecessary and extrinsic addition.

Intelligent action, on the other hand, is generally defined as one which seems to show that the creature has profited by prior experience of similar situations, that it somehow brings to bear its previous experience in the guidance of its present action. Instinct (abstractly conceived and with a capital letter) is native or inborn capacity for purposive action; Intelligence is the capacity to improve upon native tendency in the light of past experience.

The Hormic Theory

A word must be said here on the meaning of the word "teleological." I have insisted that all mental activity is purposive, that it is a striving toward a goal, however vaguely the goal may be thought of. The word "teleological," which means directed toward a goal, has often been applied to animal behavior, or to the realm of life in general, to imply that the processes of organisms are adjusted to bring about certain results which are the goal designed or willed by the Creator. If organic processes are described as teleological in this sense of the word, the view is compatible with the theory that all animals are merely machines or mechanisms. For the processes of a man-made machine are teleological in this same sense: they bring about the results designed and willed by the maker of the machine. We may, however, use the word "teleological" as equivalent to "purposive" in the sense in which it is defined in this chapter. The difference between the two meanings is very important.

[*Editor's Note:* Material has been omitted at this point.]

8B

Copyright © 1923 by Charles Scribner's Sons; copyright renewed 1951 by Anne A. McDougall

Reprinted from pages 92–93 of *An Outline of Psychology,* 2nd ed., Methuen, London, 1923, 456p., by permission of Charles Scribner's Sons

BEHAVIOR OF INSECTS

William McDougall

[*Editor's Note:* In the original, material precedes and follows this excerpt.]

Instinct and Intelligence Not Separable

The chief lesson I would have the student learn from the behavior of insects is that Instinct and Intelligence are not two diverse principles of action or of guidance of action. Instinctive behavior is indistinguishable from intelligent behavior by any outward mark. It is true that, when the same favorable conditions of the animal and of the environment are repeated, instinctive behavior is apt to be repeated with a regularity which gives it a machinelike air. But the same is true of intelligent behavior. If we have found the best way of dealing with a particular situation, then on its recurrence, if we are intelligent, we deal with it by the repetition of the same series of movements; as when we roll a cigarette or brush our hair. To do these things differently upon each occasion would add nothing to the evidence of our intelligence. When the circumstances of the creature are such that its instinctive actions do not immediately reach their goal, we see variation of direction and of mode and sequence of actions, with persistent effort. The behavior of the solitary wasps abounds in such variation and persistence in face of difficulties and failures to attain immediate success; for example, the varied modes of tugging, biting, and stinging the prey displayed by individual wasps, according to the size and shape of the prey and other circumstances of the task in hand. In some instances these variations astonish us by their effectiveness, their nice adaptation to the particular and unusual circumstances; for example, the seizing of the spider by the back, the omitting of the "ritualistic" act, the use of the pebble as hammer. These isolated acts are hard to interpret, though it is difficult to deny them the quality of intelligence.

More important are those forms of intelligent behavior which are displayed by all members of a species, and in which the past experience of the individual is unmistakably brought to bear in the guidance of present action. The return to the nest is the most striking and most abundantly illustrated of such actions. The tendency to explore the area for some days before making a nest; the tendency to make a particular study of the immediate neighborhood of the nest; the highly specialized capacity of all these species for organizing such experience in a systematic whole of knowledge of the area such as is implied by their return to the nest—all these tendencies and capacities are innate in the species. On the other hand, the knowledge of the locality which they acquire is not innate, but built up by many successive acts of perception.

How then shall we separate instinctive from intelligent behavior? We might perhaps say that the making of the locality study is instinctive, and that, when the wasp seeks her nest, the application of the knowledge so acquired is intelligent. But even this is an arbitrary distinction; for, in returning to the nest with her prey, the wasp is impelled by an innate tendency which is only awakened when the prey is secured, by an instinctive impulse of whose ultimate goal she perhaps has no foresight; yet this instinctive innate impulse can attain its end only when aided and guided by the acquired knowledge of the locality. Here, clearly, Instinct requires and implies the cooperation of Intelligence, and without its aid can achieve nothing of value to the creature or the species. And Intelligence operates only and always in the service of the instinctive impulses to action. This, then, is the relation of Instinct to Intelligence among the insects, which by common consent display Instinct in its purest and most typical forms. We shall see that the same relation obtains throughout the vertebrate realm, the human species not excepted.

8c

Copyright © 1923 by Charles Scribner's Sons; copyright renewed 1951 by Anne A. McDougall

Reprinted from pages 108-110, 113, and 118-119 of *An Outline of Psychology*, 2nd ed., Methuen, London, 1923, 456p., by permission of Charles Scribner's Sons

BEHAVIOR OF THE VERTEBRATES

William McDougall

[*Editor's Note:* In the original, material precedes and follows this excerpt.]

. . . some forms of instinctive behavior seem to involve as a rule a great, perhaps a maximal, output of energy; while others seem to be relatively feeble in this respect, and to be incapable of being excited to a degree which would involve any great output. Such a relatively feeble instinct is the "curiosity" of many species; while in very many species the mating instinct, the fighting instinct, and the instinct of escape seem capable of determining a maximal output of energy.[1] Some combination of these two views seems to be required, or something intermediate between them. We may best express the facts we are considering by saying that the excitement of "an instinct" evokes "an impulse" to action; that this impulse is variable in strength with any one instinct, according to the conditions (internal and external) of its excitement; and that the impulses of the several instincts of any animal are not capable of the same maximal strength. For we may observe how one instinctive impulse may overcome another, when they are simultaneously excited.[2]

Whether we conceive of each instinct as an independent store or spring of energy, or of all the instincts of an animal as drawing in various degrees upon a common store of energy and merely liberating and transmitting some proportion of this energy, when each in its turn comes into action, we may speak of the energy of the instinct, or of the energy derived from the instinct; and we may regard the felt strength of the impulse to action as in some sense a measure of the flow of such liberated energy.[3]

[1] Doctor W. H. R. Rivers suggests that every instinct obeys the "all or nothing" principle, *i. e.*, if it is excited at all it is maximally excited. This seems to me to be obviously untrue of many. (Cf. symposium on "Instinct and the Unconscious," *British Journal of Psychology*, 1919.) In his more recent book, "Instinct and the Unconscious," Rivers has modified his view by suggesting that the "all or nothing" principle holds good for some instincts, but not for others.

[2] Another way of expressing the same facts is to speak of the relative strengths of the several instinctive "tendencies"; but "impulse" seems the better word; impulse to action being something that we experience in ourselves as well as observe at work in others; and it denotes essentially a fact of activity; whereas "tendency" has the objectionable ambiguity, so beloved of psychologists, which enables it to be used to denote both facts of structure and facts of functioning.

[3] Though we have no guarantee that the correlation between the felt strength of the impulse and the flow of energy is a close one, our experience justifies us in assuming that it is in the main a positive correlation.

Those who like mechanical analogies may find some help in the following one. We may liken the dispositions which are the instincts of any individual organism to the following mechanical arrangement: Each instinct is represented by (1) a chamber within which a process of fermentation or other chemical change constantly liberates a gas, which accumulates under pressure. The several chambers communicate with one another by fine channels through which the gas can pass against considerable friction, when the pressure in any two chambers is different. (2) Each chamber has an outlet which branches into a complex system of pipes leading to a group of executive organs (nerves leading to muscles and glands). (3) This outlet is closed by a door or sluice-gate provided with a lock of more or less complex pattern peculiar to itself (in some cases it is a series of locks rather than a single one). This door is never quite gas-tight; gas leaks through, and leaks in larger quantity the higher its pressure in the chamber (appetite and the restlessness in which it is expressed). When the key is turned, the door swings open and the gas, issuing along the many channels, sets in action the various mechanisms to which it is led; and at the same time the relief of pressure in the chamber leads to a more rapid evolution or liberation of gas within it and to some drainage into it from other chambers. The key is the sensory-pattern presented by the specific object of the instinct (*e. g.*, the nightingale's song, the peacock's tail). The turning of the key is the act of perceiving the object. Such a mechanical model is inevitably inadequate in many respects to represent the psycho-physical disposition. It might perhaps be improved by replacing the locked door by a spring valve, which is opened by the short arm of a lever, in proportion as the long arm of the lever is depressed by a strong spring attached to its free end. Such depression is prevented, when the mechanism is at rest, by a series of stops, each of which can be drawn back by touching a key on a keyboard (like the keyboard of a piano). The complete depression of the lever and the fullest opening of the valve occur only when a certain combination of keys is struck; the striking of some of these keys will permit a partial depression of the lever. The keyboard is the array of sensory-organs; the key is any natural object which will strike the appropriate combination of keys. To complete the analogy, we should have to suppose that the mechanisms actuated by the pressure of the released gas are such as will, under favorable conditions, change the set of affairs which determines the striking of the keys, so that a new combination of keys is struck; the new combination releases the lever from its spring and so allows the spring valve to return to the closed position, and perhaps actuates another lever opening another valve (as in the case of chain-instincts), and so sets a different mechanism in action.

This rude analogy, which likens the organism to an organ whose keys and stops are played upon by Nature, is, I repeat, wofully inadequate; but certainly less inadequate than the description of instinctive behavior as a series of simple reflexes and tropisms.

When an instinctive impulse is liberated or evoked, the organism becomes absorbed in the endeavor toward the goal of the instinct; its reaction to the exciting object is a total reaction; its energies are concentrated on the task in hand and the functioning of the various organs is subordinated to and harmonized with the dominant system of activity. This absorption of the organism in any particular task or mode of activity is what we call in ourselves "attention"; and that general excitement whose indications we observe in the animals, when their instincts are strongly excited, is what we call in ourselves "emotion." We may therefore define "an instinct" as an innate disposition which determines the organism to perceive (to pay attention to) any object of a certain class, and to experience in its presence a certain emotional excitement and an impulse to action which find expression in a specific mode of behavior in relation to that object.[1]

This formula will hold good generally only if we take the word "object" in the very widest sense, namely, to include not only material things and organisms, but also various situations in which the creature may find itself, consisting in conjunctions of internal and external conditions: for we must remember always that a large part of the sensory stimuli that play upon the sense-organs come from within the organism; and these constitute an important and constantly changing part of the total complex of sense-stimuli.

I shall illustrate and further justify this definition of "an instinct" on later pages. It may be, and often is, difficult to mark off a particular action or train of behavior as the expression of one instinct. But the difficulty is not one of principle, but rather one of practice. In principle it is legitimate to ask How many instincts has this species of animal? and to endeavor to define the objects or situations which will bring each one into play, the kind of behavior which each one will determine, and the kind of goal (or change of situation) which will satisfy the impulse and bring the train of activity to a close.

[1] We might attempt to enrich the definition by placing before the word "disposition" the adjective "mental," "physiological," "neural," or "psychophysical." Of these the last is, perhaps, preferable to the others, because it clearly implies that the disposition plays a part in determining both bodily action and the course of experience. If we use the adjective "physiological" or "neural," we should do so with the explicit understanding that it is not meant to imply the mechanistic interpretation of instinctive action.

[*Editor's Note:* Material has been omitted at this point.]

This is a fact too often overlooked in discussions of biological evolution. The evolution of the animal world is primarily a process of differentiation and specialization of instincts; this much-neglected truth is obvious when we reflect on such facts as the following: Horned cattle or antlered deer do not fight because they are provided with horns or antlers; the species develops these weapons in the service of the instinct of combat. The carnivore does not prey on animals because he has large teeth; his teeth and claws have been evolved, because his food-seeking instinct has become specialized in this direction. The seal did not take to the water because his legs were flipperlike and his body fish-shaped; he acquired these peculiarities of structure in consequence of his food-seeking instinct having become specialized for the pursuit of fish. The same is true of a thousand instances of shape and form and coloration, of bodily structure and function, down to the minutest details. The evolution of the animal world may properly be conceived as primarily and essentially the differentiation of instinctive tendencies from some primordial undifferentiated capacity to strive. It is this undifferentiated capacity to strive, this primordial energy, which M. Bergson has named *l'élan vital*, which others (notably Doctor C. G. Jung) speak of as the *libido*, and which perhaps is best named *vital energy*. We may regard the instincts as so many differentiated channels through which the vital energy pours itself into or through the organism.

Two (or more) instincts may be simultaneously excited. If their tendencies are not incompatible, behavior is then a blend of the actions characteristic of the two instincts, each being modified by the other. If their tendencies are opposed, we may witness a struggle, with alternation of the movements of opposite kinds, until one gains the upper hand. Or, when one instinct is in operation, an opposed instinct may be excited (although this is not so easily excited as when the animal is at rest) and may cut short more or less suddenly and completely, may inhibit, the other mode of behavior; in which case we may assume that the impulse of the second is stronger than that of the former.[1]

[*Editor's Note:* Material has been omitted at this point.]

[1] This seems to be an instance of the general law of reciprocal inhibition of incompatible movements.

In a similar way, the first pecking of the newly hatched chick seems to be not a mere reflex in response to the optical stimulus of a small grain, but rather the actuation of this particular motor-mechanism by the impulse of the food-seeking instinct: this impulse may express itself by means of other mechanisms also; as when the chick runs to and fro, and especially when he runs to his mother as she scratches the earth and gently calls him with a specific note.

This principle, that *any one instinctive impulse may make use of a variety of motor mechanisms*, according to the circumstances of the moment, is no doubt difficult to understand or interpret in terms of the mechanistic hypothesis, *i. e.*, in terms of reflex mechanism; but it seems to be a fact, nevertheless, and unless we recognize it and take it fully into account, we meet with insuperable difficulties in attempting to interpret instinctive behavior.

How Is an Instinct Defined If Not by Its Motor Expressions?

We may raise at this stage, also, a still more difficult question. If one instinct does not always express itself in the same kind of bodily activity, but rather may actuate or make use of several different motor mechanisms in succession, according to the circumstances of the moment, how are we to recognize the goal of the instinct, and how define the instinct and mark it off from others? Obviously, if every instinct always expressed itself by setting in action some one motor mechanism peculiar to itself, or several in a given order, there would be no difficulty in defining the instinct; and also the path of the "mechanist" would be made comparatively smooth, for one of the principal distinctions between instinctive and reflex action would be lacking. Is there, in fact, any stable position between the mechanist's identification of instinctive with reflex action and the mere attribution of actions to "Instinct" written with a capital letter? I think there is. "An instinct" is to be defined and recognized, not by the kind of movements in which it finds expression, but by the kind of change of the animal's situation which its movements, whatever they be, tend to bring about and which, when it is achieved, brings the train of behavior to a close. Thus the nature of the instinct at work in an animal cannot be recognized by simple observation of its movements. You may see one pigeon pursuing another assiduously from place to place; but these varied movements of locomotion and pursuit may express either the combative instinct, or the pairing instinct, or the food-seeking impulse of the young pigeon. Yet there can be no doubt that these are distinct instincts, whose operation is attended by appetites requiring very different situations for their satisfaction. Unlike reflex action, instinctive action strives toward a goal, a change of situation of a particular kind, which alone can satisfy the impulse and allay the appetite and unrest of the organism. We must, therefore, define any instinct by the nature of the goal, the type of situation, that it seeks or tends to bring about, as well as by the type of situation or object that brings it into activity.

Part III

THE TANGLED ROOTS OF COMPARATIVE ETHOLOGY AND COMPARATIVE PSYCHOLOGY IN AMERICA

Editor's Comments on Papers 9 Through 12

In this section we consider two important figures in American psychology and biology that greatly influenced the study of animal behavior. The first of these, Charles O. Whitman, is frequently referred to as an early founder of ethology. Lorenz (Paper 23), Tinbergen (1951), Thorpe (1979) and other first-generation classical ethologists cited his work, particularly the lectures reprinted here. His widely appreciated phrase that "instinct and structure are to be studied from the common standpoint of phyletic descent" (Paper 10) helped give credence to the ethological insistence on the possibility and importance of a "comparative anatomy of behavior."

Whitman was a critical observer, suspicious of anecdotes. In this he echoed Lloyd Morgan but he was even more devastating in his criticism of anecdotal evidence. In the first reading (Paper 9) one can almost see, hear, and feel the affront to Whitman engendered by Wesley Mills' tales about squirrels. Mills was not the typical Victorian eccentric writing about animals in a romantic manner; he was America's first self-conscious "comparative psychologist," a founder of a comparative psychology society, as well as the writer of the influential suggests, he called for life-long developmental and naturalistic observation on animals (Gottlieb, 1979; Dewsbury, 1984; also cf. Morgan, Paper 7C).

Whitman's lectures (Paper 10) were more than a call for comparative analysis of behavior. They addressed a wide range of topics; it is surprising that few contemporary ethologists have read him. His use of "simple" animals such as leeches and mudpuppies was an adroit attempt to isolate basic processes for study. As the value of leeches in neurobiological research grows, Whitman is increasingly recognized for his developmental studies of them (Weisblat et al., 1984). Ornithologists are greatly in his debt for his years of work with pigeons and doves; Whitman even foster-reared different species and recognized the phenomenon of imprinting (see Paper 17). A major part of his lectures is devoted to rejecting the lapsed intelligence theory of instinct and the inheritance of acquired characteristics. The "lapsed intelligence" theory was an extreme version of the inherited habit theory of the origin of instinct. It is remarkable that forty years after Darwin presented arguments showing that reliance on inherited acquisitions was theoretically unnecessary and in some cases completely untenable, such efforts were necessary. Morgan had also been skeptical in *Habit and Instinct* but was not quite so forceful in his statements.

Because of his position at the University of Chicago and as director of the Woods Hole Marine Biology Laboratory, Whitman was at the center of much research in the United States. The year following his behavior lectures Edward L. Thorndike was invited to Woods Hole to talk on behavior.

Thorndike is frequently called the founder of experimental animal-learning psychology. Fifty years later the intellectual descendants of Whitman and Thorndike could barely communicate in a common scientific language; they had little respect or confidence in the approach, methods, and theories of their counterparts. The irony, of course, is that in 1899 there was no conflict (Burghardt, 1973); both favored rigorous objective study of both instinct and intelligence. Years later, as a leading educational psychologist, Thorndike (1913) favorably quoted Whitman's lectures at length.

The two lectures by Thorndike are on instinct (Paper 11) and associative processes (Paper 12). They demonstrate the astuteness of early animal-learning psychologists, their sympathy with evolutionary and comparative analysis, and their thoughtful appreciation of methodological rigor without rigor mortis, the fate of many to come. These Thorndike lectures are apparently little known by psychological historians. Note what Thorndike had to say about scratching in birds, reptiles, and mammals in Paper 11. Ethologists generally (e.g., Lorenz, Paper 23) credit Heinroth (Paper 16) for this compelling example of homologous behavior across vertebrate classes.

Thorndike's paper on instinct provides an attempt at careful objective definition; it is impatient with discussions of consciousness as found in Spencer and Morgan. It seems to anticipate Watsonian behaviorism by more than a decade (Watson, 1914). Thorndike was also impressed with how "indefinite and inexact," not to say "vague," were instincts. He thus presaged one basis on which experimental psychologists were eventually to discredit, then ignore, instinct. The second lecture (Paper 12) illustrates the comparison and conjunction of studies of innate and learned behavior which, with some notable exceptions (see Dewsbury, 1984*a*), were soon to largely disappear from American animal psychology. The final paragraph illustrates that the core idea of Skinner's "Phylogeny and Ontogeny" paper (1966) was known to the man who provided a prototype of the apparatus Skinner was later to develop. Thorndike, as did Whitman, took vigorous exception to writings by Wesley Mills. A circle exists (Burghardt, 1978), and it is small.

REFERENCES

Burghardt, G. M., 1973, Instinct and Innate Behavior: Toward an Ethological Psychology, in *The Study of Behavior,* J. A. Nevin, ed., Scott, Foresman, Glenview, Ill., pp. 322–400.

Burghardt, G. M., 1978, Closing the Circle: the Ethology of Mind, *Behav. Brain Sci.* **1:**562–563.

Dewsbury, D. A., 1984*a*, *Comparative Psychology in the Twentieth Century,* Hutchinson Ross, Stroudsburg, Pa., 413p.

Dewsbury, D. A., 1984*b*, *Foundations of Comparative Psychology,* Benchmark Papers in Behavior Series, Van Nostrand Reinhold, New York, 365p.

Gottlieb, G., 1979, Comparative Psychology and Ethology, in *The First Century of Experimental Psychology* (E. Hearst, ed.), Erlbaum, Hillsdale, New Jersey, pp. 147–173.

Skinner, B. F., 1966, The Phylogeny and Ontogeny of Behavior, *Science* **153:**1205–1213.

Thorndike, E. L., 1913, *Educational Psychology, Vol. 1, The Original Nature of Man,* Teachers College, Columbia University, New York, 327p.

Tinbergen, N., 1951, *The Study of Instinct,* Clarendon Press, Oxford, 228p.

Watson, J. B., 1914, *Behavior. An Introduction to Comparative Psychology,* Holt, New York, 439p.

Weisblat, D. A., S. S. Blair, A. P. Kramer, D. K. Stuart, and G. S. Stent, 1984, Cell Lineage and Cell Interaction in the Developing Leech Nervous System, *BioScience* **34:**313–317.

9

Reprinted from *Monist* **9**:524-537 (1899)

MYTHS IN ANIMAL PSYCHOLOGY

C. O. Whitman

THE life-histories of animals, from the primordial germ-cell to the end of the life-cycle; their daily, periodical, and seasonal routines; their habits, instincts, intelligence, and peculiarities of behavior under varying conditions; their geographical distribution, genetic relations and œcological interrelations; their physiological activities, individually and collectively; their variations, adaptations, breeding and crossing,—in short, the *biology* of animals, is beginning to take its place beside the more strictly morphological studies which have so long monopolised the attention of naturalists. The revival of interest in general life-phenomena, and especially in the psychical activities of animals, takes its date from Darwin's epoch-making work. The phenomenal insight which this great naturalist brought to the study of animal instinct and intelligence illuminated the whole subject and prepared the way for the development of a new science, commonly designated "Animal Intelligence; or, Comparative Psychology." That mind and body must have been evolved together and under the same natural laws was the conclusion destined to become the corner-stone, not only of biology, but also of rational psychology.

Darwin's views triumphed, as all the world knows; but while his ideas have been generally accepted, his method, the real secret of his success, has had too few followers. Darwin's method was to prepare himself for his problem by long-continued and close examination of all its details and bearings. He was no hustler on the jump for notoriety, no rapid-fire writer; but a cool, patient, indefatigable investigator, counting not the years devoted to preliminary work, but weighing rather the facts collected by his tireless industry, and testing his thoughts and inferences over and over again, until well-assured that they would stand. Such a method was altogether too laborious and searching to be imitated by students ambitious to reach the heights of comparative psychology through a few hours of parlor diversion with caged animals, or by a few experiments on domestic animals. We are too apt to measure the road and count the steps beforehand. Darwin allowed the subject itself to settle all such matters, while he forgot time in complete absorption with his theme. Neglect of Darwin's example in this respect has been unfortunate for both general animal biology and the coming science of comparative psychology. An examination of a few typical cases in recent literature may help make us more heedful of Darwin's example, and more reserved in announcing observations and conclusions which have not passed through the furnace of verification and repeated revision.

One such case[1] is furnished in a recent volume on *Animal Intelligence*, by Mr. Wesley Mills of McGill University. It is a case of

ALLEGED FEIGNING IN SQUIRRELS.

As the subject of feigning is one of great interest, as the method of treatment is especially instructive from the point of view before defined, and as the observations are presented as a contribution to comparative psychology, the case is entitled to special attention, and I shall, therefore, make it the leading subject for examination. The author stimulates interest in his communications by announcing that they give two examples in which feigning was strikingly manifested; and in another place he speaks of them as among the most typical cases of such behavior ever recorded.

After reading these observations through and through with care and in the full expectation of finding every promise fulfilled, I have to confess my inability to discover any satisfactory evidence

---

[1] The selection of this case, it may be hardly necessary to say, was due to its nature and fitness for the purpose in view. It would not be fair to judge of the book as a whole from this small part. The book contains much interesting matter and will doubtless be widely read as it deserves.

of feigning. Naturally, I am disappointed and surprised, and all the more so as it seems to me that Mr. Mills himself must be credited with all the feigning he has ascribed to his two chickarees; that is to say, the supposed feigning is a misinterpretation. Whether I am correct or not, an examination of Mr. Mills's observations cannot fail to be of interest. The subject of animal intelligence has scarcely yet emerged from the mythical state, and no part of the subject is in a more hopeless tangle of misinterpretation than the so-called feigning of animals. It must be said to the credit of Mr. Mills that he has kept his observations apart from his interpretations, and he has thus made it possible for the reader to draw his own conclusions.

A few instances to illustrate how easily people allow themselves to be misled in regard to animal intelligence and to draw conclusions from evidence supplied largely or wholly from the imagination, may put us in a more cautious frame of mind for interpreting the behavior of Mr. Mill's squirrels.

A Horse Protects His Master from the Tusks of a Savage Boar.

"George Howard, nineteen years of age, who has been employed on the farm of George Lent, about a mile outside of the city on the Buffalo road, is at the Homeopathic Hospital, suffering from injuries inflicted on him by a hog. That young Howard is not a subject for the coroner instead of the hospital surgeon is due to the fact that a horse which has been a great favorite of Howard and is greatly attached to the boy, kicked the enraged hog away as the brute was about to fasten its teeth in the boy's throat. The horse has always been looked upon by Farmer Lent as a remarkably intelligent member of the equine family, but he is now considered a wonder, and had the farmer not himself witnessed the act of the horse, he would never have believed that an animal could display such intelligence.

"The hog which made the attack on Howard was a large and particularly ugly brute. He broke out of his pen yesterday afternoon, and made a rush for the barn. The door was open and young Howard, who had just placed his favorite horse back into his stall after a careful grooming, was just starting to go out of the door when the enraged hog entered with a rush. The brute made a savage attack on the boy, and, fastening his teeth on the calf of the leg, tore and lacerated the flesh. Howard fell back into the stall and close to the feet of the horse he had just groomed.

"The hog was springing at the throat of his prostrate victim when the horse raised his hind feet and gave the hog a kick which sent him ten feet and caused him to squeal with pain. Mr. Lent, who had been attracted by the screams of the boy, was just entering the barn door as he saw the horse kick the hog off the prostrate body of the boy."

This account from the *Rochester Union and Advertiser* appears to be entirely reliable, so far as the circumstances are concerned; but these, it will be seen, do not justify the conclusion that the horse kicked the hog in order to protect the boy. The hog was probably kicked without a thought of the boy. The fright of the horse would cause it to kick in its own defence, and we are thus left without the slightest evidence of any altruistic motive in the act.

Story of the Dog-Fish (Amia Calva) and Its Young.

The following statement is taken from George Brown Goode's *Natural History of Useful Aquatic Animals* (pp. 659–660). It is a quotation from a Dr. Estes, but Mr. Goode indorses it as a part of "the best description of the habits of the fish."

Dr. Estes says:

"I come now to mention a peculiar habit of this fish, no account of which I have ever seen. It is this: While the parent still remains with the young, if the family become suddenly alarmed, the capacious mouth of the old fish will open, and *in rushes the entire host of little ones; the ugly maw is at once closed, and off she rushes to a place of security, when again the little captives are set at liberty.* If others are conversant with the above facts, I shall be very glad; if not, shall feel chagrined for not making them known long ago."

It is true that the old fish (the male) will sometimes open wide his mouth when approached, as if threatening an attack. It is also true that the swarm of young will suddenly disappear at any slight disturbance in the water, and after an interval of some minutes of quiet reappear at or near the place of disappearance.

At the moment of alarm and disappearance of the young, the old fish rushes off a short distance, stirring up the mud as he leaves. If the observer keeps perfectly quiet for some minutes, the parent fish may often be seen returning very slowly and cautiously so as not to be seen. Soon after he reaches the place in which the young are concealed at the bottom, they begin to gather about him and renew their feeding on small aquatic animals abundant in the grass along the shore.

Dr. Estes had seen the old fish open its mouth, and the young disappear as the fish dashed away. He had seen the young again with the parent fish, not far from where they were first observed. He did not take the trouble to find out how the young escaped from sight, and jumped at the conclusion that they had taken refuge in the mouth of the old fish. What a wonderful tale, and how strange that a conscientious observer could so completely humbug himself! Now this is no exceptional case; it is one of the most common occurrences, and that, too, even among men of high standing in science.

Let us now take an example from the comparative psychologist, who always has on hand an unlimited supply of this kind of material.

The Story of the Insane Pigeon.

This story, which is taken from *The Mental Evolution of Animals* (p. 173) by Mr. Romanes, has been thought worthy of translation into German by Karl Gross in his *Spiele der Thiere.* The case was reported to Mr. Romanes by a lady, and is given in her own words:

"A white fantail pigeon lived with his family in a pigeon-house in our stable-yard. He and his wife had been brought originally from Sussex, and had lived, respected and admired, to see their children of the third generation, when he suddenly became the victim of the infatuation I am about to describe.

"No eccentricity whatever was remarked in his conduct until

one day I chanced to pick up somewhere in the garden a ginger-beer bottle of the ordinary brown-stone description. I flung it into the yard, where it fell immediately below the pigeon-house. That instant down flew paterfamilias and to my no small astonishment commenced a series of genuflections, evidently doing homage to the bottle. He strutted round and round it, bowing and scraping and cooing and performing the most ludicrous antics I ever beheld on the part of an enamored pigeon. . . . Nor did he cease these performances until we removed the bottle; and, which proved that this singular aberration of instinct had become a fixed delusion, whenever the bottle was thrown or placed in the yard—no matter whether it lay horizontally or was placed upright—the same ridiculous scene was enacted; at that moment the pigeon came flying down with quite as great alacrity as when his peas were thrown out for his dinner, to continue his antics as long as the bottle remained there. Sometimes this would go on for hours, the other members of his family treating his movements with the most contemptuous indifference, and taking no notice whatever of the bottle. At last it became the regular amusement with which we entertained our visitors to see this erratic pigeon making love to the interesting object of his affections, and it was an entertainment which never failed, throughout that summer at least. Before next summer came round, he was no more."

Mr. Romanes remarks:

"It is thus evident that the pigeon was affected with some strong and persistent monomania with regard to this particular object. Although it is well known that insanity is not an uncommon thing among animals, this is the only case I have met with of a conspicuous derangement of the instinctive as distinguished from the rational faculties,—unless we so regard the exhibitions of erotomania, infanticide, mania, etc., which occur in animals perhaps more frequently than they do in man."

This pigeon, whose behavior has given it so wide fame as a case of deranged instinct, was undoubtedly a perfectly normal bird; and had Mr. Romanes been familiar with the antics of male pigeons, he would have found nothing in the performances to indicate insanity. I have seen a white fantail play in the same way to his shadow on the floor, and when his shadow fell on a crust of bread he at once adopted the bread as the object of his affection, and went through all the performances described by the lady, even to repeating the behaviour for several days afterward when I placed the same piece of bread on the floor of his pen. If one is looking for insanity in pigeons, let him first know the normal range of sanity, and pay little heed to stories of inexperienced observers who are apt to overlook circumstances essential to a correct understanding of what they report.

It is not improbable that the lady's amusing pigeon at first took the bottle for a living intruder upon his ground, and flew down to it for the purpose of driving it off. Finding it at rest, if his shadow fell upon it, or if his image was even faintly reflected from its surface, he would readily mistake it for a female pigeon, and after once getting this idea and performing before it, the bottle would be remembered and the same emotions excited the next time it was presented. The only value this suggestion can have is, that it is based on a similar case. The lady's observations were incomplete at the critical moment, i. e., at the time of the *first* performance, and it is too late to mend the failure.

The essentials to understanding any peculiar case of animal behavior are almost invariably overlooked by inexperienced observers, and the best trained biologist is liable to the same oversight, especially if the habits of the animal are not familiar. The qualification absolutely indispensable to reliable diagnosis of an animal's conduct is an intimate acquaintance with the creature's normal life, its habits and instincts. Little can be expected in this most important field of comparative psychology until investigators realise that such qualification is not furnished by parlor psychology. It means nothing less than years of close study,—the long-continued, patient observation, experiment, and reflexion, best exemplified in Darwin's work.

Let us now examine

TWO CASES OF SUPPOSED FEIGNING IN SQUIRRELS, AS REPORTED BY MR. MILLS.

Case I. (Pp. 61–62.)

"I was standing near a tree in which a red squirrel had taken up a position, when a stone thrown into the tree was followed by the fall of the squirrel. I am unable to say whether the squirrel was himself struck, whether he was merely shaken off, or how to account exactly for the creature's falling to the ground. Running to the spot as quickly as possible, I found the animal lying apparently lifeless. On taking him up, I observed not the slightest sign of external injury. He twitched a little as I carried him away and placed him in a box lined with tin, and having small wooden slats over the top, through the intervals of which food might be conveyed. After lying a considerable time on his side, but breathing regularly, and quite free from any sort of spasms such as might follow injury to the nervous centres, it was noticed that his eyes were open, and that when they were touched winking followed. Determined to watch the progress of events, I noticed that in about an hour's time the animal was upon his feet, but that he kept exceedingly quiet. The next day he was very dull—ill, as I thought, —and I was inclined to the belief, from the way he moved, that possibly one side was partially paralysed; but finding that he had eaten a good deal of what had been given him (oats), I began to be suspicious. Notwithstanding this apparent injury, that very day, when showing a friend the animal, on lifting aside one of the slats a little, he made such a rush for the opening that he all but escaped. On the third day after his capture, having left for a period of about two hours the sittingroom (usually occupied by two others besides myself) in which he was kept, I was told, on my return, by a maid-servant and a boy employed about the house, that some time previously the squirrel had escaped by the window, and, descending the wall of the house, which was 'rough-cast,' he had run off briskly along a neighboring fence, and disappeared at the root of a tree. When asked if they saw any evidence of lameness, they laughed at the idea, after his recent performances before their eyes. For several days I observed a squirrel running about, apparently quite well, in the quarter in which my animal had escaped, and I feel satisfied that it was the squirrel that I had recently had in confinement, but, of course, of this I cannot be certain.

"I believe, now, that this was a case of feigning, for if the injury had been so serious as the first symptoms would imply, or if there had been real paralysis, it could not have disappeared so suddenly. An animal even partially paralysed, could scarcely have escaped as he did and show no sign of lameness. His apparent insensibility at first may have been due to catalepsy or slight stunning. But while there are elements of doubt in this first case, there are none such in that about to be described."

Substantially the case is as follows:

1. A stone was thrown at a red squirrel in a tree, the animal fell to the ground apparently lifeless, there was no mark of external injury, but the squirrel *twitched* a little when taken up; it was placed in a box, where it lay upon its side, breathing regularly; after some time it was noticed that the eyes were open, and that winking resulted from touching.

If the squirrel was stunned, as seems probable, the behavior so far would not indicate feigning, so far as I can see.

2. In about an hour's time, the animal was found upon its feet, but it kept quiet; the next day the squirrel looked dull, but *moved as if injured in one side;* it had eaten oats.

I see nothing in all this to raise the "suspicion" that the injury was unreal and feigned.

3. This same day the squirrel tried to escape, when alarmed by the lifting of a slat.

Surely nothing surprising in a wild squirrel well enough to eat, even if it was still suffering from an injury.

4. On the third day after capture, according to testimony of servants, the squirrel escaped through an open window, ran off briskly along a fence, and disappeared at the root of a tree. Servants noticed no lameness.

An animal well enough to make a vigorous dash for liberty the day before, might well escape in the manner described. The servants' testimony as to the absence of lameness amounts to nothing. The squirrel subsequently seen by Mr. Mills, running about, "apparently quite well," may or may not have been the one he lost. Observe that Mr. Mills does not *know* whether the squirrel was injured or not. There was an appearance of injury and every reason to believe it was real, yet the cause of the injury, if real, and its nature and extent were not definitely known. Mr. Mills asserts that, *if* the injury had been as serious as the first symptoms *implied*, it *could* not have disappeared *so suddenly*. There are too many unknown elements for any positive conclusion. We do not know that the lameness had entirely disappeared at the time of escape; and if it had, there would not seem to have been any remarkable suddenness after three days' convalescence.

In this case nearly every point of critical importance was undetermined, and the author seems to be too little familiar with squirrel behavior.

The second case is claimed to be free from any element of doubt. "A more typical case of feigning than this one," says Mr. Mills, "could scarcely be found."

"A Chickaree was felled from a small tree by a gentle tap with a piece of lathing. He was so little injured that he would have escaped, had I not been on the spot where he fell and seized him at once. He was placed forthwith in the box that the other animal had occupied. He manifested no signs whatever of traumatic injury. One looking in upon him might suppose that here was a case of a lively squirrel being unwell, but events proved otherwise. He ate the food placed within the box, but only when no one was observant. He kept his head somewhat down, and seemed indifferent to everything. When a stick was placed near his mouth he savagely bit at it; but when a needle on the end of the same stick was substituted he evinced no such hostility. He made no effort to escape while we were in the room, but on our going down to dinner he must at once have commenced work, for on returning to the room in half an hour he was found free, having gnawed one of the slats sufficiently to allow him to squeeze through. With the assistance of a friend he was recaptured, but during the chase he showed fight when cornered, and finally, as he was being secured, I narrowly escaped being bitten. He was returned to his box which was then covered with a board weighted with a large stone. Notwithstanding, he gnawed his way out through the upper corner of the box during our absence on one occasion shortly afterwards.

"I think a more typical case of feigning than this one could scarcely be found."

The essentials are as follows:

1. A Chickaree, knocked from a tree with a piece of lathing, was captured and caged as before. Why, "one looking in upon him might suppose that there was a case of a lively squirrel *unwell*," is not explained. A very important point, but with no more information, we are unable to judge whether the squirrel was feigning or Mr. Mills imagining. If the animal was merely quiet through fear, as seems most probable from there being no further description, who that is familiar with squirrels would have surmised that it was feigning sick?

2. The squirrel did not eat when one was watching it. Perfectly natural. Fear would prevent.

3. It kept its head "somewhat down," and seemed indifferent, but when a stick was placed near its mouth it bit at it savagely. Mr. Mills seems to regard this as evidence of feigning indifference or sickness. If such behavior is feigning, Mr. Mills is a true discoverer.

4. The squirrel made no effort to escape while Mr. Mills was present, but did get free when left alone for half an hour at dinner-time. Such evidence of feigning has a decidedly entertaining side, to say the last. The squirrel seems to be the cleverer fellow every time, for he is serious while the observer thinks he is fooling. Who has not seen a squirrel hide behind a branch of the trunk of a tree to escape being seen by a person approaching? Is keeping *quiet* under such circumstances *feigning* quiet? If a confined squirrel, alarmed at our presence, sits still while we are watching him, but tries to get free when left alone, is there any deception in his behavior except what we ourselves invent?

5. The squirrel was recaptured, but showed fight when cornered, and Mr. Mills narrowly escaped being bitten. *Mirabile dictu!* A good bite would have been the best feint of all. Mr. Mills's good luck was an untold loss to comparative psychology.

6. "The squirrel was returned to its box, and *a board weightea with a large stone* placed over it. *Notwithstanding* he gnawed his way out through the upper corner of the box during our absence on one occasion shortly afterwards."

A large stone on a board to keep the animal in, can only be taken as another feint on the part of Mr. Mills, for of course he did not expect thus to prevent gnawing out. The size of the stone did not fool the squirrel, whoever else was taken in.

Further, on p. 71, Mr. Mills comes to the question of what is essential to feigning death or injury. "It is to be remembered," says the author, "that in these cases the animal simply remains as quiet and passive as possible. . . . It is within the observation of all that a cat watching near a rat-hole, feigns quiet. . . . A great part of the whole difficulty, it seems to me, has arisen from the use of the expression 'feigning death.' What is assumed is *inactivity* and *passivity*, more or less complete. This, of course, bears a certain degree of resemblance to death itself."

Darwin carefully compared the appearance of death-feigning insects and spiders with that of the really dead animals, and the result was, as he says, "that in no one instance was the attitude exactly the same, and in several instances the attitude of the feigners and of the really dead were as unlike as they possibly could be." (See Appendix to Romanes's *Mental Evolution in Animals*, p. 364.)

Romanes (p. 308) states this result in less cautious language: "All that 'shamming dead' amounts to in these animals is an instinct to remain motionless, and thus inconspicuous, in the presence of enemies."

Mr. Mills makes the conclusion still broader, assuming that the *essential* thing in feigning is *quiet*. That, even in the case of insects, quiet is not the distinctive character of feigning seems evident when we remember that the non-feigning state may be one of as perfect quiet as that of the feigning state. The mere passivity does not of itself discriminate between these two very different states; in other words, it does not give us the criterion of either state. The essential thing is not a non-differential element, common to the two states. The "essential" must give us the difference, and enable us to distinguish clearly between the normal state of rest and the so-called feigned condition. The quiet of an animal at rest and that of the same animal feigning death, are two very different things; otherwise we should have no use for the term "feigning" as a means of distinction. In one case the quiet is perfectly normal and signifies only a state of rest; in the other it means an *assumed* or *induced* condition, as the result of disturbance and alarm.

The cause, the conscious purpose or the blind adaptation, and the external appearances are all essentially unlike in the two cases. Look at the beetle at rest on the branch or leaf of a tree, and at the same beetle after it has dropped to the ground, alarmed by some unusual jar, lying as it fell, motionless, on its side or back. Is the quiet now the same as before? or is it as different as calm unheeding composure and the stupor of terror, or the stillness deliberately maintained to escape discovery? Whether cataleptic or voluntary, the so-called feigned quiet has no fundamental likeness with the quiet of normal rest. There is only a deceptive outward semblance, which speedily vanishes on closer comparison.

In the quiet of a cat before a rat-hole, we have quite a different phenomenon, and one to which the term feigning seems to me to have no legitimate application. There is no fear, no involuntary suspension of activity, no attempt to imitate a state of death, or to falsify appearances in order to escape enemies. The quiet is deliberately maintained, not on account of alarm, but to avoid giving alarm to her intended victim; not to elude but to capture, the rat. The cat is not surprised, but she hopes to surprise the rat. She has the same end in view when she stalks a bird, keeping behind some intervening object that hides her from view. Here the cat is in motion and glides on with manifest satisfaction in her advantage; and if she is feigning, she is certainly not feigning quiet. It must be evident, I think, that if feigning does not properly characterise the *action* of the cat in this case, it cannot properly define the *inaction* in the other.

Returning to his "feigning squirrels" (p. 72), Mr. Mills tells us more explicitly what he understands by feigning in their case.

"These little animals were naturally led, under the unwonted circumstances of their confinement, to disguise, in an extraordinary degree, their real condition, and even to imitate an unusual and unreal one. The mental process is a complex of instinct pure and simple, with higher intellectual factors added, and the cases of these squirrels, thus feigning, are among the clearest that, so far as I am aware, have ever been recorded."

This leaves no doubt that Mr. Mills believes he saw something more than feigning quiet in his squirrels. "*Disguise of the real and imitation of the unreal*," is what Mr. Mills claims to have seen, and what I have failed to find any satisfactory evidence of in the reports he has given. In fact, the observations seem to me to indicate no feigning at all on the part of the squirrels, and to show very clearly that Mr. Mills failed to get reliable data at just the most critical points. It is the old failure of anecdote psychology.

If it be true, as I think will be generally admitted, that comparative psychology is a science of the future; and if at present it is only a part of general biology, it follows that any attempt to soar to "the nature and development of animal intelligence," except through the aid of long schooling in the study of animal life, is doomed to be an Icarian flight.

10

Reprinted from *Woods Hole Mar. Biol. Lab. Biol. Lect.* (1898) **6:**285–338 (1899)

ANIMAL BEHAVIOR.

C. O. WHITMAN.

"*Natura non facit saltum*, is applicable to instincts as well as to corporeal structure." — DARWIN, *Origin of Species*, p. 231.

CONTENTS.

Animal behavior, long an attractive theme with students of natural history, has in recent times become the centre of interest to investigators in the field of pyschogenesis. The study of habits, instincts, and intelligence in the lower animals was not for a long time considered to have any fundamental relation to the study of man's mental development. Biologists were left to cultivate the field alone, and psychologists only recently discovered how vast and essential were the interests to which their science could lay claim.

The contribution which I have to offer aims at no extensive exposition of the subject, but rather to call attention to some phenomena which I have observed, and to connect therewith such interpretations and theoretical considerations as may come within the sphere of general biology.

In animal life there are many interesting modes of keeping quiet, which are instinctive and adapted to special purposes. It is a very general means of concealment and escape from enemies. For illustration, we may take first the leeches, animals relatively low in the scale. One of the lower and least active forms, occurring everywhere in ponds, lakes, and streams, is known under the generic name *Clepsine*. There are many species, varying from one-quarter inch to one inch or more in length. They are found on their regular hosts, turtles, frogs, fishes, molluscs, etc., or on the under side of stones, boards, branches, or other submerged objects near the shore. One of the larger species, found often in large numbers on turtles, will be favorable for observation.

Behavior of Clepsine.

a. *Deceptive Quiet.*

Place the animal in a shallow, flat-bottomed dish and leave it for a few hours or a day, in order to give it time to get accustomed to the place and come to rest on the bottom. Then, taking the utmost care not to jar the dish or breathe upon the surface of the water, look at the *Clepsine* through a low magnifying lens and see what happens when the surface of the water is touched with the point of a needle held vertically above the animal's back. If the experiment is properly carried out, it will be seen that the respiratory undulations (if such movements happen to be going on) suddenly cease and that the animal *slightly* expands its body and hugs the glass. Wait a few moments until the animal, recovering its normal composure, again resumes its respiratory movements. Then let the needle descend through the water until the point rests on the bottom of the dish at a little distance from the edge of the body. Again the movements will cease and the animal will hug the glass with its body somewhat expanded. Now push the needle slowly along towards the leech, and notice, as the needle comes almost in contact with the thin margin of the body, that the part nearest the needle begins to retreat slowly before it. This behavior shows a surprising keenness of tactile sensibility, the least touch of the water with a needle-point being felt at once. This delicate sensitiveness is manifested in such a quiet way that it would be generally overlooked, and an observer unfamiliar with the habits of *Clepsine*, and not realizing the necessity of extreme quiet in his own movements, would almost certainly draw false conclusions. If the dish were moved or the water carelessly disturbed in any way, the *Clepsine* would assume its motionless attitude and appear to be wholly indifferent to the disturbance. If its back were rubbed with a brush or the handle of a dissecting needle, in order to test its sensitiveness to touch, the appearance would probably still be that of insensibility and indifference to the treatment. Closer examination, however, would show that the flesh of the animal was more rigid than usual, and that the surface was covered with numerous small, stiff, conical elevations, the dermal papillæ or warts, which are so low and blunt in the normal state of rest as to be scarcely visible. It would be seen that the animal, although motionless, was in a state of active resistance to attack. Every muscle would be strained; the whole skin would be tense and rough with the stiff, pointed papillæ; and at the same time the body would be found excessively slippery and difficult to lay hold of, owing to the mucous secretion poured forth from the dermal glands. To guard still further against dislodgment, the body would be flattened out as much as possible and tightly applied to the glass. The activity of the resistance offered by this passive-looking creature would be very forcibly realized if the observer attempted to circumvent it by slipping a thin blade or spatula beneath it with a view to forcing its hold. If overcome in one part it would stick by another, and skillful manipulation would be necessary to get both ends free at the same time. With one end detached, the other will often hold against a pull strong enough to snap the body in two.

b. *Rolling into a Ball.*

Clepsine has another and entirely different method of keeping quiet. The behavior bears striking analogy to that which has been described as "feigning death" in some insects. The animal rolls itself up (head first and ventral side innermost) into a hard ball, outwardly passive, and free to roll or fall whithersoever gravity and currents in the water may direct it. The ball will bear considerable pressure and rough handling without unfolding or exhibiting any marked movements. Left in quiet for a few seconds, the animal slowly unrolls itself and creeps off. This instinct has many advantages for a slow-moving creature like *Clepsine*, as will presently be seen.

1. *Provoked by Exposure.* — The ball-like attitude is assumed under various circumstances. If a stone or board with *Clepsine* attached to its under surface be quietly turned upside down, thus bringing the leech from shade or darkness into light and exposure, it may sometimes maintain its position of rest unchanged, only hugging the stone a little more closely and not moving until all is quiet. More generally, however, it rolls itself up, and by the time the stone is turned, or before, it falls to the bottom, where it can unfold and escape without danger of discovery. If, by chance, the animal has eggs, it will not desert them to escape in this way. As soon as the eggs hatch and the young become attached to the ventral side of the parent, the latter may roll itself up with its brood inside, fall to the bottom as before, and thus escape with all its progeny.

This species, then, has two quite distinct and peculiar ways of keeping quiet and thus avoiding its enemies. If the animal has no eggs, or if it has young, it may adopt either mode of escape, while if it has eggs it has no choice but to remain quiet over them. In the species here considered, the eggs are held together in a thin, gelatinous sheet, secreted at the time of ovipositing, and of a size and form to be entirely covered by the expanded body of the parent. In some species of *Clepsine* the eggs are laid in thin membrane-like sacs, which are fastened to the under side of the parent, and in this case the rolling up into the form of a ball is the safest course of behavior and the one generally adopted.

2. *Forced by Attack.* — The same behavior will almost invariably follow when any species of *Clepsine* is closely pursued and finds itself unable to fix itself by either end, as when a spatula is repeatedly thrust under it in such a way as to break its hold and defeat its efforts to regain a footing.

3. *Induced by Gorging Blood.* — The provocations to such behavior thus far considered have all been such as might, and probably do, cause more or less alarm. It is important to note, however, that the instinct may manifest itself frequently under conditions that seem to exclude the influence of fear. I have often seen these leeches fold themselves into balls at the end of a good meal, and so roll to the edge of the shell of their host, and fall to the bottom. This mode of concluding a quiet repast, with no assignable cause for alarm, and with every reason for satisfaction and contentment, except for the desire to get out of light into darkness, under cover of a stone or some other object, will hardly pass as feigning; and cataplexy and the tropisms are equally out of question. We could not assume, for example, that *Clepsine* is positively heliotropic when hungry, and negatively heliotropic after feeding. *Shade is preferred at all times in both conditions.* If hungry, *Clepsine* leaves the shade, not because it prefers light, but because it prefers its host more than it prefers shade. If the host is not found it will again return to a shady retreat, if one is to be found, however hungry it may be. The rolling up cannot be attributed to light, as the animal takes the extended position when at rest, even if compelled to remain in the light. What, then, shall we conclude?

4. *Origin and Utility.* — Observation and inference may be stated as follows:

1. The act of rolling up into a passive ball may be performed (*a*) *under compulsion*, as when it is the last resort in self-defence; (*b*) *under a milder provocation*, as *one* of three courses of behavior, as when the resting place is turned up to light, and the choice is offered between remaining quiet in place, creeping away at leisure, or rolling into a ball and dropping to

the bottom; (*c*) or, finally, *under no special external stimulus*, but rather *from internal motive*, the normal demand for rest and shady seclusion, presumably very strong in *Clepsine* after gorging itself with the blood of its turtle host.

2. This mode of taking leave of the turtle, after a full meal, is the easiest, the quickest, and the safest way available. To drop off fully extended, as *Clepsine* sometimes does, would retard descent and increase the chances of capture by fish. To creep about on the back of the host, waiting for an opportunity to grasp a plant or stone, would be decidedly hazardous, for if it came within the stretch of its host's neck, annihilation would be almost certain, while if lucky enough to keep out of reach of its host, it would still be in danger of the same fate from other turtles.

3. This behavior is instinctive, since it is performed by the young after the first meal as perfectly as by the adult.

4. Looking more closely at the nature and origin of this instinct, it will be seen to be quite a natural performance, in keeping with the most fundamental features of the animal's organization, and only a special application of a more general act that is primary and organic as much as tasting, seeing, or sleeping.

The more general act consists simply in tucking or rolling the head under, as often happens when the animal is resting. The habit may be observed in the young as soon as they are sufficiently developed to be capable of bending the tip of the head under. The same act, carried a little further, gives the half-rolled condition, in which only the anterior of the animal is folded, while the posterior portion remains unrolled and attached by the sucker. This attitude is often assumed if the leech is sick or has been injured. It is only a step further to release the sucker and fold it over the part already rolled up, thus completing the part ball to a whole ball, which can move passively more rapidly and safely than is possible by active creeping. From beginning to end we have only one act, in different stages of completion, simply different degrees of one and the same process.

5. Having the general act to start with, it is easy to see

how it might be made of use for particular purposes; in other words, how special adaptations of a useful kind might arise. If the act is a natural concomitant of the resting condition, and is associated with a feeling of ease and security, we see how sickness, injury, fear, a heavy meal, etc., might prompt it, and in higher or lower degree, according to the nature and intensity of the inciting cause. Full and prompt action under exposure, pressure, injury, and in the event of a good meal, would carry decisive advantages, so that individuals reacting in the more favorable degrees would stand the best chance of escape and survival. Natural selection would steadily improve upon the results, and the special adaptation, in different stages of development in different species, as we find it to-day in different *Clepsines*, would lie in the direct line of progress. This view does not of course presuppose intelligence as a guiding factor, and therefore lends no support to the theory of instinct as "lapsed intelligence," or "inherited habit."

6. An instinct of the kind here considered does not depend for its development upon effort and the transmission of functionally acquired improvements in organization, but upon *the natural selection of the best qualified germs*, for that is what the survival of the fittest individuals always means. Many species of *Clepsine* require but one full meal a year, and as they seldom live more than two or three years, the number of meals is very limited. There is little room, then, for repeating the experiment often enough to affect the organization. Indeed, such a supposition would here appear to be absurd in the last degree. On the other hand, the selection of the fittest germs, provided for in the survival of the best-adapted individuals, would inevitably advance the species along the line leading to the special instinct.

If the view here taken be correct, the instinct of rolling into a ball is not a matter of deliberation at all, but merely the action of an organization more or less nicely adjusted to special conditions and stimuli. Intelligence does not precede and direct, but the indifferent organic foundation with its general activities is primary; the special behavior or instinct is built up by slowly modifying the organic basis.

c. *Sensitiveness to Light.*

The question as to how much intelligence, if any, *Clepsine* may have, I do not here undertake to settle or discuss. That the animal is endowed with keen sensibilities is evident from the behavior before described. The following simple experiment affords a striking demonstration : Pass the hand over a dish in which a number of *Clepsines* are resting quietly on the bottom, at a distance of a few inches above the animals, taking care not to make the least jar or other disturbance. If the animals are quite hungry, the slight shadow of the hand, imperceptible though it be to our eyes, will be instantly recognized by them, and a lively scene will follow, every leech rising up, supported on its posterior sucker, and swinging at full length back and forth, from side to side, round and round, as if intensely eager to reach something. Put a turtle in the dish and see what a scramble there will be for a bloody feast. The shadow of the hand was to these creatures like the shadow of a turtle swimming or floating over them in their natural haunts, and hence their quick and characteristic response. A piece of board floating over them would have the same effect. Although so sensitive to a small difference in light, the *Clepsine* eyes can give no pictures, and hence there is no power of visual discrimination between objects. They probably recognize their right host by the aid of organs of taste; at any rate they are often able to distinguish their host from closely allied species.

Instinct of Rolling into a Ball among Insects.

The following examples of the instinct of rolling into a ball among insects are from Kirby and Spence.[1]

"I possess a diminutive rove-beetle (*Aleochara complicans* K. Ms.), to which my attention was attracted as a very minute, shining, round, black pebble. This successful imitation was produced by folding its head under its breast, and turning up its abdomen over its elytra, so that the most piercing and discriminating eye would never have discovered it to be an insect. I

[1] Entomology, pp. 411, 412.

have observed that a carrion beetle (*Silpha thoracica*) when alarmed has recourse to a similar manœuvre. Its orange-colored thorax, the rest of the body being black, renders it particularly conspicuous. To obviate this inconvenience, it turns it head and tail inwards till they are parallel with the trunk and abdomen, and gives its thorax a vertical direction, when it resembles a rough stone. The species of another genus of beetles (*Agathidium*) will also bend both head and thorax under the elytra, and so assume the appearance of shining, globular pebbles.

"Related to the defensive attitude of the two last-mentioned insects, and precisely the same with that of the Armadillo (*Dasypus*) amongst quadrupeds, is that of one of the species of wood-louse (*Armadillo vulgaris*). The insect, when alarmed, rolls itself up into a little ball. In this attitude its legs and the underside of the body, which are soft, are entirely covered and defended by the hard crust that forms the upper surface of the animal. These balls are perfectly spherical, black, and shining, and belted with narrow white bands, so as to resemble beautiful beads; and could they be preserved in this form and strung, would make very ornamental necklaces and bracelets. At least so thought Swammerdam's maid, who, finding a number of these insects thus rolled up in her master's garden, mistaking them for beads, employed herself in stringing them on a thread; when, to her great surprise, the poor animals beginning to move and struggle for their liberty, crying out and running away in the utmost alarm, she threw down her prize. The golden wasp tribe also (*Chrysididæ*), all of which I suspect to be parasitic insects, roll themselves up, as I have often observed, into a little ball when alarmed, and can thus secure themselves — the upper surface of the body being remarkably hard, and impenetrable to their weapons — from the stings of those *Hymenoptera* whose nests they enter with the view of depositing their eggs in their offspring. Latreille noticed this attitude in *Parnopes carnea*, which, he tells us, *Bembex rostrata* pursues, though it attacks no other similar insect, with great fury; and, seizing it with its feet, attempts to dispatch it with its sting, from which it thus secures itself. M. Lepelletier de Saint-

Fargeau, to whom entomology is indebted for so many new facts relative to the manners of hymenopterous insects, has given us a striking account of a contest between the art of one of these parasites (*Hedychrum regium*) and the courage of one of the mason-bees in endeavoring to defend its nest from its attack. The mason-bee had partly finished one of her cells, and flown away to collect a store of pollen and honey. During her absence the female parasitic *Hedychrum*, after having examined this cell by entering it head foremost, came out again, and walking backwards, had begun to introduce the posterior part of her body into it, preparatory to depositing an egg, when the mason-bee arriving laden with her pollen-paste threw herself upon her enemy, which, availing herself of the means of defence above adverted to, rolled herself up into a compact ball, with nothing but the wings exposed, and equally invulnerable to the stings or the mandibles of her assailant. In one point, however, our little defender of her domicile saw that her insidious foe was accessible; and, accordingly, with her mandibles cut off her four wings, and let her fall to the ground, and then entering her cell with a sort of inquietude, deposited her store of food, and flew to the fields for a fresh supply; but scarcely was she gone before the *Hedychrum*, unrolling herself, and, faithful to her instinct and her object, though deprived of her wings, crept up the wall directly to the cell from whence she had been precipitated, and quietly placed her egg in it *against the side* below the level of the pollen-paste, so as to prevent the mason-bee from seeing it on her return."

Behavior of Necturus.

a. *Refusal of Food from Fear.*

Our large fresh-water salamander, popularly called mud-puppy, water-dog, hellbender, etc., is another animal that may be profitably studied with reference to its modes of quiet. The first adults which I kept in captivity in a large aquarium refused to eat pieces of raw beef or small fish, whether dead or alive. For months they went on, seemingly entirely indifferent to any proffered food, not paying the least attention, so far as I noticed, to tempting morsels dropped quietly in front of them or held in suspension before them. Living earthworms and insect larvæ were presented to them, all of which were known to be palatable to the creature in its natural habitat; but nothing availed to draw attention or elicit any evidence of hunger. Quiet and wholly indifferent in outward behavior, yet the animals were actually starving and wasting away. Were the creatures *feigning* quiet and indifference? Or was the behavior merely the expression of timidity, the animal not having the courage to perform the acts necessary to secure the food which it must have craved? I confess that I did not for a long time understand the cause of this refusal of food.

Further acquaintance with the adults, supplemented by an experience of two seasons in rearing the young, opened my eyes to the extreme timidity of these animals, which is so deep-seated and persistent that one can form only a poor idea of it without considerable actual contact with it. The outward behavior is very quiet and mild and gives little indication of fear. The animal will often submit to gentle handling without making any violent effort to escape. In short, the behavior is misleading, and one stands no chance of understanding it until he learns to keep quiet himself while observing, and discovers how to get into confidential relations with the creature. This can be done with the adults, but to better advantage with the young.

b. *Behavior of Young in Taking Food.*

The eggs may be readily hatched in a shallow dish and young thus obtained which have never learned anything from the parents. I had about fifty young hatched in this way towards the end of July. When first hatched they were loaded with food-yolk sufficient to meet their needs for about two months. By the end of September I began to get intimations of a desire for food. The method of feeding was as follows: The dish containing the young was kept on a table, where, without being moved, food could be offered in perfect quiet. I used the tiniest bits of raw beef and offered only one piece at a time, which I held in small forceps or on the point of a needle a little in front of the animal to be tested. If the meat is held closely enough to touch the head, the animal is frightened and may retreat with such haste as to alarm all its companions. If the bait is held a little to one side, an inch or so away, and very quietly for a minute or more, a slight turning of the head in that direction may be noticed, in case the animal is ready to eat and feels confidence enough to try to reach it. The turning of the head is done very cautiously and almost as slowly as the minute-hand of a clock moves, so that one may become aware of it, not by seeing the movement, but by noticing the inclination of the head to the axis of the body. If there be a decided turn of the head of this kind, the case is hopeful, as it shows an interest which may be encouraged to action by bringing the bait a little nearer, but very slowly and without any jerky movement. Halting about half an inch away, wait for further movement on the part of the animal, if you are fortunate enough not to have frightened it away. If the animal's courage holds out — in most cases it does not in the first trials — it will soon begin to move, but with a slowness that tries the observer's patience. The head at length comes up to a point a quarter of an inch away, more or less, and after making sure of the position of the bait, which seems to be done less by the aid of the eyes than by the sense of touch, the animal tries to seize it by a quick side movement of the head and a snap of the jaws. *The first attempt to take the bait corresponds in all essential points with the behavior of the adult when trying to capture a fish, a worm, or an insect larva, although the aim may not be quite so sure.*

If one is successful in getting one or more to feed, the more timid ones may be brought forward in the same way by patiently alluring them from day to day, until they are tempted to an effort. Once made, the effort becomes easier at the next trial, and in the course of a month or six weeks the bolder ones will respond fairly promptly. A few of my specimens became very familiar with me, and would come towards me when I approached the dish, looking up at me wistfully, as if knowing well the meaning of my visits.

c. *Influence of Innate Timidity.*

In the behavior above described, we see an instinctive mode of capturing prey held in check, and probably directed to some extent, by innate timidity. Fear seems to be the main factor in control at the start, holding the animal in a trance-like quiet, undecided as to what to do, waiting for confidence to attack, or for a stronger motive to flee. As fear subsides a little, the preparatory movement of attack begins, but the sly behavior is due to fear rather than the slyness of stratagem. The slow and cautious method of approach is certainly not all finesse, for the deportment bears still the stamp of hesitating timidity, and this part of the act may become much freer as the animals become tamer and more fearless. The final part of the act, that of snapping the bait, was always performed in the same characteristic way. The piece of meat seemed always to be regarded as a *living* prey, which was to be seized quickly, held firmly for a moment or two, and then swallowed. Unfortunately I did not experiment to see what could be done in modifying this part of the act.

Instinctive fear is evidently a very important element in the conduct of the lower as well as the higher animals. In *Necturus* we see how it may be just as effective as intelligence in securing a sly mode of attack. So strong is its influence that I doubt whether there is any finesse in the movement. The adaptation of acts to purposeful ends must not be accepted too quickly as proof of intelligence in the doer. Such acts are common enough in plants, and there we are under the necessity of finding some other explanation.

d. *Organization Shapes Behavior.*

Necturus appears to understand well the act of capturing its prey, and the nice adaptation of each act to the end in view naturally enough suggests forethought and refined experience.

But we see the performance executed by the young, which have never had any experience of that kind, nor any opportunity to copy from others. We cannot therefore suppose that they perform these acts understandingly. The young *Necturus*, hatched in a dish where it has never met any living thing except its companions, has nothing to guide its first effort to capture food except its organization and its simple experience in walking and swimming, which acts are again like those of the adult, not because directed by intelligence or example, but because they are performed with the same organs under similar conditions. The young has the same sensory and motor apparatus as the adult, but it has never before known the feeling of hunger, it has never experienced pain from contact with an enemy, it has never learned that a prey may escape if not approached slyly. Its movements in approaching and snatching a piece of meat, as if it were a living object, are, then, those characteristic of the species, not because they are measured and adapted to a definite end by intelligent experience, but because they are organically determined; in other words, depend essentially upon a specific organization.

The timidity of young hatched in a dish is the same as that of specimens hatched in the lake, and therefore it cannot be charged to individual experience or to parental influence. It, too, inheres in the special brand of organization, and has nothing to do with memory of pain sensations.

e. *Origin and Meaning of Behavior.*

We have taken a very important step in our study when we have ascertained that behavior, which at first sight appeared to owe its purposive character to intelligence, cannot possibly be so explained, but must depend largely, at least, upon the mechanism of organization. The origin and meaning of the behavior antedate all individual acquisitions and form part of the problem of the origin and history of the organization itself. It is the first and indispensable step, without which it would be impossible to reach sound views, either as regards the particular behavior or the difficult question of the relation of instinct to habit and intelligence. If the problem is not simplified, its nature is better defined and its perspective is relieved of many a myth that might otherwise obscure our vision. We see at once that the behavior does not stand for a simple and primary adaptation of a preëxisting mechanism to a special need. As the necessity for food did not arise for the first time in *Necturus*, the organization adapted to securing it must be traced back to foundations evolved long in advance of the species. The retrospect stretches back to the origin of the vertebrate phylum, and, indeed, to the very beginning of genealogical lines in protozoan forms. The point of special emphasis here is that instincts are evolved, not improvised, and that their genealogy may be as complex and far-reaching as the history of their organic bases.

f. *Sensibility — Sources of Error.*

Another important factor in animal behavior, namely, sensibility, is very generally underestimated and often sadly misunderstood. We are apt to gauge sensibility according to the intensity of the overt response to stimulus, forgetting that the animal has the power to inhibit such manifestations or to moderate them in a way to mislead the observer. In the struggle for existence a high premium has been placed on this power, with the result that it is well-nigh a universal attribute. The best proof of its high value to the possessor is our own readiness to accept the disguise it affords as an evidence of lack of sensibility. We are so prone to think that the exercise of such power depends upon considerable intelligence that we are incredulous of its existence in forms that give only doubtful signs of intelligence. The power is possessed to a very marked degree by *Clepsine*, and it is only when we become aware of this fact and take all necessary precautions that we can get any reliable tests of the animal's keenness of sensibility. *Necturus* is even more difficult to manage, for not until after we have won its confidence by slow degrees can we expect free responses.

Besides the great danger of being deceived by the response, or the lack of response, to stimulus, there are two other insidious sources of error to be guarded against. We habitually assume that intelligence and sensibility rise and fall together. This idea may lead to false conclusions in two directions — to overestimating sensibility at the upper end of the scale and underestimating it at the lower. That high sensibility does not imply high intelligence is clear in the case of *Clepsine* and equally so in almost any other case that might be selected among the lowest segmented animals. That high intelligence does not necessarily imply correspondingly high sensibility is shown by the well-known fact that many animals greatly surpass man in their sense powers. It can be shown, I believe, that the difference in sensibility between higher and lower animals is very much less than is generally supposed.

The second source of error is the common assumption that the grade of sensibility rises with the structural complexity of the sense organs. This view is likewise untenable. It is true that the sense organs as a rule become more complex in structure as we go up the scale, but this advance in structure is mainly confined to accessory and non-sensory parts, which are either of a protective nature or else concerned in some subsidiary function, such as muscular adjustments and regulation of the stimuli. Such improvements in the non-sensory parts may be carried to a high state of perfection and greatly raise the general efficiency of the organ (*e.g.*, the vertebrate eye), without adding much, if anything, to the sensitiveness of the individual sense cells. The sense cells may be multiplied in number and placed in a position of safety and advantage for receiving stimuli, and the stimuli may be strengthened, directed, and otherwise regulated so as to secure the best results; but all that may obviously not affect the functional power of the cells themselves. We do not know the range of variability in this power, but we do know that the sense cells often vary relatively little in structure, sometimes retaining in the higher forms the same typical features that characterize them in the lower forms. There is no known difference of structure that would warrant the assumption that the dermal sensillæ in annelids are less sensitive than those in aquatic vertebrates. We have, then, no reliable test of sensibility either in the structure of the sense organ, in the rank of the organism, or in its intelligence. We have to depend upon the response to stimuli, and, remembering that this may be deceptive, observe and experiment under conditions that insure *free behavior*.

No one who has never come into close communion with the lower animals can begin to appreciate the delicacy and efficiency of their sensory apparatus. We take up the earthworm and, as we see no eyes, we conclude that it cannot see. A little experiment shows that it is extremely sensitive to light, and further study of its structure reveals unpigmented eyes lying beneath the skin, and the whole surface thickly set with minute delicate tactile sensillæ. Even *Amphioxus*, so long reputed to have no visual organs, turns out to have such organs from end to end, imbedded in its spinal cord. I have before called attention to the highly sensitive organization of *Clepsine* and its allies. In the very lowest organisms, plant and animal alike, where special visual organs do not exist, the living protoplasm has, as has been demonstrated in many ways, a keen sensibility to light, so that one might look upon the whole organism as fulfilling the light-perceiving function.

g. *Orientation through the Dermal Sensillæ.*

Necturus, as before remarked, has a very keenly sensitive organization. The skin is richly provided with sense organs, which terminate at the surface in very short, fine hairs, invisible to the naked eye. These organs, which are of the same nature and function as the dermal sensillæ in *Clepsine* and in so many other aquatic animals, are sensitive to slight vibrations in the water that are far beyond the reach of any of our sense organs, and they are the main reliance, both in avoiding enemies and seeking prey.

It is interesting to see how little the eyes are depended upon in finding a piece of meat. A bit dropped in front of a young *Necturus* receives no attention after it reaches the bottom.

An object must be in *motion* in order to excite attention, and it is not generally the moving form that is directly perceived, but the movements of the water, travelling from the object to the sensory hairs, are felt, and in such a way as to give the direction of the disturbing centre with most surprising accuracy. If a bit of beef is taken up adhering to the point of a needle, and held in the water, the vibrations imparted to the needle by the most steady hand will be sufficient to give the animal the direction. If the meat falls to the bottom, and the needle is held in place, the animal approaches the needle and tries to capture it, without paying the slightest attention to the meat lying directly below. If, after the meat has fallen, the needle is withdrawn and touched to the surface of the water behind or at one side of *Necturus*, it turns instantly in the direction of the needle, not because it sees, but because it *feels* wave motions coming from that direction. Long experience with *Necturus* and with many of its nearer allies enables me to speak very positively on this point. When it is remembered that in the higher animals the direction of sound waves is given by the auditory sense organs, which are primarily surface sensillæ homologous with those in the skin of *Necturus*, it may not seem so strange that the animal directs its movements in the way described. *Necturus* can see, but it can feel (perhaps we should say hear) so much more efficiently that its small eyes seem almost superfluous.[1]

[1] Professor Eigenmann has kindly written the following note on the use of the tactile organs in the blind fishes:

Chologaster papilliiferus, a relative of the blind fishes living in springs, detects its prey by its tactile organs, not by its eyes. A crustacean may be crawling in plain view without exciting any interest unless it comes in close proximity to the head of the fish, when it is located with precision and secured. The action is in very strong contrast to that of a sunfish, which depends on its eyes to locate its prey. A *Gammarus* seen swimming rapidly through the water and approaching a *Chologaster* from behind and below was captured by an instantaneous movement of the *Chologaster*, when it came in contact with its head. The motion brought the head of the *Chologaster* in contact with the stem of a leaf, and instantly it tried to capture this also. Since the aquarium was well lighted, the leaf in plain sight, it must have been seen and avoided if the sense of sight and not that of touch were depended upon.

In *Amblyopsis*, the largest of the blind fishes of the American caves, the batteries of tactile organs form ridges projecting beyond the general surface of the

h. *Origin and Nature of the Behavior in Taking Food.*

1. *Some Intelligence Implied.* — Let us now return to the question of the origin and nature of the behavior of *Necturus* in capturing its food; not, however, with the expectation of reaching a complete solution, but rather in the hope of coming nearer to the problem and to the guiding principles in dealing with it. It is obvious, first of all, that automatism will not suffice to account for the whole behavior. That there is organic coördination of movements no one will dispute. But these movements must be steered in the direction of the object, and this orientation does not seem to be a purely automatic arrangement. The dermal sensillæ ("lateral-line" organs) give the impressions which enable the animal to steer its course; but action and sense impression are evidently not linked in a way to be independent of inhibitory influences. I assume that the creature is conscious, and that it has a certain intelligent appreciation of the sense impressions received. This is not saying that the young *Necturus* is a born philosopher; I assume nothing more than that it has already learned by experience how to *direct* its movements. That does not imply much, but certainly some, intelligence. I cannot otherwise understand why the same stimulus should not always evoke the same response under the same conditions. But we see that there is hesitation about starting, and this hesitation may be prolonged to any length, showing conclusively that sensation and response are not so connected as to exclude

skin. Its prey, since it lives in the dark and its eyes are mere vestiges, is located entirely by its tactile organs. This is done with as great accuracy as could be done with the best of eyes in the light, but only when the prey is in close proximity to the head. Coarser vibrations in the water are not perceived or are ignored, and apparently stationary objects are not perceived when the fish approaches them. If a rod is held in the hand, the fish always perceives it when within about half an inch of it, and backs water with its pectorals. If the head of a fish is approached with a rod, the direction from whence it comes is always perceived and the correct motion made to avoid it. This reaction is much more intense in the more active young than in the adult. One young, about 10 mm. long, determined with as great precision the direction from which a needle was coming as any fish with perfect eyes could possibly have done. It reacted properly to avoid the needle, and this without getting excited about it.

inhibitory influences. There is unmistakably a power of inhibition strong enough to counteract the strongest motive to act — the hunger of a starving animal in the presence of food.

2. *Orientation Learned by Experience.* — In assuming that the young *Necturus*, at the time of its first attempt to capture a piece of beef, has already learned to orient itself with reference to external objects, I have not gone beyond the possibilities. The animal has been out of the egg envelopes for about two months. It has been confined in a glass dish about ten inches in diameter, holding water about one inch in depth. Its life has been about as exclusively vegetative as if it had been all the while within the egg membrane, the only difference being that it has had room to straighten itself and to move about to some extent among its fellows. It has been heavily laden with food-yolk and has maintained a quiet attitude except when disturbed by change of water. Simple as the life has been, the animal has had some experience in swimming and walking, and opportunity to use to some extent its organs of orientation. The bait offered to it is something totally new in its experience, but we cannot, of course, claim that its behavior towards the bait is wholly uninfluenced by its previous experience.

3. *Deferred Instinct.* — We have to do with what Mr. Lloyd Morgan has termed a "deferred" instinct, *i.e.*, an action performed for the first time, but not until some time after birth. Mr. Morgan's remarks on the first dive of a young moor hen[1] bring out very clearly the possible influence of experience in the case of such deferred instincts.

Mr. Morgan says:

In the case of such an instinctive procedure of the deferred type as that presented by the diving of a young moor hen, though, on the first occasion of its performance, the congenital automatism predominates, *yet it is difficult to believe and is in itself improbable that the individual experience of the young bird does not, even on the first occasion, exercise some influence on the way in which the dive is performed.* If we desire to reach a true interpretation of the facts, we must realize the fact that an activity may be of mixed origin. And if we distinguish

[1] Habit and Instinct, pp. 136, 137.

— as we have endeavored clearly to distinguish — between instinct as congenital and habit as acquired, we must not lose sight of the fact that there is much interaction between instinct and habit, so that the first exhibition of a deferred instinct may well be carried out in close and inextricable association with the habits which, at the period of life in question, have already been acquired.

Although Mr. Morgan's young moor hen had undoubtedly learned far more by experience before its first dive than my young *Necturus* could have learned before its first effort to capture food, we are nevertheless well admonished to keep in mind the fact that the activity here considered may not be pure instinct. Allowing for the small though important part played by intelligence, there remains a purposive sequence of coördinated acts, which are always performed in essentially the same manner by young and old, and by the young without instruction or example or previous experience of like motive and stimulus. In so far, then, as intelligence cannot possibly be a regulating factor we must refer the activity to organization.

4. *Pause before the Bait.* — In order to exclude as much as possible the influence of experience it will be well to confine attention to the least variable part of the behavior. The concluding phase of the performance is so typical and characteristic, and so far removed from anything previously experienced, that it may be regarded as a very near approach to pure instinct. I have in mind *the pause before the bait and then the quick side-movement of the head as the jaws are opened to seize.*

If this series of acts represents an organic sequence, and if the behavior as a whole takes the form determined by the organization, as seems to me beyond reasonable doubt, we have an instinct the history of which may be coextensive with the evolution of the animal. We stand at the end of an interminable vista. The specific peculiarities of organization in *Necturus* form but an infinitesimal element of the problem. Scarcely a feature of the instinct belongs exclusively to *Necturus.* It is at least widely diffused among vertebrates, especially among fishes. The differences in the manner of execution among different forms, so far as I have observed, are of quite a superficial nature. The instinct evidently has its root in the general instinct of preying, which is doubtless coeval with animal organization. The cannibalism of our protozoan ancestors was the starting-point, and their carnal propensities were not acquired by the aid of intelligence, but given in the fundamental properties of protoplasm. The stronger ate the weaker, and made themselves stronger and more prolific by so doing. The promise of the whole animal world was contained in the act. The constitutional disposition to feed, with variable foods available, would give occasion for different appetites and various modes of getting outside of palatable victims. In primitive organisms multiplying by simple fission, structural modifications acquired during the lifetime of the individual would be carried right on from generation to generation, and hence the structural foundations for a whole animal world such as we now see could be laid in a relatively short period as compared with the time necessary to advance organization in forms limited to reproduction by germs. In fact, these fundamentals could all be established within the realm of the unicellular protozoa. Nucleus and cell-body, inner and outer layers, nerve-muscle elements, sensory and locomotor organs, mouth and stomach, respiratory and excretory mechanisms, reproductive elements, anticipating embryological development from germs — all these essentials of higher organization are presented in the protozoan.

The organic bases furnished in the protozoan world might be passed directly on to the first metazoa, or they might be reacquired in essentially the same manner as before, and in a not much longer period, as reproduction by fission would still be a condition favoring rapid organo-genesis.

To try to fill up the gaps between the protozoan and *Necturus* would lead us too far into the field of speculation, and would not contribute much to a grasp of the problem. We have to content ourselves with general facts and principles and probabilities drawn therefrom. It is enough for present purposes to know that the roots of the instinct organization we are considering run clear back to the beginnings of organo-genesis, and that they are natural products of the properties of living protoplasm. We start with known properties and get to known rudiments of organization without invoking the aid of intelligence, or finding any way in which it could be supposed to have been a guiding factor in development.

The organic basis of the preying instinct may have grown and multiplied in different phyla a long time before receiving much aid from intelligence. The rapidity and freedom of modification would be very much limited when fission ceased and reproduction by germs became the sole mode of generation. Very early in the vertebrate phylum, possibly at its dawn, the chief characters of the instinct, as we now find it, were probably fixed in structural elements differing from those in *Necturus* only in superficial details. The strikingly fish-like character of the behavior certainly suggests as much.

5. *Meaning and General Occurrence.* — If now we look more closely at the purposive character of the behavior, it will become clearer that the instinct is shared, not only by animals below *Necturus*, but also by some far above it. The pause before the final act of seizing is a well-marked feature, which means *locating the prey and fixing the aim.* The same action with the same meaning runs all through the different branches of the vertebrate phylum. It is, as I have already said, especially characteristic of the fishes and amphibia, and it is not rare among the higher branches, the reptiles, birds, and mammals. It may be seen to good advantage in the turtles, and even the common fowl halts on coming up to the insect it is pursuing in order to make sure its aim. I believe the same instinct underlies the act of *pointing* in the dog. The origin of this behavior in pointers cannot be referred to training, as was clearly seen by Darwin.[1]

It may be doubted [says Darwin] whether any one would have thought of training a dog to point had not some one dog naturally shown a tendency in this line, and this is known occasionally to happen, as I once saw in a pure terrier; the act of pointing is probably, as many have thought, *only the exaggerated pause of an animal preparing to spring on its prey.* When the first tendency to point was once displayed, methodical selection and the inherited effects of compulsory training in each successive generation would soon complete the work.

[1] Origin of Species, p. 207.

The "tendency" manifested in some one dog was regarded by Darwin as an accidental variation, the cause being unknown. May not many of the variations appearing in domestic animals, which we call "accidental," be manifestations of instinct roots of more or less remote origin?

6. *Part Played by Fear.* — We may now glance once more at the behavior as a whole, for the purpose of pointing out the part played by instinctive timidity. Gentle movements in the water, kept up with steadiness, such as are imparted by a needle in feeding as before described, may induce an attack, while less gentle or unsteady movements may lead the animal to remain quiet or to take flight. The same stimulus, according to amplitude and evenness, may then be followed either by advance, by quiet, or by retreat. In retreat, fear is manifest; in quiet it is concealed; in advance it is less concealed. There can be no doubt that fear predominates in flight and in quiet, while it certainly tempers the advance, giving the appearance of slyness deliberately acted in order to take the prey by surprise. This sly manner of advance, whatever it be due to, has a double advantage, for it is concealment against a possible foe and prevents alarming a harmless prey. If I could suppose that fear did not strongly influence the advance, I should certainly incline to think that the animal really appreciated the great advantages in quietly surprising its prey; but for reasons before given I believe the animal is quite blind to any such bearing of its action. The advantages of this manner of action, however, are just the same as if it were deliberately assumed, and the *Necturus* conducting itself in this way would certainly fare better than one reacting in a contrary way. The instincts of *Necturus* in this case coöperate to secure its welfare, while if the creature depended upon its intelligence it is difficult to see how it could escape immediate extinction.

General Considerations.

a. *Instinct Precedes Intelligence.*

The view here taken places the primary roots of instinct in the constitutional activities of protoplasm [1] and regards instinct in every stage of its evolution as action depending essentially upon organization. It places instinct before intelligence in order of development, and is thus in accord with the broad facts of the present distribution and relations of instinct and intelligence, instinct becoming more general as we descend the scale, while intelligence emerges to view more and more as we ascend to the higher orders of animal life. It relieves us of the great inconsistencies involved in the theory of instinct as "lapsed intelligence." Instincts are universal among animals, and that cannot be said of intelligence. It ill accords with any theory of evolution, or with known facts, to make instinct depend upon intelligence for its origin; for if that were so, we should expect to find the lowest animals free from instinct and possessed of pure intelligence. In the higher forms we should expect to see intelligence lapsing more and more into pure instinct. As a matter of fact, we see nothing of the kind. The lowest forms act by instinct so exclusively that we fail to get decided evidence of intelligence. In higher forms not a single case of intelligence lapsing into instinct is known. In forms that give indubitable evidence of intelligence we do not see conscious reflection crystallizing into instinct, but we do find instinct coming more and more under the sway of intelli-

[1] Professor Loeb * refers instinct back to "(1) polar differences in the chemical constitution in the egg substance, and (2) the presence of such substances in the egg as determine heliotropic, chemotropic, stereotropic, and similar phenomena of irritability." According to this view, the power to respond to stimuli lies in unorganized chemical substances, and the same powers exist in the adult as in the egg, because the same chemical substances are present. Organization serves at all stages merely as a mechanical means of giving definite directions to responses.

The view I have taken regards instinctive action as *organic* action, whatever be the stage of manifestation. The egg differs from the adult in having an organization of a very simple primary order, and correspondingly simple powers of response. Instinct and organization are, to me, two aspects of one and the same thing, hence both have ontogenetic and phylogenetic development.

* "Egg Structure and the Heredity of Instincts," *The Monist*, vol. vii, July, 1897.

gence. In the human race instinctive actions characterize the life of the savage, while they fall more and more into the background in the more intellectual races.

Every hypothesis that would derive instinctive action from teleological reflection is open to the same objections. In many cases it would be necessary to postulate an amount of prevision on the part of the animals in which the instincts arose that would simply be psychologically impossible. Conscious prevision without a possible basis in the experience of the individual, or any means of learning from others, is simply a self-contradiction. The frequently cited instance of the emperor moth puts this point in strong light. The caterpillar of this moth so constructs the upper part of its cocoon that it will resist strong pressure from without and yield to slight pressure from within. Easy egress for the imago and security against attacks from outside enemies are thus provided for. As the spinning of the cocoon happens but once in a lifetime, the caterpillar could not anticipate such needs from its own experience, nor could it learn from its parents, which were dead long before it hatched. The possibility of imitation is also excluded, as the species is not a social one.

b. *Theories of Instinct.*

1. *Pure Instinct the Point of Departure.* — The first criterion of instinct is, that it can be performed by the animal without learning by experience, instruction, or imitation. The first performance is therefore the crucial one. It is of the utmost importance in all discussion of the origin of instinct to make sure of this point, and keep clear of all ambiguous activities such as have been designated "instinct habits" (Lloyd Morgan), "acquired instincts" (Wundt), "secondary instincts" (Romanes), etc. We must not allow the question as to the relation of instinct to habit and intelligence to be obscured by confusing terminology. There may be "mixtures" and all sorts of "interactions" between habit and instinct, and these may have a far-reaching theoretical import, but they lack definiteness, and are therefore dangerous foundations for theories. A

theory of instinct must obviously make pure instinct its first concern, and keep the general course of evolution always in view.

It is not my purpose to engage in a critical examination of theories, but to indicate briefly which of the two rival theories now most in favor accords best with facts and general principles as I understand them. These two theories are the *habit theory* of Lamarck and the *selection theory* of Darwin, Wallace, and Weismann.

2. *Embryology and the Lamarckian Theory.* — The habit theory is a part of the more general theory of the transmission of acquired characters. This doctrine has never been reconciled with the teachings of embryology, the science which deals directly with the phenomena of heredity, and which is, therefore, the touchstone of every theory of inheritance. It is a fundamental tenet in embryology that all organisms reproducing exclusively by germs owe their inherited characters to the germs from which they arose, and that germs carry the primordials of adult structure, not by virtue of any mysterious transference of parental features, but by virtue of the constitution they bring with them when they arise by division of preëxisting germs. That is, I believe, a fair statement of the embryological doctrine of inheritance, which must be the final test of our theories.

The selection theory propounded by Darwin and Wallace, and further developed by Weismann, starts from the embryological law of germ continuity (Weismann), or, otherwise expressed, germ lineage, and interprets the phenomena of variation, heredity, and development, in harmony with this law and the principle of selection. This theory is incompatible with the idea that instinct is inherited habit. We could not, for example, say with Professor Wundt[1]:

"We have supposed that father can transmit to son the physiological dispositions that he has acquired by practice during his own life, and that in the course of generations these inherited dispositions are strengthened and definitized by summation."

"The occurrence [p. 405] of connate instincts renders a subsidiary hypothesis necessary. We must suppose that the physical

[1] Lectures on Human and Animal Psychology, p. 408.

changes which the nervous elements undergo can be transmitted from father to son. . . . The assumption of the inheritance of acquired dispositions or tendencies is inevitable if there is to be any continuity of evolution at all."

3. *Darwin's Refutation of Lamarck's Theory.* — Although Darwin dwelt at some length on the points of resemblance between habits and instincts, and although he thought it possible that habits could sometimes be inherited, it should be remembered that he was the first to show conclusively that "the most wonderful instincts with which we are acquainted, namely, those of the bee hive and of many ants, could not possibly have been acquired by habit" (*Origin of Species*, p. 202). Indeed, it was he who first found in the case of neuter insects a refutation of Lamarck's doctrine of inherited habit, and at the same time a demonstration of the high efficiency of the principle of natural selection. Darwin concludes his chapter on instinct with these memorable words :

"The case of neuter insects, also, is very interesting, as it proves that with animals, as with plants, any amount of modification may be effected by the accumulation of numerous slight, spontaneous variations, which are in any way profitable, *without exercise or habit having been brought into play. For peculiar habits confined to the workers or sterile females, however long they might be followed, could not possibly affect the males and fertile female, which alone leave descendants. I am surprised that no one has hitherto advanced this demonstrative case of neuter insects against the well-known doctrines of inherited habit, as advanced by Lamarck.*"

What could more forcibly illustrate the importance of crucial cases than just this work of Darwin's on the instincts of neuter insects? Here a conclusive test is reached, and no theory of the origin of instinct can stand that disregards it. If habit cannot possibly have had anything to do with the origin of such typical instincts, then we should at least be very cautious in appealing to it in any case. We certainly do not want two theories to account for the same phenomenon, if one will suffice. If the theory of inherited habit is certainly false in a single case, it must be deemed false in every case, until at least it has been shown that some cases cannot be explained without it. Is there any case where it can be clearly shown that an undoubted instinct arose from inherited habit, or any case in which it can be made clear that the theory adopted for neuter insects cannot possibly hold? Both questions, it seems to me, must be answered in the negative.

4. *Weak Points in the Habit Theory.* — The habit theory has many adherents still, and Darwin himself often found it a convenient hypothesis. But neither Darwin nor anybody else has given us a crucial test that would stand beside that furnished in neuter insects. The failure to find such a test is certainly not due to any lack of zeal or effort on the part of the advocates of the theory. The tests claimed are numerous enough, but they always fall short of the requirement. The weak points in the theory are:

1. It starts on a disputed, if not refuted assumption; namely, that habits wholly new to the individual and the species, having no hereditary basis predisposing to them, may, as the result of exercise frequently repeated, and continued in successive generations, eventually become hereditary instincts.

2. It appeals to the less typical rather than to the more typical cases — to cases in which the critical points are undetermined or doubtful, or open to a different interpretation.

3. Its definition of habit and instinct verges towards a *petitio principii.* Two or more classes of instincts are set up so as to facilitate a nearer approach to habit, *e.g.*, acquired and connate (Wundt); primary and secondary (Romanes). Habit is used indiscriminately for an action originating in some congenital variation and an action forced upon the individual by special circumstances. A fundamental distinction, on which the validity of the theory must be tested, is thus ignored.

c. *Two Demonstrations of the Habit Theory Claimed by Romanes.*

The evidence adduced to show that habit may pass into instinct cannot here be examined in detail. Romanes brings forward two cases — the instincts of *tumbling* and *pouting* in pigeons — which he declares are alone sufficient to demonstrate the theory. We may, therefore, take these as fair samples of the arguments generally appealed to.

After quoting Darwin's remarks on this subject, Romanes adds:

"This case of the tumblers and pouters is singularly interesting and very apposite to the proposition before us; for not only are the actions utterly useless to the animals themselves, but they have now become so ingrained into their psychology as to have become severally distinctive of different breeds, and so not distinguishable from true instincts. This extension of an hereditary and useless habit into a distinction of race or type is most important in the present connection. *If these cases stood alone, they would be enough to show that useless habits may become hereditary*, and this to an extent which renders them indistinguishable from true instincts."[1]

Granting that we have here true instincts, — and I do not doubt that, — what proof have we that they originated in habits? Did there preëxist in the ancestors of these breeds organized instinct bases, which, through the fancier's art of selective breeding, were gradually strengthened until they attained the development which now characterizes the tumblers and pouters? Or was there no such basis to start with, but only a new mode of behavior, accidentally acquired by some one or more individuals, and then perpetuated by transmission to their offspring, and further developed by artificial selection? The original action in either species is called a "habit," and this so-called habit must have been inherited; ergo, habit can become instinct. Obviously, argument of that kind can have weight only with those who overlook the test-point, namely, the real nature and origin of the initial action.

If the instinct had its inception in a true habit, *i.e.*, in an action reduced to habit by repetition in the individual, and not determined in any already existing hereditary activity, is it at all credible that it could have been transmitted from parent to progeny? Does not our general experience contradict such an assumption in the most positive manner? But may not the habit have originated a great many times, and by repetition in successive generations, gradually have become "stereotyped

[1] Mental Evolution in Animals, p. 189.

into a permanent instinct"? To suppose that such *utterly useless* action originated a great many times without compelling conditions or any organic predisposition is not at all admissible.

Darwin saw at once from the nature of the actions that they could not have been taught, but "*must have appeared naturally*, though probably afterwards vastly improved by the continued selection of those birds which showed *the strongest propensity*." Darwin, then, postulates as the foundation of each instinct a "propensity" — something given in the constitution. That view of the matter is in entire accord with the theory adopted in the case of neuter insects and quite incompatible with the habit theory.

1. *The Instinct of Pouting.* — I believe the case is much stronger than Darwin suspected, and that it shows, not the genesis of instinct from habit, but from a prëexisting congenital basis. Such a basis of the pouting instinct exists in every dovecot pigeon, and is already an organized instinct, differing from the instinct displayed in the typical pouter only in degree. I could show that the same instinct is widely spread, if not universal, among pigeons. It will suffice here to call attention to the instinct as exhibited in the common pigeon. Observe a male pigeon while cooing to his mate or his neighbors. Notice that he inflates his throat and crop, and that this feature is an invariable feature in the act, often continued for some moments after the cooing ceases. Compare the pouter and notice how he increases the inflation whenever he begins cooing. The pouter's behavior is nothing but the universal instinct enormously exaggerated, as any attentive observer may readily see under favorable circumstances.

2. *The Instinct of Tumbling.* — The origin of the tumbling instinct cannot be fixed by the same direct mode of identification; but I believe that here also it is possible to point to a more general action, instinctively performed by the dovecot pigeon, as the probable source of origin. I have noticed a great many times that common pigeons, when on the point of being overtaken and seized by a hawk, suddenly flirt themselves directly downward in a manner suggestive of tumbling, and thus elude the hawk's swoop. The hawk is carried on by

its momentum, and often gives up the chase on the first failure. In one case I saw the chase renewed three times, and eluded with success each time. The pigeon was a white dovecot pigeon with a trace of fantail blood. I saw this same pigeon repeatedly pursued by a swift hawk during one winter, and invariably escaping in the same way. I have seen the same performance in other dovecot pigeons under similar circumstances.

But this is not all. It is well known that dovecot pigeons delight in quite extended daily flights, circling about their home. I once raised two pairs of these birds by hand, in a place several miles from any other pigeons. Soon after they were able to fly about they began these flights, usually in the morning. I frequently saw one or more of the flock, while in the middle of a high flight, and sweeping along swiftly, suddenly plunge downwards, often zigzagging with a quick, helter-skelter flirting of the wings. The behavior often looked like play, and probably it was that in most cases. I incline to think, however, that it was sometimes prompted by some degree of alarm. In such flights the birds would frequently get separated, and one thus falling behind would hasten its flight to the utmost speed in order to overtake its companions. Under such circumstances the stray bird coming from the rear might be mistaken for the moment for a hawk in pursuit, and one or more of the birds about to be overtaken be thus induced to resort to this method of throwing themselves out of reach of danger.

The same act is often performed at the very start, as the pigeon leaves its stand. The movement is so quick and crazy in its aimlessness that the bird often seems to be in danger of dashing against the ground, but it always clears every object.

As this act is performed by young and old alike, and by young that have never learned it by example, it must be regarded as instinctive, and I venture to suggest that it probably represents the foundation of the more highly developed tumbling instinct.

The behavior of the Abyssinian pigeon, which, when "fired at, plunges downwards so as to almost touch the sportsman, and then mounts to an immoderate height," may well be due to the same instinct. The noise of the gun, even if the bird were not hit, would surprise and alarm it, and the impulse to save itself from danger would naturally take the form determined by the instinct, if the instinct existed. This seems to me more probable than Darwin's suggestion of a mere trick or play.

d. *The Habit Theory Losing Ground.*

The two instincts of pouting and tumbling, claimed as demonstrations of the habit theory, thus turn out to be explicable only on the selection theory. It is significant that this theory is fast losing ground even among the psychologists. A. Forel's conversion illustrates the trend of opinion. "I, too," he says, "used to believe that instincts were hereditary habits, but I am now convinced that this is an error, and have adopted Weismann's view. It is really impossible to suppose that acquired habits, like piano playing and bicycle riding, for instance (these are certainly acquired), could hand over their mechanism to the germ-plasm of the offspring." [1]

In his latest work, *Habit and Instinct*, Lloyd Morgan has also abandoned the theory. On the same side stand James, Baldwin, Ziehn, Flügel, and others. The following, from Karl Groos, pp. 60, 61, will show how the difficulties with the theory are multiplying.

"As regards instinct," says Groos, "there is, further, the *a priori* argument that it is inconceivable how acquired connections among the brain cells could so affect the inner structure of the reproductive substance as to produce inherited brain tracts in later generations. And, finally, there is this consideration mentioned by Ziegler as a suggestion of Meynert's: 'It is well known that in the higher vertebrates acquired associations are located in the cortex of the hemispheres. As an acquired act becomes habitual, it may be assumed that the corresponding combination of nervous elements will become more dense and strong, and the tract proportionally more fixed. This being the case, it follows that the tracts of acquired and habitual association, as well as those of acquired movement, pass

[1] Gehirn und Seele, 1894, p. 21. Taken from The Play of Animals, by Groos, p. 56. (Translated by Elizabeth L. Baldwin.)

through the cerebrum. Instincts and reflexes, however, have their seat for the most part elsewhere. The tracts of very few of them are found in the cortex of the hemispheres. It is chiefly in the lower parts of the brain and spinal cord that the associations and coördinations corresponding to instincts and reflexes have their seat. When the comparative anatomist investigates the relative size of the hemispheres in vertebrates (especially in amphibians, reptiles, birds, and mammals), a very evident increase in size is observed which apparently goes hand in hand with the gradual gain in intelligence. In the course of long phylogenetic development, during which the hemispheres have gradually attained their greatest dimensions, they have constantly been the organ of reason and the seat of acquired association. If, then, habit could become instinct through heredity, it is probable that the cerebrum would, in much greater degree than is the fact, be the seat of instinct.' "

The stronghold of the Lamarckian view is Paleontology. It is here that the doctrine of acquired characters, or ctetology as Professor Hyatt calls it, has been nearly as unyielding as the fossils to which it adheres. But a new light seems to be penetrating even here under the name of "organic selection." This idea, first formulated by Professor Baldwin, but almost simultaneously and independently reached by Lloyd Morgan and Professor Osborn, is, that adaptive modifications are not transmissible, but that they have, nevertheless, acted as *the fostering nurses of congenital variations,* since organisms surviving through them would carry forward to the next generation such congenital variations as happened to be coincident with them. It may be, perhaps, a fine question to determine whether so-called "adaptative modifications," which really have selection value, are not themselves the coincidents of congenital bases. Be that as it may, the conversion of so eminent a paleontologist as Professor Osborn to the selection theory is all the more significant on account of the prominent part he has taken in defending the Lamarckian idea.

e. *Hyatt on Acquired Characters.*

Professor Hyatt was the first to demonstrate a wonderfully complete parallelism between the ontogenetic and the phylogenetic series, and he has presented the paleontological argument in terms that seem, to many at least, to be beyond controversy. With all respect to Professor Hyatt's monumental work, I must say that I find nothing in the evidence that compels one to take his view of acquired characters.

"We have been unable" [says Professor Hyatt] "to find any characters which were not inheritable in some series. The behavior of all characteristics which have been introduced into any series of species shows them to be subject to the law of acceleration, in whatever way they have originated, whether primarily as adaptive characters, according to our hypothesis, or by natural selection and through the combination of the sexual variations, as supposed by Weismann."[1]

This is a very sweeping statement, at least in implication. I can hardly believe that the author would have us understand that acquired characters are just as readily and invariably transmitted as congenital characters; and yet, if that is not the argument, there is no argument there. Nothing is more certain than that, in living forms accessible to direct experimental test, acquired characters are not invariably, if at all, transmissible. Demonstrations have been sought for, but so far without avail. Unless the *Arietidæ* are a wholly exceptional group, we must conclude from the above statement that all the characters found were of congenital origin, and that no acquired characters were recognized. It is easier to believe that such characters were overlooked than to believe a miracle.

The *law of acceleration* established by Professor Hyatt is complemental to *the biogenetic law* formulated by Fr. Müller and Haeckel, and both laws rest on the theory of germ continuity, as formulated by Weismann. Logically, neither of these laws implies the transmission of acquired characters. That is an assumption which has never been reconciled with the fundamental law of the genetic continuity of germs. The

[1] "Genesis of the Arietidæ," p. 43, *Mem. Mus. Comp. Zoöl.*, Cambridge, 1889.

pangenesis theory of Darwin was an attempt in this direction, but that theory has no scientific basis and it stands as a theoretical failure, rejected because it could not possibly be reconciled with what we know about the genesis of germs. That is the inevitable fate of every view which fails to adjust itself to the primary law of germ continuity.

Sense impressions and physical impressions or modifications stand on the same footing. Repetition may become habit and produce marked effects on the nervous mechanism or other organs; but the individual structure so affected is not continued from generation to generation, so that the effects are cancelled with each term of life, and there is no conceivable way by which they could be stamped upon the germs and so carried on cumulatively. If they reappear in the offspring, it cannot be because they were inherited, but because they are reproduced in the same way as they were acquired in the parent.

f. *Preformation the Essence of the Doctrine.*

This doctrine of the transmission of acquired characters is a species of preformation that eclipses the old creation hypothesis, for the miracle of stamping the germ with the form it is to present in the adult has to be repeated at each generation.

It may be objected that "stamping" is not the method by which parental characters are given to the germ. They are commonly said to be inherited. But it is too late to juggle with the term "heredity." That term either means something or nothing. If it means that characters acquired by the parent can be transmitted to the offspring, then the transmitted characters, which *ex hypothesi* are not originally determined in the germ, must in some way be determined for it by the parent. What better term than "stamp" or "impress" can be suggested? Whatever the *modus operandi*, the determining influence or impress must be imparted, at least in the great majority of cases, before development begins. Is it conceivable that perfectly definite form features can be in any way reflected back upon the ovum? Can we think of the germ as vibrating sympathetically with each acquired peculiarity of the parent organism? What vibration could there be between germ and passive structures, such as shell configurations? Could an indentation, groove, ridge, or protuberance forced upon a shell by environmental action be at the same time wrought into the germs in such a definite way as to reappear in the offspring without the aid of the same environmental causes? Or could the repetition of the same environmental action on a long line of parents gradually modify the germs in the same direction? In whatever way we turn the question, we are confronted with the same miracle of preformation. *The character arises in the parent organism by epigenesis, but it is thrown back on the germ, nobody knows or can conceive how, in such a way that it becomes a preformation capable of unfolding without the aid of its epigenetic causes.*

On the other hand, the hypothesis that all hereditary characters in organisms exclusively gamogenetic originate in spontaneous or induced (by *direct* action of environment) germ variations, appeals only to known facts and principles, and provides for the same amount of preformation as before without any miraculous transfer of characters from one organism to another. We know that germ variations are transmissible; we do not know that individually acquired modifications can be transferred to germs; we know the principle of selection to be rational and verifiable; we know of no substitute for it.

The Genetic Standpoint in the Study of Instinct.

a. *The Genealogical History Neglected.*

The problem of psychogenesis requires a more definite genetic standpoint than that of general evolution. It is not enough to recognize that instincts have had a natural origin; for the fact of their connected genealogical history is of paramount importance. From the standpoint of evolution as held by Romanes and others, instincts are too often viewed as disconnected phenomena of independent origin. The special and more superficial characteristics have been emphasized to the exclusion of the more fundamental phylogenetic characters.

Biologists and psychologists alike have very generally clung tenaciously to the idea that instincts, in part at least, have been derived from habits and intelligence; and the main effort has been to discover how an instinct could become gradually stamped into organization by long-continued uniform reactions to environmental influences. The central question has been: How can intelligence and natural selection, or natural selection alone, initiate action and convert it successively into habit, automatism, and congenital instinct? In other words, the genealogical history of the structural basis being completely ignored, how can the instinct be mechanically rubbed into the ready-made organism? Involution instead of evolution; mechanization instead of organization; improvisation rather than organic growth; specific *versus* phyletic origin.

This inversion, or rather perversion, of the genealogical order leads to a very short-focussed vision. The pouting instinct is supposed to have arisen *de novo*, as an anomalous behavior, and with it a new race of pigeons. The tumbling instinct was a sort of *lusus naturæ*, with which came the fancier's opportunity for another race. The pointing instinct was another accident that had no meaning except as an individual idiosyncrasy. The incubation instinct was supposed to have arisen *after* the birds had arrived and laid their eggs, which would have been left to rot had not some birds just blundered into "cuddling" over them and thus rescued the line from sudden extinction. How long this blunder-miracle had to be repeated before it happened all the time does not matter. Purely imaginary things can happen on demand.

b. *The Incubation Instinct.*

1. *Meaning to be Sought in Phyletic Roots.* — It seems quite natural to think of incubation merely as a means of providing the heat needed for the development of the egg, and to assume that the need was felt before the means was found to meet it. Birds and eggs are thus presupposed, and as the birds could not have foreseen the need, they could not have hit upon the means except by accident. Then, what an

infinite amount of chancing must have followed before the first "cuddling" became a habit, and the habit a perfect instinct! We are driven to such preposterous extremities as the result of taking a purely casual feature to start with. Incubation supplies the needed heat, but that is an incidental utility that has nothing to do with the nature and origin of the instinct. It enables us to see how natural selection has added some minor adjustments, but explains nothing more. For the real meaning of the instinct we must look to its phyletic roots.

If we go back to animals standing near the remote ancestors of birds, to the amphibia and fishes, we find the same instinct stripped of its later disguises. Here one or both parents simply remain over or near the eggs and keep a watchful guard against enemies. Sometimes the movements of the parent serve to keep the eggs supplied with fresh water, but aëration is not the purpose for which the instinct exists.

2. *Means Rest and Incidental Protection to Offspring.* — The instinct is a part of the reproductive cycle of activities, and always holds the same relation in all forms that exhibit it, whether high or low. It follows the production of eggs or young, and means primarily, as I believe, *rest* with incidental *protection to offspring.* That meaning is always manifest, no less in worms, molluscs, crustacea, spiders, and insects, than in fishes, amphibia, reptiles, and birds. The instinct makes no distinction between eggs and young, and that is true all along the line up to birds which extend the same blind interest to one as to the other.

3. *Essential Elements of the Instinct.* — Every essential element in the instinct of incubation was present long before the birds and eggs arrived. These elements are: (1) the disposition to remain with or over the eggs; (2) the disposition to resist and to drive away enemies; and (3) periodicity. The birds brought all these elements along in their congenital equipment, and added a few minor adaptations, such as cutting the period of incubation to the need of normal development, and thus avoiding indefinite waste of time in case of sterile or abortive eggs.

(1) **Disposition to Remain over the Eggs.** — The disposition to remain over the eggs is certainly very old, and is probably bound up with the physiological necessity for rest after a series of activities tending to exhaust the whole system. If this suggestion seems far-fetched, when thinking of birds, it will seem less so as we go back to simpler conditions, as we find them among some of the lower invertebrate forms, which are relatively very inactive and predisposed to remain quiet until impelled by hunger to move. Here we find animals remaining over their eggs, and thus shielding them from harm, from sheer inability or indisposition to move. That is the case with certain molluscs (*Crepidula*), the habits and development of which have been recently studied by Professor Conklin.[1] Here full protection to offspring is afforded without any exertion on the part of the parent, in a strictly passive way that excludes even any instinctive care. In *Clepsine* there is a manifest unwillingness to leave the eggs, showing that the disposition to remain over them is instinctive. If we start with forms of similar sedentary mode of life, it is easy to see that remaining over the eggs would be the most likely thing to happen, even if no instinctive regard for them existed. The protection afforded would, however, be quite sufficient to insure the development of the instinct, natural selection favoring those individuals which kept their position unchanged long enough for the eggs to hatch.

(2) **Disposition to Resist Enemies.** — The disposition to keep intruders from the vicinity of the nest I have spoken of as an element of the instinct of incubation. At first sight it seems to be inseparably connected with the act of covering the eggs, but there are good reasons for regarding it as a distinct element of behavior. In birds this element manifests itself before the eggs are laid, and even before the nest is built; and in the lower animals the disposition to cover the eggs is not always accompanied by an aggressive attitude. This attitude is one of many forms and degrees. A mild self-defensive state, in which the animal merely strives to hold its position without trying to rout intruders, would perhaps

[1] *Journ. of Morph.*, vol. xiii, No. 1, 1897.

be the first stage of development. In some of the lower vertebrates the attitude remains defensive and is aggressive only in a very low degree, while in others pugnacity is more or less strongly manifested. Among fishes the little Stickleback is especially noted for its fiery pugnacity, which seems to develop suddenly and simultaneously with the appearance of the dark color of the male at the spawning season.

In pigeons, as in many other birds, this disposition shows itself as soon as a place for a nest is found. While showing a passionate fondness for each other, both male and female become very quarrelsome towards their neighbors. The white-winged pigeon (*Melopelia leucoptera*) of the West Indies and the southern border of the United States is one of the most interesting pigeons I have observed in this respect. At the approach of an intruder the birds show their displeasure in both tone and behavior. The tail is jerked up and down spitefully, the feathers of the back are raised as a threatening dog "bristles up," the neck is shortened, drawing the head somewhat below the level of the raised feathers, and the whole figure and action are as fierce as the bird can make them. To the fierce look, the erect feathers, the ill-tempered jerks of the tail, is added a decidedly spiteful note of warning. If these manifestations are not sufficient, the birds jump toward the offender, and if that fails to cause retreat, wings are raised and the matter settled by vigorous blows.

This pugnacious mood is periodical, recurring with each reproductive cycle, and subsiding like a fever when its course is run. The birds behave as if from intelligent motive, but every need is anticipated blindly; for the young pair, without experience, example, or tradition, behave like the parents.

It seems to me that this mood or disposition, although in some ways appearing to be independent of the disposition to cover the eggs, can best be understood as having developed in connection with the latter. It has primarily the same meaning, — protection to the eggs, — but the safety of the eggs and young depends upon the safety of the nest, and this accounts for the extension of its period to cover all three stages — building, sitting, and rearing.

(3) **Periodicity.** — The periodicity of the disposition to sit coincides in the main with that of the recuperative stage. Its length, however, at least in birds, is nicely correlated with, though not exactly coinciding with, the time required for hatching. It may exceed or fall short of the time between laying and hatching. The wild passenger pigeon (*Ectopistes*) begins to incubate a day or two in advance of laying, and the male takes his turn on the nest just as if the eggs were already there. In the common pigeon the sitting usually begins with the first egg, but the birds do not sit steadily or closely until the second egg is laid. The birds do not, in fact, really sit on the first egg, but merely stand over it, stooping just enough to touch the egg with the feathers. This peculiarity has an advantage in that the development of the first egg is delayed so that both eggs may hatch more nearly together. The bird acts just as blindly to this advantage as *Ectopistes* does to the mistake of sitting before an egg is laid. *Ectopistes* is very accurate in closing the period, for if the egg fails to hatch within twelve to twenty hours of its normal time, it is deserted, and that too if, as may sometimes happen, the egg contains a perfect young, about ready to hatch. Pigeons, like fowls, will often sit on empty nests, filling up the period prescribed in instinct, leaving the nest only as the impulse to sit runs down. It happens not infrequently that pigeons will go right on with the regular sequence of activities, even though nature fails in the most important stage. Mating is followed by nest-making, and at the appointed time the bird goes to the nest to lay, and after going through the usual preliminaries, brings forth no egg. But the impulse to sit comes on as if everything in the normal course had been fulfilled, and the bird incubates the empty nest, and exchanges with her mate as punctiliously as if she actually expected to hatch something out of nothing. This may happen in any species under the most favorable conditions. It is possible by giving an abundance of rich food to wind up the instinctive machinery more rapidly than would normally happen, so that recuperation may end in about a week's time, when incubation will stop and a new cycle begin, leading to the production of a second set of eggs in the same nest.

This has happened several times with the crested pigeon of Australia (*Ocyphaps lophotes*).

Schneider[1] says: "The impulse to sit arises, as a rule, when a bird sees a certain number of eggs in her nest." Although recognizing a *bodily disposition* as present in some cases, sitting is regarded as a *pure perception impulse*. I hold, on the contrary, that the *bodily disposition* is the universal and essential element, and that sight of the eggs has nothing to do primarily with sitting. It comes in only secondarily and as an adaptation in correlation with the inability in some species to rear more than one or two broods in a season. In such species the advantage would lie with birds beginning to incubate with a full nest.

The suggestions here offered on the origin of the incubation instinct, incomplete and doubtful as they may appear, may suffice to indicate roughly the general direction in which we are to look for light on the genesis of instincts. The incubation instinct, as we now find it perfected in birds, is a nicely timed and adjusted part of a periodical sequence of acts. If we try to explain it without reference to its physiological connections in the individual, and independently of its developmental phases in animals below birds, we miss the more interesting relations, and build on a purely conjectural chance act that calls for a further and incredible concatenation of the right acts at the right time and place, and is not even then completed until its perpetuation is secured by a miracle of transmission.

A Few General Statements.

1. Instinct and structure are to be studied from the common standpoint of phyletic descent, and that not the less because we may seldom, if ever, be able to trace the whole development of an instinct. Instincts are evolved rather than involved (stereotyped by repetition and transmission), and the key to their genetic history is to be sought in their more general rather than in their later and incidental uses.

[1] Der Thierische Wille, pp. 282, 283. As cited in Professor James's Psychology, p. 388.

2. The primary roots of instincts reach back to the constitutional properties of protoplasm, and their evolution runs, in general, parallel with organogeny. As the genesis of organs takes its departure from the elementary structure of protoplasm, so does the genesis of instincts proceed from the fundamental functions of protoplasm. Primordial organs and instincts are alike few in number and generally persistent. As an instinct may sometimes run through a whole group of organisms with little or no modification, so may an organ sometimes be carried on through one or more phyla without undergoing much change. The dermal sensillæ of annelids and aquatic vertebrates are an example.

3. Remembering that structural bases are relatively few and permanent as compared with external morphological characters, we can readily understand why, for example, five hundred different species of wild pigeons should all have a few common undifferentiated instincts, such as, drinking without raising the head, the cock's time of incubating from about 10 A.M. to about 4 P.M., etc.

4. Although instincts, like corporeal structures, may be said to have a phylogeny, their manifestation depends upon differentiated organs. We could not, therefore, expect to see phyletic stages repeated in direct ontogenetic development, as are the more fundamental morphological features, according to the biogenetic law. The main reliance in getting at the phyletic history must be comparative study.

5. Instinct precedes intelligence both in ontogeny and phylogeny, and it has furnished all the structural foundations employed by intelligence. In social development also instinct predominates in the earlier, intelligence in the later stages.

6. Since instinct supplied at least the earlier rudiments of brain and nerve, since instinct and mind work with the same mechanisms and in the same channels, and since instinctive action is *gradually* superseded by intelligent action, we are compelled to regard instinct as the actual germ of mind.

7. The automatism, into which habit and intelligence may lapse, seems explicable, in a general way, as due more to the preorganization of instinct than to mechanical repetition. The

habit that becomes automatic, from this point of view, is not an action on the way to becoming an instinct, but action preceded and rendered possible by instinct. Habits appear as the uses of instinct organization which have been learned by experience.

8. The suggestion that intelligence emerges from blind instinct, although nothing new, will appear to some as a complete *reductio ad absurdum*. But evolution points unmistakably to instinct as nascent mind, and we discover no other source of psychogenetic continuity. As far back as we can go in the history of organisms, in the simplest forms of living protoplasm, we find the sensory element along with the other fundamental properties, and this element is the central factor in the evolution of instinct, and it remains the central factor in all higher psychic development. It would be strange if, with this factor remaining one and the same throughout, organizing itself in sense organs of the keenest powers and in the most complex nerve mechanisms known in the animal world—it would be strange, I say, if, with such continuity on the side of structure, there should be discontinuity in the psychic activities. Such discontinuity would be nothing less than the negation of evolution.

9. We are apt to contrast the extremes of instinct and intelligence—to emphasize the blindness and inflexibility of the one and the consciousness and freedom of the other. It is like contrasting the extremes of light and dark and forgetting all the transitional degrees of twilight. In so doing we make the hiatus so wide that derivation of one extreme from the other seems about as hopeless as the evolution of something from nothing. That is the last pit of self-confounding philosophy.

Instinct is blind; so is the highest human wisdom blind. The distinction is one of degree. There is no absolute blindness on the one side, and no absolute wisdom on the other. Instinct is a dim sphere of light, but its dimness and outer boundary are certainly variable; intelligence is only the same dimness improved in various degrees.

When we say instinct is blind, we really mean nothing more than that *it is blind to certain utilities* which we can see. But we ourselves are born blind to these utilities, and only discover them after a period of experience and education. The discovery may seem to be instantaneous, but really it is a matter of growth and development, the earlier stages of which consciousness does not reveal.

Blindness to the utilities of action no more implies unconsciousness in animals than in man. It is the worst form of anthropomorphism to claim that animal automatism is devoid of consciousness, for the claim rests on nothing but the assumption that there are no degrees of consciousness below the human. If human organization is of animal origin, then the presumption would be in favor of the same origin for consciousness and intelligence. Automatism could not exclude every degree of consciousness without excluding every form of organic adaptation.

10. The clock-like regularity and inflexibility of instinct, like the once popular notion of the "fixity" of species, have been greatly exaggerated. They imply nothing more than a low degree of variability under normal conditions. Discrimination and choice cannot be wholly excluded in every degree, even in the most rigid uniformity of instinctive action. Close study and experiment with the most machine-like instincts always reveal some degree of adaptability to new conditions. This was made clear by Darwin's studies on instincts, and it has been demonstrated over and over again by later investigators, and by none more thoroughly than by the Peckhams in the case of spiders and wasps.[1] Intelligence implies varying degrees of freedom of choice, but never complete emancipation from automatism. The fundamental identity of instincts and intelligence is shown in *their dependence upon the same structural mechanisms and in their responsive adaptability.*

Instinct and Intelligence.

In order to see how instinctive action may graduate into intelligent action it is well to study closely animals in which the instincts have attained a high degree of complexity, and

[1] Wisconsin Geological and Natural History Survey, Bulletin No. 2, 1898.

in which there can be no doubt about the automatic character of the activities. These conditions are perfectly fulfilled in the pigeons, a group in which we have the further advantage that wild and domestic species can be studied comparatively.

It is quite certain that pigeons are totally blind to the meanings which we discover in incubation. They follow the impulse to sit without a thought of consequences; and no matter how many times the act has been performed, no idea of young pigeons ever enters into the act.[1] They sit because they feel like it, begin when they feel impelled to do so, and stop when the feeling is satisfied. Their time is generally correct, but they measure it as blindly as a child measures its hours of sleep. A bird that sits after failing to lay an egg, or after its eggs have been removed, is not acting from "expectation," but because she finds it agreeable to do so and disagreeable not to do so. The same holds true of the feeding instinct. The young are not fed from any desire to do them any good, but solely for the relief of the parent. The evidence on this point cannot be given here, but I believe it is conclusive.

But if all this be true, where does the graduation towards intelligence manifest itself. Certainly not in a comprehension of utilities which are discoverable only by human intelligence. Whatever the pigeon instinct-mind contains, it is safe to say that the intelligence is hardly more than a grain hidden in bushels of instinct, and one may search more than a day and not find it.

a. *Experiment with Pigeons.*

Among many tests, take the simple one of removing the eggs to one side of the nest, leaving them in full sight and within a few inches of the bird on the nest. The bird sees the uncovered eggs, but shows no interest in them; she keeps

[1] Professor James, Psychology, II, p. 390, thinks such an idea may arise and that it may encourage the bird to sit. "*Every instinctive act in an animal with memory*," says James, "*must cease to be 'blind' after being once repeated.*" That must depend on the kind of memory the animal has. It is possible to have memory of a certain kind in some things, while having absolutely none of any kind in other things. That is the case in pigeons, as I feel very sure.

her position, if she is a tame bird, and after some moments begins to act as if the current of her feelings had been slightly disturbed. At the most she only acts as if a little puzzled, as if she realized dimly a change in feeling. She is accustomed to the eggs, and now misses them, or, rather, misses something, she knows not what. Although she does not know or show any care for the eggs out of the nest, she does appear to sense a difference between having and not having.

There is, then, something akin to memory and discrimination, and little as this implies it cannot mean less than some faint adumbration of intelligence. Now this inkling of intelligence, or, if you prefer, this nadir of stupidity, so remote from the zenith of intelligence, is not something independent of and foreign to instinct. It is instinct itself just moved by a ripple of change in the environment. The usual adjustment is slightly disturbed, and a little confusion in the currents of feeling arises, which manifests itself in quasi-mental perplexity. That is about as near as I can get to the contents of the pigeon mind without being able, by a sort of metempsychosis suggested by Bonnet, to live some time in the head of the bird.

In this feeble perplexity of the pigeon's instinct-mind, in this "nethermost abyss" of stupidity, there is a glimmer of light, and nature's least is always suggestive of more. The pigeon has no hope of graduating into a *homo sapiens*, but her little light may flicker a little higher, and all we need to know is, how instinct behavior can take one step toward mind behavior. This is the dark point on which I have nothing really new to offer, although I hope not to make it darker.

b. *The Step from Instinct to Intelligence.*

Some notion of what is involved in the step may be gathered by comparing wild with semi-domesticated and fully domesticated species. These grades differ from each other in respects that are highly suggestive. In the wild species the instincts are kept up to the higher degrees of rigid invariability, while in species under domestication they are reduced to various degrees of flexibility, and there is a correspondingly greater freedom of action, with, of course, greater liability to irregularities and so-called "faults." These faults of instinct, so far from indicating psychical retrogression, are, I believe, the first signs of greater plasticity in the congenital coördinations and, consequently, of greater facility in forming those new combinations implied in choice of action.

If we place the three grades of pigeons under the same conditions and test each in turn in precisely the same way, we can best see how domestication lets down the bars to choice and at the same time gives more opportunities for free action. The simplest experiment is always the best. Let us take three species at the time of incubation and repeat with each the experiment of removing the eggs to a distance of two inches outside the edge of the nest. The three grades are well represented in the wild passenger pigeon (*Ectopistes*), the little ring-neck (*Turtur risorius*), and the common dovecot pigeon (*Columba livia domestica*). The results will not, of course, always be the same, but the average will be about as follows:

1. *The Passenger Pigeon.* — The passenger pigeon leaves the nest when approached, but returns soon after you leave. On returning she looks at the nest, steps into it, and sits down as if nothing had happened. She soon finds out, not by sight, but by feeling, that something is missing. Her instinct is keenly attuned and she acts quite promptly, leaving the nest after a few minutes without heeding the egg. The conduct varies relatively little in different individuals.

2. *The Ring-neck Pigeon.* — The ring-neck is tame and sits on while you remove the eggs. After a few moments she moves a little and perhaps puts her head down, as if to feel the missing eggs with her beak. Then she may glance at the eggs and appear as if half consciously recognizing them, but make no move to replace them, and after ten to twenty minutes or more leave the nest with a contented air, as if her duty were done; or, she may stretch her neck toward the eggs and try to roll one back into the nest. If she succeeds in recovering *one*, she is satisfied and again sinks into her

usual restful state, with no further concern for the second egg. The conduct varies considerably with different individuals.

3. *The Dovecot Pigeon.* — The dovecot pigeon behaves in a similar way, but will generally try to get *both* eggs back; and, failing in this, she resigns the nest with more hesitation than does the ring-neck.

4. *Results Considered.* — The passenger pigeon's instinct is wound up to a high point of uniformity and promptness, and her conduct is almost too blindly regular to be credited even with that stupidity which implies a grain of intelligence. The ring-neck's stupidity is satisfied with one egg. The dovecot pigeon's stupidity may claim both eggs, but it is not always up to that mark.

In these three grades the advance is from extreme blind uniformity of action, with little or no choice, to a stage of less rigid uniformity, with the least bit of perplexity and a very feeble, uncertain, dreamy sense of sameness between eggs *in* and eggs *out* of the nest, which prompts the action of rolling the eggs back into the nest. That is the instinctive way of placing the eggs when in the nest, and the neck is only a little further extended in drawing the eggs in from the outside. How very narrow is the difference between the ordinary and the extraordinary act! How little does the pendulum of normal action have to swing beyond its usual limit![1]

But this little is in a forward direction, and we are in no doubt as to the general character of the changes and the modifying influences through which it has been made possible. Under conditions of domestication the action of natural selection has been relaxed, with the result that the rigor of instinctive coördinations which bars alternative action is more or less

[1] We come to equally surprising results in many different ways. Change the position of the nest-box of the ring-neck, without otherwise disturbing bird, nest, or contents, and the birds will have great difficulty in recognizing their nest, for they know it only as something in a definite position in a fixed environment. If a pair of these birds have a nest in a cage, and the cage be moved from one room to another, or even a few feet from its original position in the same room, the nest ceases to be the same thing to them, and they walk over the eggs or young as if completely devoid of any acquaintance with or interest in them. Return the cage to its original place and the birds know the nest and return to it at once.

reduced. Not only is the door to choice thus unlocked, but more varied opportunities and provocations arise, and thus the internal mechanism and the external conditions and stimuli work both in the same direction to favor greater freedom of action.

When choice thus enters, no new factor is introduced. There is greater plasticity within and more provocation without, and hence the same bird, without the addition or loss of a single nerve-cell, becomes capable of higher action and is encouraged and even constrained by circumstances to *learn* to use its privilege of choice.

Choice, as I conceive, is not introduced as a little deity, encapsuled in the brain. Instinct has supplied the teleological mechanism, and stimulus must continue to set it in motion. But increased plasticity invites greater interaction of stimuli and gives more even chances for conflicting impulses. Choice runs on blindly at first, and ceases to be blind only in proportion as the animal learns through nature's system of compulsory education. The teleological alternatives are organically provided; one is taken and fails to give satisfaction; another is tried and gives contentment. This little freedom is the dawning grace of a new dispensation, in which education by experience comes in as an amelioration of the law of elimination. This slight amenability to natural educational influences cannot, of course, work any great miracles of transformation in a pigeon's brain; but it shows the way to the open door of a freer commerce with the eternal world, through which a brain with richer instinctive endowments might rise to higher achievement.

The conditions of amelioration under domestication do not differ in kind from those presented in nature. Domestication merely bunches nature's opportunities and thus concentrates results in forms accessible to observation. Natural conditions are certainly working in the same direction, only more slowly. The direction and the method of progress must, in the nature of things, remain essentially the same.

Nature works to the same ends as intelligence, and to the natural course of events I should look for just such results as

Lloyd Morgan[1] so clearly pictures and ascribes to intelligence. "Suppose," says Mr. Morgan, "the modifications are of various kinds and in various directions, and that, associated with the instinctive activity, a tendency to modify it *indefinitely* be inherited. Under such circumstances *intelligence would have a tendency to break up and render plastic a previously stereotyped instinct.* For the instinctive character of the activities is maintained through the constancy and uniformity of their performance. But if the normal activities were thus caused to vary in different directions in different individuals, the offspring arising from the union of these differing individuals would not inherit the instinct in the same purity. The instincts would be imperfect, and there would be an inherited tendency to vary. *And this, if continued, would tend to convert what had been a stereotyped instinct into innate capacity; that is, a general tendency to certain activities (mental or bodily), the exact form and direction of which are not fixed, until by training, from imitation or through the guidance of individual intelligence, it became habitual. Thus it may be that it has come about that man, with his enormous store of innate capacity, has so small a number of stereotyped instincts.*"

The following from Professor James[2] is suggestive:

"Nature implants contrary impulses to act on many classes of things, and leaves it to slight alterations in the conditions of the individual case to decide which impulse shall carry the day. Thus, greediness and suspicion, curiosity and timidity, coyness and desire, bashfulness and vanity, sociability and pugnacity seem to shoot over into each other as quickly, and to remain in as unstable equilibrium, in the higher birds and mammals as in man. They are all impulses, congenital, blind at first, and productive of motor reactions of a rigorously determinate sort. Each one of them, then, is an instinct, as instinct is commonly defined. *But they contradict each other; experience, in each particular opportunity of application, usually deciding the issue. The animal that exhibits them loses the 'instinctive' demeanor and appears to lead a life of hesitation and choice, an intellec-*

[1] Animal Life and Intelligence, pp. 452, 453.
[2] Psychology, II, pp. 392, 393.

tual life; not, however, because he has no instincts — rather because he has so many that they block each other's path."

Looking only to the more salient points of direction and method in nature's advance towards intelligence, the general course of events may be briefly adumbrated. Organic mechanisms capable of doing teleological work through blindly determined adjustments, reproduced congenitally, and carried to various degrees of complexity and inflexibility of action, were first evolved. With the organization of instinctive propensities, liable to antagonistic stimulation, came both the possibility and the provocation to choice. In the absence of intelligent motive, choice would stand for the outcome of conflicting impulses. The power of blind choice could be transmitted, and that is what man himself begins with.

Superiority in instinct endowments and concurring advantages of environment would tend to liberate the possessors from the severities of natural selection; and thus nature, like domestication, would furnish conditions inviting to greater freedom of action, and with the same result, namely, that the instincts would become more plastic and tractable. Plasticity of instinct is not intelligence, but it is the open door through which the great educator, experience, comes in and works every wonder of intelligence.

Spencer[1] has shown clearly that this plasticity must inevitably result from the progressive complication of the instincts. "*That progressive complication of the instincts,*" he says, "*which, as we have found, involves a progressive diminution of their purely automatic character, likewise involves a simultaneous commencement of memory and reason.*"

[1] Psychology, I, pp. 443 and 454, 455.

11

Reprinted from *Woods Hole Mar. Biol. Lab. Biol. Lect.* (1899) **7**:57-67 (1900)

INSTINCT.

EDWARD THORNDIKE.

I HAVE first of all a request to make of the reader. It is that, in receiving and estimating what I say, he temporarily discard the definitions or formulæ for psychological phenomena now in his mind. Many of the discussions and quarrels of comparative psychology are about mere words, and are therefore fruitless. We can avoid all such if for the time being I am allowed to *name* the phenomena, the facts about which we are to think, as I please. I hope to make clear what facts I am referring to in every case, and you will be at liberty to rename them after your own preferences as soon as we part company.

The facts to which I refer by the words "instinctive reactions" or "instinctive activities" or "instincts" are any activities which do not have to be learned, which the animal is capable of without experience. Let us not mind whether the act be accompanied by consciousness or not, whether it represent the inheritance of some ability acquired by the animal's ancestors or not, whether it involve emotional feeling or not. I shall, for the sake of keeping in line with customary usage, deal with such instinctive activities as physiology generally leaves out, *e.g.*, instinctive fears, food preferences, motor control in running, jumping, flying, etc., though breathing, defecation, sleeping, etc., really deserve treatment of just the same sort as these more complex activities. Under instincts, then, let us study all the abilities to respond to different situations (more particularly certain complicated external situations) which the animal has to start with, has apart from parental or self tuition. An instinctive act, then, will be the opposite of an act due to experience. The sum of an animal's instinctive abilities plus the habits taught him by his life struggle will be the total of his store of ability.

It is clear that what an animal has to start with, due to the organic structure it inherits, is a matter of prime importance in the case of any animal whose consciously directed activities you are studying. If we do not know its instinctive equipment, we are likely to credit it with intelligent thinking for doing something it really could n't have helped doing. Suppose I say to you, "I have an intelligent chick who has learned so to adapt his movements in the water that he can swim to shore," and tossing him into the water, demonstrate that fact to you. It is probable that many a one would say, "How remarkable! Do you suppose he reasoned out the way to act? How *did* he learn to do it?" The real fact would be that he did not *learn* to do it at all; that all chicks react to water by swimming out of it the very first time they ever get in it; that chicks swim instinctively. How long error may persist about any animal activity is well shown by the recency of our knowledge that walking is instinctive in the human infant. So the first task of comparative psychology is to find out the instinctive equipment of any animal studied. Instincts are, however, well worth study for their own sake. An instinctive fear of a certain enemy may be as truly useful to an animal as sharp teeth or protective coloration. Instincts are the expressions of structures and functions of the nervous system, and are as real and as important matters for the biologist as are bones and blood vessels.

It is outside the province of this lecture to enumerate or describe the particular instincts of any group of animals, but we should note that in spite of the tremendous number of instincts that have been observed, the story has not been half told. A rich field awaits the investigator. The humble chicken has been under everybody's observation, and has been specially studied by Spalding, Eimer, Preyer, Lloyd Morgan, and others, yet I was able to record the following additional instinctive activities: swimming to shore when thrown into the water, fighting (observed as early as the sixth day), appropriate reactions to distance in jumping down from heights, avoidance of open places, reacting to honey-bees by seizing, knocking them against the ground, and then eating them (see *Psychological Review*, May, 1899).

Turning to our proper task, the statement of some characteristics of instinctive activities in general, we should first recall the familiar fact that to say that a certain ability or tendency is in an animal apart from teaching or experience, is born in him, is not to say that at birth he possesses it or that all through life he keeps it. Scratching the ground in the case of the chick, walking in the case of the human infant, are obvious examples of true instincts which yet do not appear for some time after birth. The instinct to follow and the instinctive avoidance of loneliness which the chick manifests so markedly in the earliest days of his life disappear to a large extent if given no chance to harden into habits. Instincts, that is, may be *delayed* and may be *transitory.* Care must always be taken by the student, however, not to interpret as delayed instincts cases where the reason for delayed activity is really lack of strength in some organ, or as transitory instincts cases where the reason for loss of the activity is loss of strength or inhibition of the activity by unpleasant consequences.

Perhaps the most important general statement one can make about instinctive reactions is that they are often indefinite and inexact. The same situation may be reacted to differently by different individuals of a species, or by the same individual on different occasions. The Peckhams (Wisconsin Geological Survey, *Bulletin No. 2*) have shown decisively this inexactness, this vagueness of reaction in the case of the nest-building and stinging habits of the solitary wasps. Among vertebrates one sees it almost everywhere. If you slam a door in a room where you have half a dozen chicks, one may run and crouch for twenty seconds, another may squat where he is for an equal time, another may also chirp, while the others may keep on feeding undisturbed. The old theory of instinct, that it was a sort of God-given power bestowed on animals to make up for their lack of reason, which in mysterious ways directed the animal's footsteps and aroused in him by direct inspiration the appropriate act in any set of circumstances, left us an unfortunate legacy in the shape of a tendency to expect an accuracy and infallibility and unchangeableness in animals' reactions such as supernatural inspiration might well give, but which are usually signs of death in the natural world. The common human interest in the marvelous and unusual coöperated with this tendency by selecting for observation such instinctive activities as web-spinning, honeycomb construction, etc., and neglecting the more ordinary activities. Thus arose such amusing expressions as "Failure of instinct." Furthermore, so long as there survived vestiges of this teleological notion of an entity, "instinct" which fitted acts to situations so as to get desirable results, there was the less interest in the real intermediary between situation and act, namely, the nervous system, and less likelihood of men looking for the variations in response to be expected in the results of the activity of any living organ.

The recognition of the vague in instinctive activity not only brings such activity into line with other biological phenomena, and lends a healthier tone to investigation, but also provides a useful warning to the observer. The naturalist who studies a single case of such activity is almost sure to be misled. "In a multitude of witnesses there is strength."

The utility in the struggle for existence of an animal's equipment of instinctive reactions is much increased by reason of their ability to harden into habits. It is a wide if not a general law that any act which has been performed in a certain situation and resulted in pleasure, or even indifferently, is the more likely to be performed again in that same situation, the reverse happening to any act resulting in discomfort.

The general manifestations of this law will be dealt with in a later lecture, but its influence on instinctive reactions deserves mention here. First of all, transitory instincts may gain thereby equal value with permanent ones.

The chick that has followed the mother-hen for six or eight days does not thenceforth need any instinctive impulse to follow, for the act has become habitual and will remain in the absence of any such impulse. Again vague responses may be hardened into more definite and more successful forms. A cat with a general instinct to jump at small birds might with practice jump straighter and more quickly because the quick, straight jumps would result in the pleasure of capture. One is likely, however, to be misled if he argues from the presupposition that acts always tend to assume a "perfect" form. If the animal gets along very well with its instinct still "imperfect," there may be no change. Lloyd Morgan, for instance, has chosen a dubious example of perfecting through habit in the seizing of bits of food by chicks. They often do fail to seize in their first experiences, as he observed, but they often, perhaps just as often, fail even after long experience. I took nine chickens from 10 *to* 14 *days old* and placed them one at a time on a level surface over which were scattered bits of cracked wheat (the food they had been eating in this same way for a week) and watched their pecking. Out of 214 objects pecked at, 159 were seized, 55 *were not.* Out of the 159, *only* 116 were seized on the first peck, 25 on the second, 16 on the third, and 2 on the fourth. This is far from a perfect record.

In the growth of the chick's discrimination between objects as food we find a sure manifestation of our law. The chick instinctively pecks at all sorts of objects of suitable size, *e.g.*, tacks, match ends, printed letters, the eyes and toes of his mates, his own excrement, etc. The pecks at bits of food and small stones bring satisfaction, and the chick that when first confronted by the situation, "grain of wheat, match end, and excrement," was as likely to peck at one as another, becomes a chick who almost inevitably pecks at the wheat. Thus the vague instinctive response may be educated into a lot of definite food preferences and avoidances. Not only thus directly, but also in indirect and complex ways, instincts may furnish the foundation of habits or, as we may better call them, associations; associations, that is, between certain situations or circumstances and certain acts. A young animal instinctively follows or keeps near its parent and thereby forms associations which later will lead him to go independently to certain feeding grounds, to eat certain sorts of animals or plants, to run away from certain enemies, to sleep in certain lairs. A kitten when confined in a cage reacts instinctively with squeezings, clawings, bitings, etc., some one of which may happen to open a thumb-latch on the door and give it freedom. The pleasure consequent may so stamp in that particular act, in connection with that situation, that after enough experiences the animal will, when put in that situation, manifest nothing of all the instinctive activities first observed save the one particular poke at the thumb-piece. Its activity now would seem far enough from instinct to many people, but it is really a consequence of instinctive activity. It has come to neglect the unsuccessful squeezings and bitings, to choose the successful clawing, in just the same way that the chick comes to neglect the peckings at excrement and match ends, and choose the food.

I have already implied that instinctive activities may be inhibited just as truly as they may be confirmed and reinforced. If the conditions in which an animal lives become such that an instinctive act brings discomfort, that activity tends to disappear; or if the animal has, prior to the appearance of a certain instinct, learned to meet otherwise a situation which would normally call forth the instinct, he may continue to meet it in that rather than the instinctive way. It would even be fairly reasonable to interpret the transitory instincts as instincts which were inhibited by mere lack of exercise. A convenient account of the inhibition of instincts may be found in James's *Psychology*, Vol. II, pp. 394–397. I may quote in addition examples of inhibition (1) by virtue of the previous formation of a habit, (2) by (1) plus actual abolition through resulting discomfort.

An instance of the former sort is found in the history of a cat which learns to pull a loop and so escape from a box whose top is covered by a board nailed over it. If, after enough trials, you remove a piece of the board covering the box, the cat, when put in, will still pull the loop instead of crawling out through the opening thus made. But, at any time, if she happens to notice the hole, she *may* make use of it. An instance of the second sort is that of a chick which has been put on a box with a wire screen at its edge, preventing it from jumping directly down, as it would instinctively do, and forcing it to jump to another box on one side of it and thence down. In the experiments which I made, the chick was prevented by a second screen from jumping directly from the second box also. That is, if in the accompanying figure, A is a box 34 inches high, B a box 25 inches high, C a box 16 inches high, and D the pen with the food and other chicks, the subject had to go A–B–C–D. The chick tried at first to get through the screen, pecked at it and ran up and down along it, looking at the chicks below and seeking for a hole to get through. Finally it jumped to B and, after a similar process, to C. After enough trials it forms the habit and when put on A goes immediately to B, then to C and down. Now if, after 75 or 80 trials, you take away the screens, giving the chick a free chance to go to D from either A or B, and then put it on A, the following phenomenon appears. The chick goes up to the edge, looks over, walks up and down it for a while, still looking down at the chicks below, and then goes and jumps to B as habit has taught it to do. The same actions take place on B. No matter how clearly the chick sees the chance to jump to D, it does not do so. The impulse has been truly inhibited. It is not the mere habit of going the other way, but the impossibility of going *that* way. In one case I observed a chick in which the instinct was all but, yet not quite, inhibited. When tried without the screen, it went up to the edge to look over *nine times*, and at last, after seven minutes, did jump straight down. — *Animal Intelligence, an Experimental Study*, etc., pp. 99–100. New York, 1898.

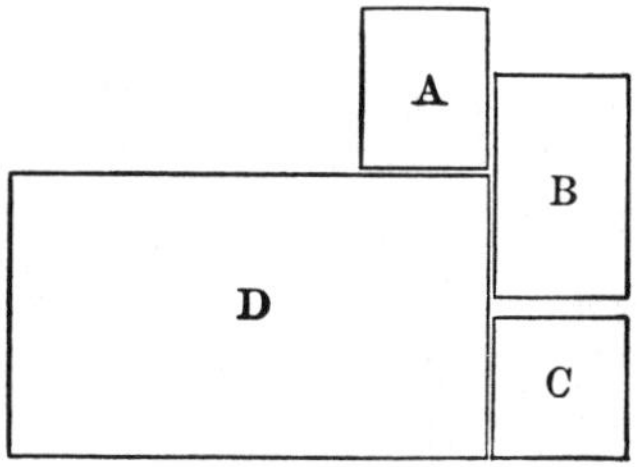

It remains to consider the so-called "perversions" of instinct, that is, the cases where an instinctive activity appears in response to a situation which it does not fit, or does not ordinarily go with. The cat that nurses rats or puppies, the pigeon that capers before a bottle, are stock examples. Most of such are due to the essential indefiniteness of the instincts in question. The cat does not have an impulse to nurse a particular, definite thing, to wit, a young kitten, and to abhor all else as subjects for nursing, and if her own young are only taken away and other young nursing things happen to be around, they may call forth the activity quite as well. Other "perversions" may be brought about by naturally or artificially caused inhibitions. There may also, of course, be in animals, as there may be in men, abnormalities in the nervous system which may occasion abnormal instinctive manifestations. These would need separate treatment. The interest of the ordinary "perversions" is simply their witness, first, to the subordination of instincts to the laws of nervous activity in general, *e.g.*, variation, association, inhibition, and, secondly, to the coarse, rough-hewn nature of many instincts.

So much for the important general facts about how animals meet situations apart from experience. Before entering on the tasks of justifying the definition of instinct, which I asked you to accept provisionally, and of saying somewhat about the origin and evolution of instincts, I wish to take this opportunity to advocate the study of instinctive activities by students of elementary biology. Their observation requires no knowledge of psychological terminology, no ability to make psychological interpretations. The observer has simply to answer the question, "How does an animal react to a certain situation the first time he is in it?" The demonstration of every matter discussed so far in this paper may be effected at practically no expense in any quiet place where the student can find twenty square feet of space in which to keep a few chicks. Such study will serve as a partial corrective to the tendency to make biology a study, not of life, but of form. It may give the student a taste of and for fact in this field of biology, which will keep him from getting lost when he meets with the speculations about instinct. It is, moreover, worth while for its own sake.

Returning now to the question of definition, it is clear that any one has a logical right to name things to suit himself and to determine the particular things which he shall denote by a certain word at his own sweet will, so long as he is clear and consistent and does not reason from any other meaning of the word than that he has given to it. One might, for example, in a study of fission call Paramœcium "Hylobates McKinleyensis" and commit no logical error, provided his figures showed clearly the animal he meant to talk about, and provided he did not use the name to argue, for instance, that we might, having found fission in Hylobates McKinleyensis, expect it in the forms ordinarily called "Hylobates." Names are not facts, and quarrels about them are generally quarrels about rhetorical expediency. So if any one chooses to call instinctive only such unlearned acts as have consciousness accompanying them, leaving other such acts the name "reflex action" and lumping in still other such acts under the words "processes of respiration, digestion," etc., I, personally, have no theoretical objection. It seems to me *practically* unwise to separate the discussion of unlearned activities with consciousness from that of unlearned activities without consciousness, so long as you are thinking about the external form of the act; and I have even chosen to so define the word that even digestion and respiration could be studied under the name. There is really no need for *any* name and no cause for worry if we cannot invent any one definition which will consistently apply to all the facts we treat. For what is important is concrete information about particular facts. My use of a name or definition is to point to certain actual facts, and I trust that I have made clear by my illustrations to what sort of facts my statements refer. To one who should say, "By your definition the knee-jerk or the formation of saliva is an instinct, for they are 'unlearned reactions,' — may even be considered reactions to a 'complex external situation,' — yet your statements about the formation of habits and inhibition are not true of them," I should reply, "I plead guilty, though I did describe further my facts as those which 'physiology generally leaves out,' but you are a more stupid reader than I expected to address."

The fact is that unlearned activities range from direct physical and chemical activities of single cells to most complicated activities involving the nervous system; that the series seems to be continuous; that if you try to separate off one lot "instincts," another lot "reflexes," another lot "non-nervous reactions," you find your classes running into each other. It is because of my confidence in this continuity of nature that I used my definition only to point to what I was talking about —

not to describe authoritatively the attributes of a fixed class of facts. Practically it seemed worth while to bring out this continuity in the definition.

This rather tiresome explanation serves an additional purpose by introducing us to a real question—the question of the origin of instincts. For some thinkers have cheerfully begged this question, by defining instincts as inherited habits, and by studying those unlearned abilities which looked like inherited habits, out of connection with all other unlearned abilities. In his lecture on "Animal Behavior" (Woods Holl *Biological Lectures*, pp. 285–338, 1898) Prof. C. O. Whitman has, in opposition to these thinkers, defended with great care and force the theory that such unlearned abilities as the spider's web-spinning, chick's scratching, dog's pointing, etc., — such activities, in short, as have been discussed in this paper, — (1) have the same origin as the unlearned activities of reflex action, digestion, circulation, etc., (2) are due, in fact, to organic features, which again (3) are due to germ variations, and that (4), therefore, we should look for and expect to find, so far as traces have been left for us to follow, as continuous a development of such activities, as true an evolution of "instincts" as of organs.

All that I have to say about Professor Whitman's first two contentions is that they seem to me so true that I cannot understand any one's doubting them. As one runs through animals' unlearned activities, from those most to those least like products of learning, he can nowhere stop and say, "hitherto inherited habits, from hence chance variations." These first two contentions would remain true whatever might happen to the third. As to the third, I can only add my mite to the evidence. The failure of chicks to avoid their own excrement and their ability to swim seem hard to explain on Lamarckian principles. For the avoidance of excrement must have been formed as a habit in every individual for many generations, while the swimming instinct has been unused for even more perhaps. Again, the vagueness of response discovered the more we study animals' unlearned activities is just what one would expect if these responses were due to germ changes selected by reason of their success in procuring survival, while it is not what we would expect from the inheritance of habits acquired in reference to definite objects. If the instinct to follow in chicks came from the habit of following (acquired, of course, with a hen as the object followed), we would expect it to require as its object something at least like a hen. As a matter of fact it does not.

Professor Whitman's fourth contention, that, since instinctive activities are the results of gradual development, they should be, not merely enumerated, described, and explained as to their utility, but also explained as to their development and relationships, comes as a timely piece of advice. Even if students of instinct should never succeed in working out the genealogy of one instinct out of a hundred, the genetic method of study would be valuable by preventing mythologizing and reminding the student that instinctive activities are expressions of organic structure as truly as are the activities of digestion or excretion. And, in closing this lecture, I wish to give some samples of development among instincts. Professor Whitman has traced the ancestry of some particular acts. Let us look at some instinctive activities which persist, with modifications of course, over a wide range of forms, which correspond in a way to the notochord, or brain-eye, or arthropod appendages among physical organs. The frog, lizard, chick, and cat all react to irritation of the head by scratching with the hind leg with a quick, repetitive motion that is startlingly alike in the last three. Here we have an instinct which apparently ranges nearly over a subkingdom. In the primates it is modified, the monkeys (at least some of them) using *either* hind or front limb for the purpose, while man uses only the front. Distaste at confinement is another widely prevalent instinct, which may be a foundation-stone laid by germinal variation once for all. The instinct to follow is another. It might be that the gregarious instincts originated as exaggerations of it by the selection of individuals in whom the impulse to follow varied in the line of greater intensity.

12

Reprinted from pages 69-77 and 91 of *Woods Hole Mar. Biol. Lect.* (1899)
7:69-91 (1900)

THE ASSOCIATIVE PROCESSES IN ANIMALS.

EDWARD THORNDIKE.

IN the previous lecture we studied the general nature of those reactions which animals make to various situations instinctively, or apart from experience. It is obvious that with many animals, in many situations, acts are observable which cannot be so explained. The cat that comes when we call "Kitty, kitty," is not provided by her organic inheritance with any tendency to respond to the situation, "hearing the sound *kitty, kitty,*" by the act, "running toward the source of that sound." The ten-days-old chick that, on coming out from the brooder, turns round a corner and goes straight to the dish of water kept always in a certain place, is not guided by any innate tendency to meet the situation, "feeling thirsty while in brooder," by that particular act. So, too, with the dog that sneezes when you say, "Sneeze, Bowser!" the chick that avoids pecking at excrement, and with millions of animals performing all sorts of acts. We ordinarily distinguish such activities from those purely instinctive by saying that the animal has *learned* them. They are the results, not of organic inheritance, but of individual experience. The object of the present lecture will be to explain such activities, to show at least one of the ways in which animals learn or profit by experience.

Some hints of how this happens were given in our discussion of how instincts led to habits and how they were inhibited. But the matter is so important that I may be pardoned for beginning at the beginning. Let us watch some animals as they learn to meet certain situations in appropriate ways. If we make a pen, as shown in Fig. 1, and put a chick, say six days old, in at *A*, it is confronted by a situation which is, briefly, "the sense-impression or feeling of the confining surfaces, an uncomfortable feeling due to the absence of other chicks and of food, and perhaps the sense-impressions of the chirping of the chicks outside." It reacts to this situation by running around, making loud sounds, and jumping at the walls. When it jumps at the walls, it has uncomfortable feelings of effort; when it runs to *B*, or *C*, or *D*, it has a continuation of the feelings of the situation just described; when it runs to *E*, it gets out, feels the pleasure of being with the other chicks, of the taste of food, of being in its usual habitat. If from time to time you put it in

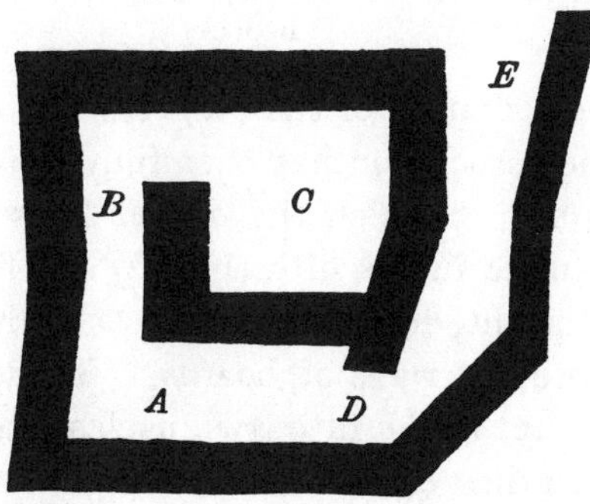

FIG. 1.

again, you find that it jumps and runs to *B*, *C*, and *D* less and less often, until finally its only act is to run to *D*, *E*, and out. It has, to use technical psychological terms, formed an association between the sense-impression or situation due to its presence at *A* and the act of going to *E*. In common language it has *learned* to go to *E* when put at *A* — has learned the way out. The decrease in the useless runnings and jumping and standing still finds a representative in the decreasing amount of time taken by the chick to escape. The two chicks that formed this particular association, for example, averaged one about three and the other about four minutes for their first five trials, but came finally to escape invariably within five or six seconds.

The following schemes represent the animal's behavior (1) during an early trial and (2) after the association has been fully formed — after it has learned perfectly the way out.

(1)

SITUATION.	IMPULSES.	ACTS.	RESULTING FEELINGS.
As described above.	To chirp, etc. To jump at various places. To run to *B*. " " " *C*. " " " *D*.	Corresponding to impulses.	Continuation of situation. Fatigue.
	" " " *E*.		Pleasure of company. " " food. " " surroundings.

(2)

SITUATION.	IMPULSES.	ACTS.	RESULTING FEELINGS.
Same as (1).	To run to *E*.	Corresponding to impulse.	Pleasurable as above.

A graphic representation of the progress from an early trial to a trial after the association has been fully formed is given in the following figures, in which the dotted lines represent the path taken by a turtle in his fifth (Fig. 2) and fiftieth (Fig. 3) experiences in learning the way from *A* to his nest. The straight lines represent walls of boards. Besides the useless traveling, there were, in the fifth trial, useless stoppings. The time taken to reach the nest in the fifth trial was seven minutes; in the fiftieth, thirty-five seconds. The figures represent typical early and late trials, chosen from a number of experiments on different individuals in different situations, carried on at Woods Holl by Mr. R. M. Yerkes, of Harvard University, to whom I am indebted for permission to use these figures.

Now the process of learning here consists of the selection, from among a number, of a certain impulse and act in connection with a certain situation. And our first business is to discover the cause of that selection. The result of such discovery was dogmatically stated in the previous lecture: "Any impulse to an act which, in a given situation, leads to pleasurable feelings, tends to be connected more firmly with that situation; and any impulse to an act which, in a given situation, leads to discomfort, tends to become weakened in connection with that situation." I say *tends* because the pleasurable feelings must follow the act within certain limits of time — must be important enough to outweigh possible instinctive opposition to the act, or possible discomfort in its performance. Any one may dem-

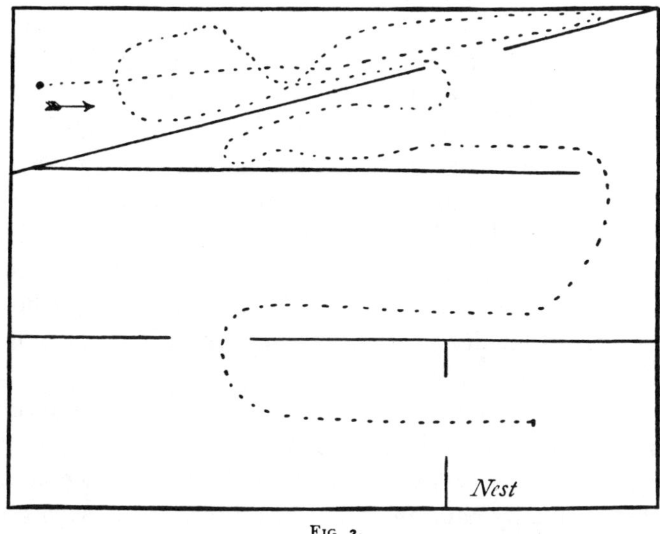

Fig. 2.

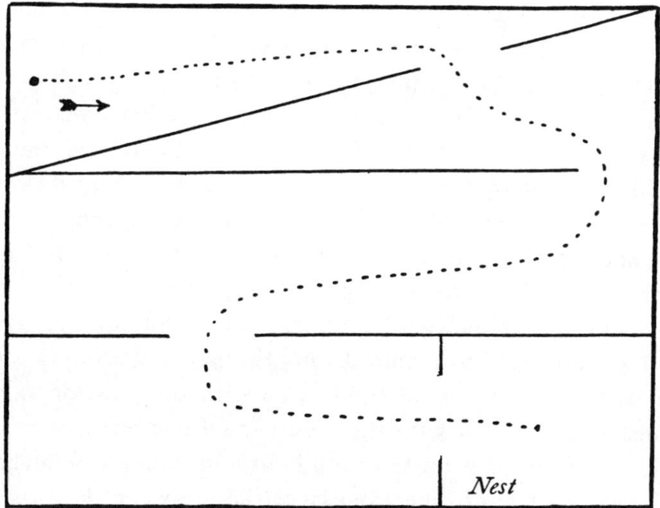

Fig. 3.

onstrate to himself the truth of this law over a wide range of animal life by observing the nature of the acts selected and of those eliminated by experience. He will find that the common element of the former is resultant satisfaction, and of the latter the opposite. Later on it will be worth our while to examine more carefully the way in which pleasurable and painful consequences respectively stamp in and stamp out the impulses which lead to them. For the present we had better return to actual specimens of animal behavior.

In the first place, the selection may be of one impulse and act from only one. If the chick had happened to go to *E* the first thing he did, the resulting satisfaction might have so stamped in that impulse that on the second, third, and later trials he would never have done otherwise. If accident or instinct furnishes the right impulse, it will naturally be confirmed just as readily as if picked out by its success from any number of inappropriate ones.

It will be well now to examine a more ambitious performance than the mere discovery of the proper path by a chick. If we take a box twenty by fifteen by twelve inches, replace its cover and front side by bars an inch apart, and make in this front side a door arranged so as to fall open when a wooden button inside is turned from a vertical to a horizontal position, we shall have means to observe such. A kitten, three to six months old, if put in this box when hungry, a bit of fish being left outside, reacts as follows [1]: It tries to squeeze through between the bars, claws at the bars and at loose things in and out of the box, reaches its paws out between the bars, and bites at its confining walls. Some one of all these promiscuous clawings, squeezings, and bitings turns round the wooden button, and the kitten gains freedom and food. By repeating the experience again and again, the animal gradually comes to omit all the useless clawings, etc., and to manifest only the particular impulse (*e.g.*, to claw hard at the top of the button with the paw, or to push against one side of it with the nose) which has resulted successfully. It turns the button round without delay whenever put in the box. It has formed an association between the situation, "confinement in a box of a certain appearance," and the impulse to the act of clawing at a certain part of that box in a certain

[1] Confinement alone, apart from hunger, causes similar reactions, though not so pronounced.

definite way. Popularly speaking, it has learned to open a door by turning a button. To the uninitiated observer the behavior of the six kittens that thus freed themselves from such a box would seem wonderful and quite unlike their ordinary accomplishments of finding their way to their food, beds, etc., but the reader will realize that the activity is of just the same sort as that displayed by the chick in the pen. A certain situation arouses, by virtue of accident or, more often, instinctive equipment, certain impulses and corresponding acts. One of these happens to be an act appropriate to secure freedom. It is stamped in in connection with that situation. Here the act is "clawing at a certain spot" instead of "running to *E*," and is selected from a far greater number of useless acts.

In the examples so far given there is a certain congruity between the impulse associated with the situation and the result. The act which lets the cat out is hit upon by the cat while trying to get out, and is, so to speak, a likely means of release. But there need be no such congruity between act and result. If we confine a cat and open the door and let it out to get food only when it scratches itself, we shall, after enough trials, find the cat scratching itself the moment it is put into the box. Yet in the first trials it did not scratch itself in order to get out, or indeed until after it had given up the unavailing clawings, etc., and stopped to rest. The association is formed in the same way with such an "unlikely" or incongruous impulse as that to scratch, or lick, or, in the case of chicks, to peck at the wing to dress it.

The examples chosen so far show the animal forming a single association, but such may be combined into series. For instance, a chick learns to get out of a pen by climbing up an inclined plane. You then so arrange a second pen that the chick can, say by walking up a slat and through a hole in the wall, get from it into pen No. 1. After a number of trials the chick will, when put in pen No. 2, go at once to pen No. 1 and thence out. You then arrange a third pen so that the chick, by forming another association, can get from it to pen No. 2, and so on. In such a series of associations the "act" of one brings the animal into the "situation" of the next, thus arousing its act, and so on to the end. Three chicks thus learned to go through a sort of long labyrinth without mistakes, the "learning" representing twenty-three associations.

The next matter to examine is the set of feelings in the animal which we have called "feelings of the situation," or, more simply, "the situation." The important thing about them is their vagueness, indefiniteness. The kitten did not have in mind every bar of the box confining it, and feel the impulse to turn the button in connection with such a precise sense-impression. It felt the whole environment in an extremely hazy way, and would still have turned the button if you had planed off one-quarter of each bar, or painted the button, or put the box in a different place, or sprinkled the bottom with sawdust, though any of these acts would have vastly altered the situation from our point of view. Cats that climbed up a certain screen when I whistled three times and said, "I must feed those cats," would climb up just as surely if I whistled only or spoke the words only, or even used any short sentence with the same voice and manner. If I took the button off in the box referred to above, the animals that had formed the association would often claw as before, though the (to us) essential part of the situation was absent. Cats that had often obtained food in consequence of being made to go into a certain box, from which they liberated themselves by pulling down a wire loop, would go in and pull the loop and then come out, though no food was there and the door was open all the time. All these facts witness to the vague, indefinite character of the animal's feelings of any of these situations, and to the consequent fact that much may be added to or subtracted from the external situation without essentially modifying the animal's reaction to it. Thus the kitten that in the first place responded with the act of approach to the situation, "smell and sight of milk in a certain dish," might later respond similarly to the sight of the dish, though it was empty; thus the kitten that in the first place responded to the situation, "sight or smell of food plus hearing of certain sounds," might later respond similarly to the sounds alone, might come, that is, at our call. It is by virtue of this power of a single element in a situation to set off the reaction associated with the vague total situation, that horses are trained to stop at "Whoa!" and to turn at the slightest pull of the rein, etc. It is in any particular case a concrete problem how far you can change the external situation and still secure the reaction associated with it. By such *gradual* modification as we make use of in training domestic animals you can of course change it without limit. For our present purposes we need not bother with any instances of this problem, remembering only that the animal does not react to a well-defined, hard-and-fast feeling or notion of its surroundings, nor with an immutable act, much less does it react to a well-defined feeling of the essentials (for its purpose) in the situation, and that therein lies the explanation of a host of animal activities.

So much for the feelings which are the first element in the association. Somewhat must now be said about the "impulses." I have all along spoken of impulse and act, because for various reasons out of place in this discussion I believe that there is in one of these associations, besides the feeling of the situation and the resulting act, a feeling preparatory to that act, such as we ourselves feel in connection with any non-automatic muscular performance. Such a feeling is meant by the word "impulse." It does not, in my usage, mean "motive" or "desire," or even the idea of the act to be performed, but merely the, in our own case, rather evanescent feeling of doing. Whoever wishes may discard this feeling from consideration, and substitute for "impulse and act" the word "act" alone, and refer what is said in the rest of this paragraph about the former to the latter. In the cases so far considered, the causes of the impulses and acts manifested at the first experience of a certain situation have been, apart from accident, the instinctive tendencies of the animals in question. That was because the particular situations chosen (the pen for the chick, box for the kitten, etc.) were situations for which the animals' previous experiences had given them no preparation. But suppose us to take the kitten that had formed an association between confinement in that box and the act of clawing at a certain place and put it in another box. We would witness less biting and squeezing and more clawing than we did in our previous experiment. For its previous experience would have strengthened the association between clawing and confinement, or at least weakened it less than it had weakened the impulses to bite and squeeze. That is, in any situation the impulses and acts which appear are due not only to the animal's instinctive tendencies, but also to any modifications of them which have been wrought by associations already formed. The question of how an animal will act in any situation may thus require, for correct prophecy, knowledge not only of its inborn constitution, but also of its entire previous history.

Let us now return to the simple cases of animal behavior with which we started and interpret them in accordance with the facts we have found in our observation of the formation of associations under test conditions. The cat's coming when we call, "Kitty, kitty," we have already explained as a case of an act associated with a certain total situation first (because, of course, the act brought the pleasure of eating), then with one element of that situation. The chick coming out of the brooder, turning round a corner, and going straight to the dish of water, leads back to a history somewhat like this:

Situation.	Impulses.	Consequent Acts.
Being in yard. Feeling thirsty.	To walk around and to peck at things, *e.g.*, at stones, seeds, worms, specks of all sorts, *e.g.*,	
	specks of dirt in the water of water dish, the edge of the dish.	These lead to the further acts of drinking.

The last part of this piece of experience leads by repetition to the formation of an association between the situation, (*A*) "presence near dish when thirsty," and the act, "drinking." But as soon as it is partly formed there is also being formed an association between the situation, (*B*) "presence near brooder when thirsty," and the act, "walking around the corner"; for whenever in that situation the chick by accident does stroll around the corner, it comes to feel situation *A*, and so to be led to the pleasurable drinking.

[*Editor's Note:* Material has been omitted at this point.]

The degree of permanence of these associations, once they are formed, vastly increases their utility. It allows experiences rarely met with to still, little by little, build up habits. It allows the experience of certain localities, for instance, to be useful again, even if the locality is not revisited for a considerable time. It lets the animal's past influence its present conduct in all sorts of ways.

It seems hardly necessary to make any statement about the general usefulness of the power of association in securing survival. If our view of the process is correct, it is a process of selection among reactions, not by eliminating the animal that does not react suitably and so developing a stock with certain instincts, but by eliminating the unsuitable reaction directly by discomfort, and also by positively selecting the suitable one by pleasure, and so developing certain associations in the individual. It is, then, selection *within the individual* that is the great case of plasticity, and is of tremendous usefulness, in that it definitely enables the animal to modify his acts and so meet varieties and modifications of environment. New feeding grounds, new foods, new friends, new enemies are dealt with by virtue of it. "He who learns and runs away, *will live* to learn another day."

Part IV

CORE ETHOLOGY: PRECURSORS OF THE CLASSICAL SYNTHESIS

Editor's Comments on Papers 13 Through 16

We now turn to European scientists that greatly influenced the development of classical ethology. Jacob von Uexküll, a non-establishment German biologist who could not accept evolution as a completely materialistic phenomenon, provided the inspiration for the ethological concepts of sign stimulus, releaser, Innate Releasing Mechanism, and the functional circle. His writings are known in America today primarily through his popular charming book produced late in life (Uexküll, 1934; translation in Schiller, 1957). Its many illustrations by Kriszat, attempted to depict the environment as perceived by various species of animals.

The excerpts included here (Paper 13) are from the first edition of his book, which was cited heavily by Lorenz and other early ethologists. In this edition the focus is entirely on invertebrates. The claim at the end that all "higher" animals possess the perceptual world of all lesser creatures is an echo of the recapitulationist philosophy of Haeckel. Neither this work nor the second edition of 1921 has previously been translated. Schleidt (1962) has written (in German) an important historical and theoretical analysis of the Innate Releasing Mechanism that discusses von Uexküll at length.

In ethology, the term *Umwelt* is often used as it is thought von

Uexküll did: to refer not to the *real* environment but to what was both noticeable and important to the animal (see Lorenz, 1958). In contrast to this usage (which was encouraged by the translation in Schiller, 1957), the present translators have here adopted the following approach to represent more clearly the fact that von Uexküll took German words with a well established meaning still followed today and redefined them somewhat for his own purposes. Thus in the present translation: *Umwelt* = environment, *Umgeburg* = surroundings, *Gegenwelt* = counterworld, *Aussenwelt* = external world, *Innenwelt* = inner world, *Bauplan* = construction plan. English words were chosen for several reasons despite the currency 'umwelt' has achieved in ethological writings in English. In German *Umwelt* generally is equivalent to environment. When von Uexküll first used the term *Umwelt*, he took an existing term meaning "milieu" or "environment," narrowed its semantic scope for his immediate purpose, and established this new usage in the material translated here. Thus, rendering *Umwelt* as "environment" most closely approximates a German's reading of the original and avoids the mistaken impression of English speakers that *Umwelt* is a specialized term in German that means something other than 'environment' in its English sense.

Ethology, Lorenz (in K. Heinroth, 1971) has written, is the field of study founded by Oskar Heinroth. Heinroth's contributions were indeed great, especially in their comparative and descriptive emphases. His groundbreaking 1911 paper, large sections of which are published here in English for the first time (Paper 14), was one of the first to use the term ethology in its modern sense. Primarily an ornithologist and zoo curator, Heinroth was most widely known by others in those circles. He was not isolated from current controversies in animal behavior, however. He was, for example, a member of one of the scientific commissions that investigated Clever Hans, the notorious horse that some claimed could count and read (Pfungst, 1911). Lorenz (in K. Heinroth, 1971) noted that Heinroth was not much interested in philosophical issues; this contrast with von Uexküll is unmistakable in the readings included here.

In Paper 14, an overall comparative look at geese, swans, and ducks, the great debt that Lorenz felt he owed Heinroth is apparent. Not only is the comparative phylogenetic perspective in evidence, but releasers, imprinting, intention movements, inciting, and ritualization are discussed, although in a rather discursive manner embedded in discussions of specific species. The discussion of ritualization is of particular interest since Julian Huxley is usually credited as its discoverer (see Paper 15). Heinroth's observations and discussion on forced copulation in mallards anticipated recent sociobiological writings and controversy (e.g. McKinney, Derrickson, and Mineau, 1983). [Translators' note: Heinroth wrote "triumph call" where today "triumph ceremony" would be used.]

Of further interest in this paper is its running commentary on humans, which Heinroth used largely as analogies and as a means to help the reader appreciate the birds' behavior. But the last paragraph indicates that not all of the resemblances were to be considered superficial, a view that Lorenz was much later to extend in his most controversial book (Lorenz, 1966).

Julian Huxley is well known for his work in a variety of areas in biology, particularly the synthetic theory of evolution, comparative anatomy, and the philosophical, religious, and human implications of evolution. Early in his career he used his well-developed ornithological talents to observe the behavior of bird displays. His most important contribution to ethology is the 1914 paper excerpted here (Paper 15), on the ritualization of courtship displays in Great Crested Grebes. His ability to extract so much information from a two-week "vacation" is dispiriting to lesser mortals, particularly those of us who have spent months in the field with little to show for our time, efforts, and discomfort. However, Huxley's contribution is compromised by his not being able to recognize individual birds and by the paucity of data on other aspects of the species' life history. Neither of these limitations existed for Heinroth, who had plenty of contemporaneous comparative data on related species. Still, Huxley had the great advantage of working in the natural environment, and leaning upon an excellent classical education, current biological knowledge, and the academic skills of writing a paper as an organized, extended argument; compared to Heinroth and von Uexküll, Huxley's paper appears more contemporary in style. Portions of the paper's appendix (not included here) present quantitative data. Many years later Huxley (1966) organized a symposium that evaluated the current status of the ritualization concept and paid homage to his contributions.

In 1930 Heinroth briefly addressed the possibility of evolutionarily derived movements extending beyond closely related species (Paper 16). His discussion of behavioral homologies across and within vertebrate classes helped give ethologists the confidence to generalize, although whether such homologies are comparable to morphological ones is still controversial.

REFERENCES

Heinroth, K., 1971, *Oskar Heinroth: Vater der Verhaltensforschung 1871–1945*, Wissenschaftliche Verlagsgesellschaft MBH, Stuttgart, 257p.

Huxley, J. S., ed., 1966, Ritualization of Behavior in Animals and Men, *Philos. Trans. Roy. Soc. London,* **251**(7), entire issue.

Lorenz, K., 1958, Methods of Approach to the Problems of Behaviour, in his *Studies in Animal and Human Behaviour,* vol. 2, 1970, Harvard, Cambridge, Mass., pp. 246–280.

Lorenz, K., 1966, *On Aggression,* Harcourt, Brace & World, New York, 306p.

McKinney, F., S. R. Derrickson, and P. Mineau, 1983, Forced Copulation in Waterfowl, *Behaviour* **86:**250–294.

Pfungst, O., 1911, *Clever Hans, the Horse of Mr. Von Osten,* Holt, New York, 274p.

Schiller, C. H., ed., 1957, *Instinctive Behavior,* International Universities Press, New York, 328p.

Schleidt, W., 1962, Die historische Entwicklung der Begriffe "Angeborenes auslösendes Schema" und "Angeborenes Auslösemechanismus" in der Ethologie, *Z. Tierpsychol.* **19:**697–722.

13

ENVIRONMENT [UMWELT] AND INNER WORLD OF ANIMALS

J. von Uexküll, M.D. (honorary)

These excerpts were translated expressly for this Benchmark *volume by Chauncey J. Mellor and Doris Gove, The University of Tennessee, Knoxville, from the book* Umwelt und Innenwelt der Tiere, *Julius Springer Verlag, Berlin (1909), pp. 4–253. Page numbers are indicated in parentheses in the text.*

Translation © 1985 by Chauncey J. Mellor and Doris Gove.

INTRODUCTION

[*Editor's Note*: In the original, material precedes this excerpt.]

(4) The fate of biology in lower animals has taken peculiar shape. Here, anatomy and physiology did not go separate ways, but physiology was for a time completely suppressed. This occurred through *Darwinism*. Darwinism, not Darwin himself, considered the achievements of anatomical structure to be immaterial in respect to the main problem—how the structure of the higher animals had developed from that of the lower ones.

In the animal kingdom, one saw evidence for the stepwise increase of perfection from the most simple to the most complex structure. But along with this, one thing was forgotten: that the perfection of the structure cannot be judged from its complexity. No one will assert that a battleship is more perfect than modern rowboats of the international rowing club. Also, a battleship in a rowing regatta would make a poor showing. In the same way, a horse could only very imperfectly fulfill the role of an earthworm.

The question about the higher or lower degree in the perfection of organisms can be asked if one correlates each construction plan with its execution and tests the cases in which the execution has been best realized. There is no doubt that in asking this question, the lower animals, because they are among the oldest groups, will be the winners because the following rule seems to hold: the older the family, the more thorough the construction.

In addition, one attempts to discuss the perfection problem by comparing the needs of the organism with the construction plan and by asking to what degree the construction plan corresponds to (5) the need. That was also the question that Darwinism asked, and only from this question does the assertion that higher animals are the more perfect make any sense.

For example, if one looks at the needs of people, as a standard by which all construction plans of animals are to be measured, then of course, the higher animals are the most perfect. But this is too palpable an error. For the investigation of the needs of an animal, we have, after all, no aids whatsoever at hand other than, simply, its construction plan. It alone gives us information about the active and passive role that the animal is called upon to play in its environment. For that reason, the whole question is senseless.

But even the assertion that the varying individuals of one species are more or less well-suited to their environment is taken out of thin air. Every varying individual is different according to its differing construction plan, but equally perfectly adapted to its surroundings, for, *within broad limits, the construction plan by itself creates the environment of the animal.*

This fact, which I intend to prove step-by-step, can alone be seen to be the lasting basis of biology. Only through it do we gain the proper understanding of how organisms order and rule the chaos of the inorganic world. Every animal in a different place and in a different manner seeks out from the vast complexity of the inorganic world precisely that which fits it; that is, it creates its needs itself corresponding to its own construction type.

Only to the superficial eye does it seem as if all sea animals were living in a homogeneous world common to all of them. Closer study teaches us that each of these myriad life forms possesses an environment peculiar to it that is mutually determined by the construction plan and the animal itself.

It cannot amaze anyone that the environment of an animal also includes other animals. Then the mutual determination appears as well among animals themselves and brings about the remarkable phenomenon that the pursuer fits just as well to the pursued as vice-versa. (6) Thus, not only is the parasite adapted to the host, but also the host to the parasite.

Experiments to explain this mutual co-evolution of sympatric animals by gradual adaptation have failed miserably. In addition, they have turned interest away from the most pressing task, which consists first of all in establishing the environment of every single animal.

This task is not as simple as the inexperienced might think. It is, of course, not difficult to observe any animal in its surroundings. But the task is not solved in the least by doing this. The experimenter must attempt to establish which parts of the surroundings affect the animal and how this occurs.

Our anthropocentric way of looking at things must retreat further and further, and the standpoint of the animal must be the only decisive one.

When this occurs, everything that we hold to be self-evident disappears: all of nature, earth, heaven, stars, indeed, all the objects that surround us. Only those effects remain as factors of the world that, corresponding to the construction plan, exert an effect on the animal. Their number, and the way they fit together, is determined by the construction plan of the animal. Once this correspondence between the construction plan and the external factors has been carefully researched, then a new world takes shape around every animal, completely different from our world: its *environment*.

The effects evoked in the nervous system by the factors of the environment have been, and indeed must be, conceived of just as objectively as those factors in the environment itself. These effects are likewise sifted out and regulated by the construction plan; together they constitute the *inner world* of the animal.

This inner world is the unfalsified fruit of objective research and should not be obscured by psychological speculations. One might perhaps raise the question in order to bring the impression of such an inner world to life: What would our psyche do with such a confined inner world? But to elaborate and decorate this inner world with psychic qualities that we can prove just as well as deny is not an activity of serious researchers.

(7) The construction plan stands over the inner world and the environment, dominating everything. Only research on the construction plan can, in my estimation, give a healthy and certain basis for biology. It also brings anatomy and physiology together again in a productive mutuality.

If the organization of the construction plan is placed at the focus of research for every species, then each newly discovered fact finds its natural place, and only thus does it gain sense and significance.

The content of the book here before us is to serve that purpose—to bring the significance of the construction plan as vividly and as penetratingly as possible before our eyes and to show in individual examples how the environment and the inner world are connected by the construction plan. We are not offering here a text of special biology, but rather are only showing the way to arrive at one.

Above all, my wish to give the most systematic images possible determined my choice of the examples presented. Of course, there are gaps everywhere, and, to be sure, not just in the physiological, but also in the anatomical material. Since I was only able to use such anatomical material as was physiologically enriched, the great mass of anatomical and zoological knowledge had to be ignored. Likewise, all the physiological results that only offered physical or chemical interest had to be neglected. But even those structures whose achievements are well researched had to remain unconsidered if their complication placed excessive demands on the imaginative ability of the reader.

Finally, I have limited myself to invertebrates because I myself am at home there, and I leave the higher animals to those more qualified. The bees and the ants are not considered because we have detailed books on them already.

I could now proceed to the content of the book because the point of view from which it should be viewed has been adequately presented. But it is still necessary to examine further those opinions that attempt to provide biology with a different foundation.

(8) A special biology of all animal species can come into being in the manner just now presented. Such a biology would be very one-sided if one were to do without comparison. All animals carry out their achievements with the help of tissues that remain very similar throughout the whole animal kingdom. Muscle and nerve tissues all show analogous achievements, even when found in very diverse organs. This is of great significance for special biology, because the generally valid properties of the muscles and

nerves can be assumed to hold even in those animals whose bodily configuration permits no physiological analysis of the individual tissues. Therefore, comparative physiology of tissue will always remain a very necessary component of specific biology, and it cannot be denied that comparative cytology should precede the discussion of the individual animals. I have distanced myself from this because I wanted to show which animal species most easily permit us to make broader conclusions for general cytology.

Biology cuts a quite different figure when one chooses comparison as the basis for the whole field of study. This was done by Loeb in an extraordinarily original and interesting way.

Most animal movements proceed in the following manner. An external stimulus impinges on a receptive organ. This sends the nervous system an impulse. Conducted by the nervous system, the impulse eventually reaches the muscle, which then contracts. One calls this process a *reflex*. Now Loeb found that a large number of animals, if they are subjected to quite simple stimuli, such as light, gravity, or simple chemical substances, always respond with a systematic movement in which they turn either toward or away from the source of the stimulus. In this he saw an elementary process that he termed *tropism*, and, according to the direction of the movement, it was positive or negative.

Loeb conceded the possibility that in many tropisms we might be (9)
dealing with reflexes that are not yet sufficiently analysed. But he wants to explain certain tropisms, for example, *phototropism*, which occurs in the case of one-sided lighting, as an elementary phenomenon considered equal to the physical phenomena. The light rays are said to be capable, in passing through the animal body, of turning it, as, for example, a magnet turns iron filings. Animals that react this way to light are called *photopathic*.

But no doubt remains that, in many cases, the light simply releases a reflex on the illuminated side of the animal that must lead to a one-sided movement, since no reflex occurs on the shadow side. Animals that react this way toward light are called *phototactic*.

Photopathic tropism is a physical process; phototactic, however, is a reflex.

Now F. Lee has been able to establish in one-celled animals that the photopathic explanation of movement can quite easily be replaced by a phototactic one. Recently Radl has attempted to demonstrate that light has just as much directional influence on insects as gravitation has on a floating body. By contrast, G. Bohn has found that the unquestionable directional effect of illuminated objects on snails and crabs is dependent on the physiological condition of the animals.

From this one sees how uncertain the interpretation of these processes is.

To be sure, it is tempting to attribute all movements of animals to tropisms because that frees us of the task of treating the seemingly simple processes as achievements of a structure that is difficult to determine, but one obtains a stable basis only by the study of the structure and the construction plan.

This view appears to be gaining ground even now. But only some of the researchers are turning toward the study of the construction plan. Another group follows a different theory that rejects the study of the construction plan and wants to observe the animals free of any analogy with machines.

(10) There is of course no doubt that the establishment of the construction plan of the animals only makes sense when the structure of the animals is equated with the structure of machines.

With this we approach the foundation of all biology, which cannot be decided through speculation, but only by observation of the living substance out of which all organisms are built, whereas machines are built of dead material—the *protoplasm issue*.

[*Editor's Note*: Material has been omitted at this point.]

(54) THE REFLEX

[*Editor's Note:* Material has been omitted at this point.]

People attempt to attribute all actions of animals to *reflexes*. The reflex is therefore the basic element of all actions. This basic element, however, already unites various factors into one common function. Every reflex is actually the *response* of one part of the animal body to an incident effect of the external world. Amoebas are able to make the response with a single protoplasmic organ. Multicellular creatures use three different organs for the same purpose: a receiving, conducting, and an executing organ—a *receptor*, the *nerves*, and an *effector*. The connection of these three organs that execute the reflex is called *the reflex arc*. (55) In the three organs constituting the reflex arc, we must recognize the three elementary organs that are found in all animals, since all animals are machines that respond to effects from the external world.

No matter how differently the receptors may be constructed, and no matter how various the effects of the external world they serve, they still have only one single job to do: to transform the effect of the external world to the reflex arc. Every receptor is set up for a certain segment of the effects of the external world. The segment can be larger or smaller and can involve strong or very weak effects that can in cases be very specialized. The effects may be of a chemical or physical nature, but we always, when wishing to express their connections to the receptors of the animals, term them *stimuli*. I have given a detailed classification of the receptors in my *Leitfaden*. It will be enough to note here that the manner of construction of the receptors of every single animal determines absolutely which effects of the external world will be connected with that animal and which effects will not. The sum of all the stimuli that an animal receives, thanks to the construction of its receptors, constitutes its *environment*. This is the outwardly directed side of every receptor. Its inner side is directed towards the animal's body, and its job is only fulfilled when it has made the external stimuli available to the reflex arc. The organ connected to the receptor is the nerve. The nerve does not possess the ability to accept and transmit any sort of physical or chemical process. It can transmit only one quite specific factor that we call the *excitation*. The excitation passes along the nerve like a wave. The wave has quite specific electrical properties that permit us to establish with certainty the form and speed of propagation.

[*Editor's Note:* Material has been omitted at this point.]

(58) The truth lies directly before us in the reality surrounding us. However, we cannot use it as is. An unbroken description of reality would be simultaneously the truest and most useless thing in the world, and it would certainly not be science. If we want to make reality and therefore truth useful to science, we must do violence to reality. We must introduce the distinction, which does not exist in nature, between *essential* and *inessential*. In nature, everything is equally essential. By seeking out the relationships that seem essential to us, we order the material in a surveyable way at the same time. Then we are doing science. Now, many of these essential connections remain hidden from us. To find them, we use microscopes, galvanometers, dyeing, etc., etc. For those connections that we cannot find despite all aids, but of whose existence we are convinced, we use temporary models. We use these models just like any other hand tool; if one is not useful, another is made.

(59) Just take a look at physics to convince yourself how that science operates with its models. Now electricity is a fluid, now a motion, now consisting of tiny particles of matter that *are* material, but have no characteristic of mass. Chemistry operates the same way. What is all of stereochemistry but working with models? Such an excellent chemist as Emil Fischer casually talks about key and keyhole to characterize purely chemical properties. Now why should biology continue to be sworn to the hallowed, but often outmoded, manner of expression of physics and chemistry, particularly since these models generally are not at all appropriate for biology? Since biology is seeking mechanical relationships, it needs mechanical models. The more visual these models are, and the more intimately they fit the observed processes, the better. For that reason, I stand in complete agreement with the term fluid, with its volume, its pressure, and other trappings, since fluid seems to me to provide the most visual model for the unknown connections of the nervous system.

The coming and going of excitations are the only objective processes that we can use to reconstruct the inner life of the animal. The inner world, in contrast to the variable and manifold surroundings, knows no change in quality. Therefore, one can consider the dynamic excitations only as tokens for the fact that something is going on outside, since they do not bear the least similarity to the events in the surroundings.

From the effects of the surroundings, the receptors choose those stimuli that, according to the construction plan of the animal, are appropriate to be noticed. The receptors thus give the nervous system a sign that the stimulus in question is making itself felt in the surroundings. Consequently, one can establish how many signs an animal receives from its environment. So many stimuli—so many signs. If an animal has received a sign of its environment in this way, it must respond to it. Now the nervous system is built such that it uses the sign itself to cause the response to be given by the muscle. For the dynamic excitation is (60) channeled by the dominating factors of the nervous system so that it gets to those muscles whose activity constitutes the correct response of the animal to the stimulus.

The dominant factors that channel the dynamic excitation to the muscles are, above all, the structures in the nervous system. When only certain muscles are connected to certain receptors by a specific nervous network, then the answer to the central question of biology, "How does the excitation find the right muscle?" is an easy one. But this changes even in the case of the sea urchin, which, in spite of its richly elaborated nervous system, strictly adheres to the principle of *coordination* of the reflex arcs. In bilateral animals, possessing a definite front end that houses the higher receptors, a progressive *subordination* of the centers naturally occurs, and the simplicity of the reflex arc is lost. The structure then becomes extremely complicated, and the determination of it becomes exceedingly difficult, but it is at least constant. Its changes in the adult animal can, for our purposes, be set equal to zero.

The other factors are more variable and therefore more difficult to survey. This holds true especially for the central reservoirs of static excitation whose degree of fullness changes and therefore exerts a decisive influence on the condition of the entire nervous system. But the greatest influence on the pathway taken by the dynamic excitation comes from the numerous small reservoirs standing in direct connection with the muscles. These reservoirs possess the ability to deflect or attract the dynamic excitation according to their degree of fullness, or, to use Jordan's term, according to their potential. This degree of fullness is dependent, on the one hand, on the central reservoirs, and, on the other hand, on the muscles themselves, whose demands on the excitation of the central network they must represent. Therefore, I have called them *"representatives."*

Three organs, therefore, are situated on the central nervous network (which may have split up into various networks and can have assumed all kinds of structures): the *receptors*, the *central reservoirs*, and the *representatives*. Among them (61), the inner life of the animal is played out in three forms:

1. As a *reflex*. This is the pure form of the dynamic excitation beginning in the receptor and diffusing in the muscle.

2. As a *rhythm*, which can be a uniformly repeated reflex or can be produced by the oscillation of the static excitation.

3. As an *automatic action*, which presupposes a high degree of elaboration of the central equipment, since, while the rhythm depends on the active intervention of the muscles, the often very complex automatic excitations are guided solely by the activity of the central equipment. One has the impression that the entire static excitation begins to flow, and instead of dominating the overall pressure from the individual reservoirs, this automatic action courses through its entire nervous domain.

In the various construction plans, one finds all possible combinations of the two structural construction principles: *coordination* and *subordination* with the three *movement principles* of excitation.

And yet, even with this, the variety is not exhausted. Up to this point, we have only observed animals whose movements are caused by the influx of excitations, be they static or dynamic. One might call them *power machines*. But there are also such animals whose actions are regulated by the discharge of excess excitation. One could call these animals *braking machines*, and it is again Jordan's contribution that made us more clearly familiar with such oddly built animals. Most prominent among these braking machines are the snails.

[*Editor's Note:* Material has been omitted at this point.]

ANEMONIA SULCATA

[*Editor's Note:* Material has been omitted at this point.]

(73) It is interesting to measure both findings by Bohn's theories, which, as he says, split American researchers into two camps. These experiments doubtlessly show that, in the phototropism of Loeb, we are dealing not merely with unresolved reflexes, but that a direct effect of light on protoplasm (74) must be assumed. However, this effect is not mechanical, since the protoplasm has the ability to use all vital stimuli in a purposeful manner. Bohn's experiments, on the other hand, prove Jennings to be right when he speaks of the effects of inner processes in the establishment of habits. However, they contradict the theory of "trial and error," since purpose is not first sought and then found, but it is itself the fundamental property of protoplasm, and is present before all trials. As we see, the theories dealing with protoplasm are correct for just as long as they do not attempt to give a mechanical and physical interpretation. The mechanical interpretation comes into its own when we are no longer dealing with protoplasm, but with structures.

[*Editor's Note:* Material has been omitted at this point.]

JELLYFISH

2. Carmarina and Gonionemus

[*Editor's Note:* Material has been omitted at this point.]

(88) Comparing Rhizostoma [a large jellyfish without tentacles] with Gonionemus [a small jellyfish with tentacles], we are most struck by the fact that two organisms with such similar structure can live in such completely different environments. Rhizostoma perceives only the beat of

its own body; Gonionemus, on the other hand, is touched and moved by light and darkness, gravitation, and mechanical and chemical stimuli. For both, the external world is the same, but Rhizostoma constantly closes it out, while Gonionemus admits (89) a rich stream of effects of the external world through the portals of its receptors. The organism is like a magic world, closed off to all effects of the external world, opening only to the right key. If no lock is present, no key can be found. And so it is with Rhizostoma. Gonionemus has many doors, each equipped with its own special lock. Like house doors, these are put in those places where an appropriate entrance is located, corresponding to the construction plan of the whole. Who could claim that a house with many doors is more perfect than a house with few entrances? Thus, one may not place a lower value on the exclusion of stimuli, which allows Rhizostoma its great uniformity and tightness, than on the acceptance of stimuli in Gonionemus, which, thanks to the many stimuli, performs many feats. Rhizostoma has no need for these feats; they would be of no value to it. And yet, Rhizostoma is as artfully constructed as Gonionemus. No jellyfish species can be substituted for another. Gonionemus cannot live on the pelagic meadow; nor can Rhizostoma catch its own prey.

Although these animals have the same construction type, with the same nervous equipment and central nervous system regulating the beat rhythm, they are nevertheless not at all comparable if one examines their way of life. The inclination to classify all animals into more or less perfect ones in order to demonstrate thereby an upward development proceeding from the inferior to the higher is nowhere more forcefully taken ad absurdum than in those cases where animals of the same type, that are only differentiated in different directions, possess quite different environments. We should no longer speak of different degrees of adaptation, but only of equally perfect adaptations to different environments. It would also be better to refrain from an evaluation of the environments, since the environment is, for its part, only comprehensible from its connections to the actions of the animal. The environment consists only of those questions that the animal can answer. And finally, the construction type of the central nervous system that gives those answers is nothing more than the part of the (90) answer given by the construction type of the entire animal to the question of life. In this, sometimes, the emphasis rests on the development of a specific organ. Putting a special value on the central nervous system is entirely unjustified, since nature can answer its own questions with any organ.

THE SEA URCHINS

[*Editor's Note:* Material has been omitted at this point.]

(117) The Environment [Umwelt]

The treatment of stimulus combinations by sea urchins is so important because it solves the question of the nature of the environment. The objects we notice in the surroundings of the sea urchin have no means at all to exert an effect as independent entities except by the generation of stimulus combinations characteristic of them alone. Or, said another way, a sea urchin can achieve no knowledge of its surroundings if it is incapable of transforming stimulus combinations into excitations. The excitations generated by stimulus combinations must, moreover, be capable of exerting separate effects in the sea urchin so that one can speak of a true effect of an object. Otherwise, there are only disjunct stimuli, and the environment of the animals contains properties, to be sure, but no objects.

The surroundings of the sea urchin, as they appear to us, can be easily enumerated: water, rocky bottom, small stones, algae, light for certain species, and also shadow; also prey animals such as crabs and worms, and finally, enemies such as starfish and sea slugs. For the nervous system of the sea urchin, these objects, one and all, do not exist. For the sea urchin, there are only weak and strong stimuli that release only weak and strong excitations, and occasionally a (118) combination of weak and strong stimuli that are not further distinguished. The only stimulus worthy of special treatment is shadow. All other stimuli always generate only excitations that must make their way in the general nervous network in an undifferentiated manner.

Even if we wanted to make things pleasant for ourselves, and quite consciously insert our mind into the core of the central nervous system of the sea urchin (something that comparative psychologists do unconsciously), then we can still never receive from such a nervous system anything but isolated perceptions. Our mind would receive two linked perceptions only on the stalk of poison pinchers. But what would be most perplexing for our mind would be the impossibility of giving the body a unified impulse.

To be sure, there are centrally located reservoirs regulating the general excitation pressure, but the individual reflexes run their course completely independently. Not only each organ, but even each muscle fiber with its nerve center, acts completely on its own. That in this situation something sensible nevertheless results is only the contribution of the plan by which the independent individual parts fit together such that always and in every case the benefit of the entire animal is maintained. We can therefore call the sea urchin a republic of reflexes and distinguish it from higher animals in the following vivid manner: "When a dog runs, the animal moves its legs. When a sea urchin runs, the legs move the animal."

To emphasize what is essential once again, it is not the unitary impulse that is dominant in the sea urchin, but the unitary plan, which draws the entire surroundings of the sea urchin into its organization. Of the useful and hostile objects in its surroundings, it chooses those effects appropriate as stimuli for sea urchins. Corresponding to these stimuli are graded reception organs and centers that react differently to different stimuli, and, in so doing, excite the muscles that must carry out the motions foreseen by the plan.

So even the sea urchin is not abandoned to a hostile outer world in which it leads a brutish struggle for existence (119). On the contrary, it lives in an environment that harbors harmful and beneficial aspects; an environment that so well fits its capabilities down to the last detail, as if there were only one world and only one sea urchin.

[*Editor's Note:* Material has been omitted at this point.]

THE COUNTERWORLD

[*Editor's Note:* Material has been omitted at this point.]

(192) As we know, all receptors have the same task: to change the stimuli of the external world into excitations. Therefore, in the nervous system, the stimulus itself does not really appear. Instead, a quite different process appears that has nothing to do with the events in the environment. It can only serve as a *token* for the fact that a stimulus that has struck a receptor is found in the environment. It says nothing about the quality of the stimulus. The stimuli of the external world are fully and completely translated into a nervous sign language. Remarkably, for all types of external stimuli, the same sign appears again and again, which changes in intensity in correspondence with the strength of the stimulus. The strength of the stimulus must first have exceeded a certain threshold before a sign of excitation appears, but then the strength of the excitation grows with the strength of the stimulus.

The introduction of a threshold is a very effective means permitting the organism to block or select the stimuli of the environment. But when the nervous system receives only the same signs for the same stimuli, how is it then possible to distinguish the types of stimuli? This occurs through the use of special nervous pathways for the specially differentiated types of stimuli. Every receptive organ has a very great number of centripetal pathways at its disposal and is thereby enabled to distinguish even very fine differences in the type of stimulus just as certainly as the coarsest differences are differentiated by reserving a special nerve pathway for each type of stimulus.

Even in the lower animals, the use (193) of special pathways for the various receptors is evident, but as soon as these pathways flow into the general nervous network, the differentiation is lost again, and the nervous system differentiates the stimuli of the external world not by type but only corresponding to their strength. If the centripetal pathways remain isolated from one another, then the possibility arises of also evaluating stimulus types separately by their effect on the organism.

In the higher organisms, various centripetal pathways, which correspond to specific, frequently occurring stimulus combinations, join in isolated networks and serve as a collection point for the corresponding combinations of excitations. The possibility is therefore offered to the organism to treat even combinations of stimuli in a differentiated manner. One could just cut things short and term such combinations of stimuli 'objects' and, correspondingly, consider the nervous system of an animal that reacts differently to different combinations of stimuli as capable of distinguishing objects.

Up to this point, this conclusion had seemed to me inescapable. But the more I pondered the question of what mechanical arrangements the nervous system must possess so that it treats different objects of its environment differently, the more I concluded that simple combinations of excitation do not suffice for this. Each and every object is primarily characterized by its *spatial extension*.

For the lower animals, it is certain that they do not use this characteristic. The combination of a mechanical stimulus with a chemical stimulus is completely adequate for the sea urchins to distinguish with certainty the enemy starfish from all other aspects of the environment. But in the higher organisms, this is no longer the case. They are no longer content with this primitive machinery of classification, thanks to their higher organization. They also distinguish spatial delimitations of objects. The mere earthworm supplied the first test for this.

Here, all of a sudden, the *problem of space* confronts us in all its difficulty. Each individual stimulus quality can be registered in an isolated fashion through a special sign in the central nervous system (194) by use of an isolated nerve pathway regardless of which route the nerve pathway may take. But the spatial arrangement of the stimuli is lost if it is not registered by a similar arrangement of the nerve pathways. Now we see how significant it is for the organizational plan of the central nervous system that the types of stimuli are not reproduced by various types of excitation in the same nerve fiber, but are registered by the use of different nerve fibers: One could not order the types of excitation at all in a spatial manner corresponding to the form of objects. But one can do this very well with nerve fibers.

One can order the nerve fibers by laying them side-by-side on a surface and in this way creating a spatial arrangement corresponding to the arrangement of stimuli in the environment. Through this, the central nervous system achieves the possibility of arriving at quite new and much more intimate connections with its surroundings than was possible through the mere combination of stimuli. It does not matter at all whether a circle in the environment should correspond to a circular or triangular arrangement of nerve pathways or vice versa. The main thing is that the differentiation of the spatial delimitations of objects by higher central nervous systems and brains requires a fixed spatial arrangement of nerve pathways. One can assert that the higher brains know their environments not only through sign language, but also reflect a piece of reality in the spatial relationships of their parts.

Through the introduction of this reflection of the world into the organization of the central nervous system, even though it may be very simplified, the motor portion of the nervous system has lost its earlier connections to the environment. No external stimulus transformed into excitation signs penetrates any longer directly to the motor networks. These networks receive all their excitations only second hand—from a new world of excitation originating in the central nervous system that is erected between the environment and the motor nervous system. All actions of the muscular equipment can now only be related to them and can only be understood through them. (195) The animal no longer flees from the stimuli that the enemy sends to him, but rather from the mirrored image of the enemy that originates in a mirrored world.

But so as not to cause misunderstanding with the use of the words "mirrored world," because a mirror does much more than merely reflect some spatial relationships in a very simplified form, I call this new individual world that has arisen in the central nervous system of the higher animals the *counterworld* of the animals.

In the counterworld, the objects of the environment are represented by schemata that, according to the organizational plan of the animal, can be kept very general and can subsume very many types of objects. However, the schemata can also be very exclusive and refer only to quite particular objects. The schemata are not a product of the environment, but rather are individual tools of the brain determined by its plan of organization. These tools always stand ready to become active in response to appropriate stimuli from the external world. Their number and selection cannot be inferred from the surroundings that we see; they can only be deduced from the needs of the animal. If the schemata also represent spatially reflected images of the objects, then the form and number of these images is nevertheless the peculiarity of the mirror and not of that which is mirrored.

The schemata change with the construction plans of the animals. A great variety of counterworlds thereby arises representing the surroundings. For it is not nature, as one is accustomed to saying, that compels animals to adapt. On the contrary, the animals construct nature for themselves according to their special needs.

If we possessed the capability of holding the brains of animals before our mind's eye, as we can hold a glass prism before our physical eye, then our environment would appear just as changed before us. There might not be anything more charming or interesting than such a world through the medium of various counterworlds. Unfortunately, this vision is denied us, and we must make do with a tedious and imprecise reconstruction of the counterworlds as they become probable to us by detailed and difficult series of experiments. (196) One guiding thought gives us hope of constructing something useful from this uncertain material, and that is the certainty that nature and the animal are not, as it might appear, two separate things, but that they together constitute a higher organism. The surroundings that we see spread out around the animal are obviously distinct from the animals themselves, but all the same it is not *their* environment that we see, but *ours*. The environment as reflected in the counterworld of the animal is always a part of the animal itself, constructed by *its* organization, and processed into an indissoluble whole with the animal itself. One can easily imagine an animal devoid of the surroundings we see and conceive of the animal as isolated, but one cannot think of an animal as isolated from its environment, for the environment is properly understood only as a projection of *its* counterworld. And the counterworld is a part of its most special organization.

After we have obtained a general impression of the significance of the counterworld, we want to try to give an account of which view, according to our present knowledge, best corresponds to the counterworld. This can happen only by way of suggestion and must necessarily remain very incomplete until more observational material is collected. But in any case, a visually clear conception is useful because, on the one hand, it aids in the clear statement of questions, and, on the other hand, it gives us a feeling for an overall context. Once the counterworld has come into being, it exercises a significant attraction on all receptors, which gradually drop their direct connections to the general nervous system and connect up with the receptive network of the counterworld.

The earthworm can serve as a starting point for our observations. In it, a definite discrimination of form arises for the first time. The central network of the earthworm enters at the front end and goes into the upper pharyngeal ganglia. The upper pharyngeal ganglia must house at least two separate centers to make the simplest distinction between left and right. Both these centers must stand in a fixed connection with one another (197) when they cause a certain muscular movement in response to a certain object shape that stimulates heavily on the left, but little on the right. Deviating somewhat from my expression in *Leitfaden*, I want to call each of these centers a "*nucleus of excitation*." Both centers working together constitute a common schema; consequently, the earthworm would possess the simplest form of a schema consisting of two nuclei of excitation and their conductive connection. This schema can be considered the first step towards a counterworld.

We meet the next highest level of the counterworld in animals whose eyes transmit a movement, or to use Nuel's expression, serve *motoreception*. In this case, we must begin to conceive of a surface containing many excitation nuclei. These nuclei only release a well-defined muscular activity when, as a group, they are successively excited, so that a wave of excitation passes over them like a wave in a field of grain. Fixed neural connections leading to the establishment of schemata do not yet exist between the individual nuclei. In their environment, an object is something "that moves together" without regard to the form.

We find the next higher counterworld where the eye already distinguishes images, and where the simplest *iconoreception* appears. Here, in the field of the nuclei of excitation, the first schemata appear, which are like rough outline drawings of the images traced out on the retina. In this case, we can begin to speak of spatial schemata. These are aroused as soon as an object corresponding to the schemata approaches the animal. Firmly delineated objects in the environment correspond to spatial schemata in the counterworld.

The counterworld of *chromoreception*, which makes possible the distinction of colored objects without regard to their form, interposes itself between the counterworld of motoreception and iconoreception. In chromoreception, groups of differentially, strongly aroused nuclei of excitation must come into play. In the environment of such animals, the (198) definition of an object is: "An object is something that is the same color."

As we see, in this way, the three characteristics we ascribe to each object seen in the animals' surroundings are separated. The individual parts constituting a seen object all have a *common outline*, as a rule a *common color*, and a *common motion*. How great the advance here with respect to the lower animals that learn something of the unity of objects only because these objects have a *uniform aroma*, cast a *uniform shadow*, or give a *uniform impulse*!

[*Editor's Note: Material has been omitted at this point.*]

THE CEPHALOPODS

[*Editor's Note: Material has been omitted at this point.*]

(231) The main characteristic of higher organization was seen in the appearance of the counterworld, that is, a re-formation in the receptive portion of the central nervous system. By observation of octopods it has been adequately established that they react to the form of photoreceived objects. But as we saw, the form of an object can only act as a stimulus when a corresponding form in the construction of the nervous pathways and ganglia is pre-formed in the brain. The form of the ordering of the nervous pathways can be considered a transformer for the form of the object in the broadest sense, and it must therefore be assigned to the receptive part of the central nervous system. The octopods certainly possess a counterworld, and where would it be more appropriate for it to take up residence than in the cerebral ganglia? They are situated such that they are equidistant from all receptors and receive all external impressions along the shortest pathway. Furthermore, they are situated over the central ganglia that house the highest motor stations from where the complete actions are directed. If we pursue this thought further, then the entire environment of the animal lies stored in the cerebral ganglia in the form of neural schemata. And each schema is ready to use its connections with the highest motor centers as soon as excitation is sent to it in the form uniquely suited to it. In this way alone, do we succeed in achieving a vivid image of the processes in the brain halfway corresponding to our general experiences with the animal.

Unfortunately, with our raw stimuli, we cannot have the individual schemata called up. Only then, might one hope for success. All experiments with intricate locks that open only to a specific password are in vain if one does not know the word. One achieves nothing at all with all the variations. On the other hand, it is very easy to achieve movements of the bolt if one manipulates the gears themselves in the workings of the lock. And so it is with the stimuli in the receptor (232) and motor areas. The former yield no effects at all; the others always yield effects that, to be sure, are often quite aberrant.

The notion of the counterworld also opens up for the octopods entirely new questions. One knows that the octopus recognizes by form the crab dangled in front of it on a string. As soon as the octopus sees the crab, it changes color and rushes toward it. The eye supplies a flawless image of the external objects in the retina and can even accommodate in an excellent way. But how precise the schemata are has not been examined at all. Nothing is yet known about whether, say, a brittle starfish would be treated just like a shrimp or whether the artificial coloration of the shrimp neutralizes the effect of the form.

Removal of the cerebral ganglia is just as ineffectual as artifical stimulation. At least, all the coordinated movements are preserved and can be reflexively released by stimulation of the receptor nerves. Therefore the central inner life remains untouched by this operation. The highest motor centers have remained unharmed and bring the complicated movement apparatus into play with the same certainty regardless of from where they receive their nervous impulse. Since in normal life it can become necessary at any time to bring one of the motor mechanisms into action as quickly as possible, then it cannot be astonishing that direct pathways run from the receptors to the highest motor centers. From this point of view, the continued existence of the capability of movement as a whole after removal of both of the cerebral ganglia is not so striking. However, this fact, taken together with the complete incapability of reacting to artificial stimuli, will easily lead to the belief that the cerebral ganglia possess merely blocking properties. And even if, in fact, blocking effects appear to come from the ganglia, nevertheless, this scarcely hints at their significance. From experiments on snails, whose activity is of a quite different kind, we now know how blocking centers work.

Quite recently, the analysis of reactions in higher animals has made the assumption of the counterworld in the central nervous system necessary, and we have been able to make plausible the inexcitability (233) of the schemata by artificial stimuli. The fact that dysfunction does not occur after the removal of the counterworld is just as plausible if the receptors send direct neural pathways to the motor centers. The importance of cerebral ganglia with their counterworld is not affected in the least by this. The inner life of octopods simply falls into two main portions: a *central* inner life and a *cerebral* inner life.

The central inner life, which forms a completely closed entity, is derived directly from the inner life of the lower animal forms. Here, the path of the reflexes is likewise receptor-network-effector with nothing more than a higher elaboration in the motor portion. The environment associated with the central inner life consists not of objects but of individual physical or chemical effects, which perhaps undergo a certain grouping in the receptor network portion. The octopods can still live with the central inner world because no indispensable machine part has dropped out. The organism still functions as a whole.

Even in the cerebral inner life, nothing can happen other than that movement reaction takes place in response to external stimuli. The receptors and effectors remain the same, and only the receptive pathways undergo a restructuring. This restructuring, however, does not change the organism itself so much as, above all, the environment, which is restructured from the ground up by the introduction of spatial forms and the generation of real objects. What objects they are, and to what extent they are congruent with the objects recognized by us in the animal's surroundings, suitable experiments must yet clarify for us. Octopus vulgaris builds itself a house of stones and blocks of rock, and this, after all, demands a certain knowledge of the forms of the building blocks used.

At this moment in America, interesting experiments are being carried out on various animal species that relate to the development of habits (Yerkes). In doing this, one hopes to find evidence for the operation of a psyche. To the extent that a new formation of habits (234) permits one to posit new formations in the brain itself, it is, to be sure, to be concluded from these experiments that a supramechanical factor is at work in the brain. But I see no cause to call this factor a psyche or a psychic entity. For structure formation is a mechanically inseparable property of unformed protoplasm that differs precisely by this property from all other formed and unformed materials. To what extent a new formation is to be assumed in an octopod's brain has not yet been established with certainty. Undoubtedly, an Octopus vulgaris who has pounced on a torpedo fish, been chased away by its electric discharges, and is sitting again on the edge will leave the torpedo fish in peace for a while, but whether a lasting habit will emerge from this has not been investigated.

I have found in hungry specimens of Eledone moschata [another octopus] that they like to pounce on hermit crabs. But if the hermit crab shell possesses a sea anemone that stings the Eledone, it promptly gives up the vain attempts, soon gives up eating altogether, no longer accepts the favored shrimp, but perishes miserably. This experiment shows that the so-called plasticity of the brain of Eledone is meager, for the new experience brings on no new habits, but tears its counterworld to shreds.

In contrast to Yerkes and Driesch, who seek proof for the psyche in the protoplasmic achievements of the brain, Loeb and, most recently, Bohn believe they see proof of a psyche in an associative memory. Now an associative memory, if one uses this term to designate an objective achievement of an animal as such, is by no means a supramechanical capability. You can very well imagine machines in which the release of a certain placement of the gears constantly influences the operation of the machine. This attempt to prove the existence of the psyche objectively, which, after all, is identical with the life of feelings, seems to me therefore even less successful.

The existence of an associative memory, which also plays a great role in the origin of habits, has surely been indicated for octopods, but has not been strictly proven. In general, the organization of our (235) knowledge in this direction is completely absent. But I believe that the way to a completely acceptable ordering of our experience can only take place on the basis of the concept of the environment and counterworld.

DRAGONFLIES

[*Editor's Note: Material has been omitted at this point.*]

Through Radl, we have become aware, above all, of the central environment of insects, and he has emphatically described the environment of insects as a *field of light* with respect to which a flying insect finds itself in a *light equation*. This light field acts on the animal only through the eyes, as Parker was able to show, and the light equation is maintained only in a reflexive way.

[*Editor's Note: Material has been omitted at this point.*]

(236) For the orientation of insects in their environment, neither the intensity of the light nor the forms of the outlines, nor the color of the objects comes into consideration, but merely the size and distribution of darknesses on a light background. Parker found the simplest type of this orientation in the mourning cloak butterfly, which always orients itself when alighting such that both eyes are equally strongly lit by the sun. If, however, a shadow falls on it, it leaves its place and flies to the largest illuminated surface, but never towards the sun. These observations were checked and verified in every detail by Parker experiments.

[*Editor's Note: Material has been omitted at this point.*]

(240) Odonates seem to be more independent of the light equation. To be sure, I have been able to observe an Aeschna that had a quite definite circuit flyway, and tirelessly, for more than a half an hour, flew around the same bushes at the same height. But for the damselflies, sitting still, lurking for their prey, the proof that they owe their position solely to the light equation might be difficult to provide. Nevertheless, their body executes quite definite compensation movements. Every passive shifting of the body up or down, right or left, is equalized by an opposite motion of the neck muscles. It is not necessary to attribute this phenomenon to the light equation.

[*Editor's Note: Material has been omitted at this point.*]

The biological significance of the compensatory motions is very great because they provide the animal, even when it is resting on a fluttering leaf, a still background against which the moving prey animals most certainly stand out.

Just as the totality of the outlines of the objects depicted on the retina can release a reflex when this totality begins to move, so an individual outline can achieve this also. Here is where motoreception begins in the strict sense as Nuel (241) termed the effect of the motion of the surroundings on the eye of the animal. The movement of all outlines on the retina only occurs when the dragonfly itself is moved. The motion of one outline alone is always brought about by a process independent of the animal. When in spite of this, the motion of a single outline is answered with a compensatory motion, this has the advantage that the dragonfly can fix a prey in flight on a certain spot on the retina.

Generally, however, another reflex sets in. The dragonfly pounces on the moving object, and, if it is prey, grasps it. I have often had occasion to observe Aeschna's pouncing on a slowly falling small leaf. As soon as it approached the leaf, it swerved off without touching it. In agreement with Exner's results, I have also succeeded in deceiving Aeschna by flicking little bits of paper, something which was unsuccessful in the common damselfly Caleopleryx.

The observations on Aeschna show us directly that there are two reflexes here: one reflex elicited by a motion on the retina and a second generated by the form of the outline. The first we call a motoreflex; the second, iconoreflex. In a normal prey capture, both reflexes, which so clearly differ from each other in the experiments with paper bits, must complement each other and elicit a unified action. The motoreflex produces the pouncing; the iconoreflex produces the seizure. Together, both constitute the prey capture.

I assume that the iconoreflex and the motoreflex originate similarly. Every outline that is traced out on the retina produces a neural excitation in all those nerve endings that the outline covers with its surface, and that excitation is carried on to the central terminus of the optic bundle. The excited surface and the central plane of the bundle, thanks to its electrical properties, may exert an induction effect on the central network provided that an ordering of fibers corresponding to the form of the excited surface is located there. I call this (242) central pathway complex, formed in correspondence with the images of the objects on the retina, a *schema* and assert that the exact number of types of objects in the surroundings is distinguished by the animal as there are schemata present in its counterworld.

[*Editor's Note: Material has been omitted at this point.*]

(248) THE OBSERVER

We are approaching the end. The introduced images of individual animal species already indicate the great lines of direction that lead to the ultimate conclusions of which biology is at all capable.

If we first cast a glance back at the environments of the various animals that we have observed, then we recognize that our own surroundings give the common basis in all cases for all observations. The world surrounding us is the only objective reality that we deal with when we pursue objective natural research. It consists of numerous colored objects that have a manifold structure, and it is full of sounds and scent. It would seem that all other animals also live in the same world. But none of them possesses anything even approaching such a rich mutual interaction with all the objects with which our receptive organs constantly come in contact.

Each animal possesses its own environment, which displays ever greater differences from ours the farther distant it is from us in its organization. We have seen that the higher insects live in a world that is still somewhat similar to that of our own: bushes, trees and water surfaces also appear in their world as effective factors. But as soon as we get to the lower insects and crabs, the world changes significantly. Outlines and colors of the background disappear, and only the size of the illuminated surfaces serves to guide them. Even the number of the outlines with which the closest objects are distinguished, which is still significant in the cephalopods, decreases more and more. Some crabs seem to be surrounded only by colors and no longer by forms.

The further one descends the animal kingdom, the more the world of the eye, with its colored and formed objects, disappears; the more the environment transforms itself into a world of smells and mechanical resistances that, according to the construction plan, exert attractive or aversive effects, until finally, in the environment of the tunicates, only some harmful stimuli are still present. In the Rhizostoma, (249) the entire environment is filled only by its own stimulus producing movements, whereas all the worms and sea urchins still possess an environment that has an effect upon their entire stimulatable surface. These animals can therefore simultaneously receive stimuli that impinge upon them at more than one place. The tunicates rely upon a single gate of introduction through which the stimuli can be effective. Their environment therefore consists merely of a sequence of harmful stimuli. Everything spatial has disappeared from their environment.

In this respect, the environment of the tunicates is even simpler than that of the amoebas, which consist of a clump of homogeneously stimulatable substance and which therefore can simultaneously accept stimuli at more than one place.

One can say that the more simply an animal is built, the more simple its environment will be also, the one exception being the tunicates. This can be explained by the fact that their nervous system plays a subordinate role in their construction plan. The environment is always only that portion of the surroundings that has an effect on the excitable substance of the animal body, and, with the simplification of the entire constructional type, the constructional type of the excitable substance becomes simpler as well.

Whereas the environment changes, the surroundings remain essentially unchanged simply because they represent the environment of the observer and not of the animal. The animal body is exposed to the effects of the surroundings even where it harbors no stimulatable substance. Sometimes, the non-stimulatable part of the animal body makes up the greater portion of the surface, as is the case in the tunicates and the Rhizostoma. The non-stimulatable parts of the animal body are equivalent to inorganic bodies. However, unlike them, they can be changed. Both their mechanical and their chemical structure are influenced by the agents of the surroundings. It seems reasonable to look at even these connections between the object and the surroundings under the image of environment, although in this case, the organism confronts the external world no differently than does any lifeless stone. The factors of the surroundings can have an effect even on a stone, both from all sides and successively. Therefore, the stone's environment will possess both spatial and temporal extent. (250) Since the stone can undergo chemical, thermal and mechanical changes, one will ascribe corresponding effects to its environment. If one now succeeds in attributing all these changes to the movements of its molecular structure, then one will be able to demonstrate, even in the environment, nothing other than moved particles in space and time. And this is also the point of view that the science of inorganic matter has arrived at. It is applicable to all organic matter in the same way, to the extent that the material is not stimulatable, that is, is not alive or (something that means the same thing) does not consist of protoplasm.

But it is too much to swallow when the materialist demagogues want to convince us that that "environment of stones," which is only a conceptual abstraction of the reality surrounding us, is more real than this one surrounding us. Despite the fact that in the environment of stones no objects exist at all, but only a chaos of dancing points, this feeblest of all conceptualizations is more substantive than everything that surrounds us as real objects. It seems high time to once and for all get rid of this obscurantism that is worthy of an alchemist.

The comparison of inner worlds is just as instructive as the comparison of environments. Whereas our own environment, which constitutes the surroundings for all animals, is full of colorful, resounding, scented objects, our counterworld is limited to the succession of excitations in the prestructured nervous tissue schemata of our brain. In their form they resemble those objects. In addition, however, every dependent quality of objects must be represented by a particular individual nerve. Only the intensity of the stimulus is transmitted into the intensity of the excitation. By the way, stimulus and excitation are something completely different. Therefore, the excitation must be seen only as a token of an external occurrence.

The counterworld, which is so rich in our brain that it can serve as a mirror to our environment, quickly diminishes in the animals in scope and wealth. Whereas the cephalopods still possess great cerebral ganglia in the brain whose task it is to harbor the counterworld, and exhibit correspondingly special physiological properties, (251) special cerebral ganglia are not demonstrable for the arthropods. In the simpler animals the counterworld is totally absent, and the entire internal life plays itself out only in the central networks. These can offer, as we have seen, an abundance of variety, from the sea anemones to the leeches. In Rhizostoma, the inner life is reduced to a simple back and forth flow of excitation. In the amoebas, the course of excitation is no longer as regular, since no established pathways are present, and the excitation must accommodate itself to the changing shape of the organism.

The relationship between the inner world and the environment is an unchangeable one in all animals since they are mutually determined. All the stimuli of the surroundings are subject, as we well know, first to a choice by the receptors. A great portion of the effects is a priori excluded; the other portion is transformed into excitation. If all excitations are directly fed into the general network, then all the qualitative differences that are present in the surroundings are lost. Only when definite nerves are present for definite qualities are these qualities preserved for the environment. Likewise, the spatial relationships that we recognize in the objects of the surroundings are transmitted into the environment of the animal when its counterworld harbors corresponding schemata.

The activity of the receptors is always a threefold one. First, the selection from the effects of the surroundings takes place, whereby the greatest part is discarded. Then, the analysis of the portion to be accepted takes place; that is, the entire amount of stimuli is split down into groups that correspond to the structure of the receptor. Thus, in one animal, light can only have an effect as bright or dark, whereas in another animal, an entire color scale is differentiated. Thirdly, the transformation of the individual stimulus groups into nerve excitation takes place.

The synthesis in the nervous system only takes place when the various neural personae, each corresponding to a stimulus quality in the environment, come together in highly complicated structures. In what way these structures can fit together as schemata of objects has been set down in great detail.

The more the inner world is (252) enriched by the elaboration of such structures, the greater and richer will be the environment of the animals. Therefore, the environment of the next higher one includes again and again the environment of the next lower one. And when one as an observer imagines the animals, then each time the environment of the higher animal can be considered to be the surroundings of the lower animal, in which it is observed by this one (the lower one). The lower animal, together with its environment, presents itself to the observer as a closed unit, whereas the unity of the higher animal with its environment can never be grasped by the lower animal. This conception of the animal kingdoms creates the notion of ever greater circles each enclosing the next smaller one.

Even we people each live in the environment of the other. Doubtless, there are people in whose environment we, along with our entire environment, live enclosed as if by alien surroundings. One merely need observe the pictures of a Holbein to convince oneself of this fact, that the world in which he lived was one of much greater richness than ours. When he paints the simplest objects, they possess such an incomparably high reality that the objects surrounding us pale by contrast. If Holbein wanted to investigate the connections of our environment to our inner world, as we have done with animals, then he would find in the counterworld of our brains schemata to which the objects in our environment correspond. He would correspondingly conclude "for these realities, this human object is still barely receptive." The gaps of our counterworld, however, would not remain concealed to him, and he would say, "for those higher realities, this object is not created to perceive them."

Thus each higher environment grows with the increasing number of effects that it contains, and approaches more and more the surroundings that enclose it. It does not matter at all whether we want to imagine these surroundings in turn as the environment of a higher being or not. The fact remains that we are surrounded by higher realities that we are incapable of surveying. We must acknowledge, as Keyserling penetratingly demonstrates, that that thing which runs from the egg to the hen and extends its purposeful construction through time with no gaps, forming a chain of objects (253), without itself becoming an object, exists without our being able to recognize it. We are simply surrounded by countless realities that our perceptive abilities cannot reach. They remain imperceptible because they transcend perception. All organisms, plants as well as animals, belong here. Of them, we possess only the image of their momentary appearance. We can make no image for ourselves of their being that reaches in an unbroken chain from the seed to the adult, and of which we know that it harbors a unitary lawfulness. All plant and animal species, which we manipulate as if they were known quantities, are realities transcending perception. Indeed, we ourselves constitute such a reality that we cannot survey, since we can only observe ourselves from moment to moment. All nations, all states, reach with their reality beyond our ability to perceive. Whoever has at any time cast his glance on these things soon comes to the conviction that we live not in an "environment of stones," but are enclosed on all sides by higher surroundings that we cannot survey, and by which we ourselves are guided in an unrecognizable fashion. And since we subsume these higher surroundings in the word "life," the problem of life slips away again and again from our short-sighted eyes.

LITERATURE CITED

Bauer: Einführung in die Physiologie der Kephalopoden. *Mitteil. der Zool.* Station Neapel, Bd. 19, 1909.

Bohn: Introduction à la psychologie des Animaux à symétrie rayonée. *Bull. Instit. général Psychol.* 1907.

———: Attractions et oscillations des animaux marins sous l'influence de la lumière. *Mém. Instit. psychol.* Paris, 1905.

———: La Naissance de l'Intelligence. Paris, Flammarion, 1909.

Exner: *Physiologie der facettierten Augen.* Leipzig-Wien, 1891.

Jordan: *Die Physiologie der Lokomotion bei Aplysia limacina.* Inaugural-Dissert. Oldenbourg, München, 1901.

———: Untersuchungen zur Physiologie des Nervensystems bei Pulmonaten. Teil I. *Pflügers Arch.*, Bd. 106, 1905. Teil II. *Pflügers Arch.*, Bd. 110, 1905.

———: Über reflexarme Tiere. Teil I. *Verworns Zeitschr.*, Bd. 7, 1907. Teil II. *Verworns Zeitschr.*, Bd. 8, 1908.

H. Graf Keiserling: *Unsterblichkeit.*

Loeb: Beiträge zur Gehirnphysiologie der Würmer. *Pflügers Archiv,* Bd. 56, 1894.
———: Concerning the Theory of Tropisms. *Journ. Exp. Zool.*, Vol. IV, 1907.
———: *Einleitung in die vergl.* Gehirnphysiologie. Leipzig, 1899.
———: *Untersuchungen z. physiol.* Morphologie der Tiere. Teil II. Würzburg, 1891.
———: Zur Physiologie und Psychologie der Aktinien. *Pflügers Archiv*, Bd. 59, 1894.
Nuel: La Vision. *Bibl. internat. de Psychol.* expér. Paris, 1904.
———: Les fonctions spatiales etc. *Arch. Internat. de Physiologie*, Vol. I, 1904.
Parker: *The phototropism of the mourning-cloak Butterfly*. Mark Anniversary 1903.
Radl: *Untersuchungen über den Phototropismus der Tiere*. Leipzig, Engelmann, 1903.
Uexküll: *Leitfaden*. Wiesbaden, Bergmann.
Yerkes: A contribution to the Physiology of the Nervous System of Medusa Gonionemus Murbachii, I. *Amer. Journ.* Phys., Bd. 6, 1902.
———: A Study of the reaction Time of the Medusa. *Amer. Journ. Phys.,* Bd. 9, 1903.
——— and Huggins: *Habit formation in the crawfish*. Harvard Psychol. Studies, Vol. I.

14

CONTRIBUTIONS TO THE BIOLOGY, ESPECIALLY THE ETHOLOGY AND PSYCHOLOGY OF THE ANATIDAE

O. Heinroth

These excerpts were translated expressly for this Benchmark *volume by Doris Gove and Chauncey J. Mellor, The University of Tennessee, Knoxville, from Beiträge zur Biologie, namentlich Ethologie und Psychologie der Anatiden,* Verhand-lungen des V. Internationalen Ornithologischen Kongresses, Berlin, 1910, *Deutsche Ornithologische Gesellschaft (1911), pp. 589–702. Page numbers are indicated in parentheses in the text.*
Translation © 1985 by Doris Gove and Chauncey J. Mellor.

(589) Since my earliest youth, I have observed the world of birds, especially the larger birds and particularly the ducks in the broadest sense. This avocation earned me, even as a child and later as a schoolboy, many a harsh word and nickname and much doubtful headshaking at the university from others who did not understand. I had always made it my goal not to consider that which one generally terms biology, that is, those things that we all know from Brehm or Naumann; rather I chose to consider the more subtle habits of life, the customs and practices—that which one calls ritual in the college students' understanding of the term. I soon found that there is next to nothing of this sort of ethology in the literature and that, with my observations, I was plowing somewhat untilled ground in which it was first of all necessary to draw guide furrows.

If one wants to uncover the secrets of nature, then one must ask questions of it—here more than anywhere one can rightly say: no questions, no answers. But it is precisely the asking of these questions that is so often totally missing in our handbooks. Observed facts are usually placed side by side without connection and frequently are generalized in the crudest fashion. At present we are so far behind in explaining the purpose of nuptial plumage, mating calls, and other such things precisely because we do not know the habits of the individual species. For example, we have no idea which characteristics are most important to animals in the evaluation of a conspecific, etc. Certainly this is often quite different in rather closely related genera and species. It is difficult and extremely time consuming to observe these most subtle habits of life. Perhaps it seems unbelievable to the uninitiated how little knowledge exists, often about crude biological things, even in otherwise well-informed ornithological circles. Just two examples: in all our handbooks, the vocalizations of male and female ducks are almost always confused, despite the fact that basic

(590) vocal differences are usually more prevalent in ducks than in any other birds. The very existence of the tracheal bulla *(bulla osses)*, which is completely unknown to quite a number of field ornithologists, substantiates this difference. Or: if I look up the illustration of the whooper swan in the new Naumann, I scarcely believe my eyes, for there is a picture of a bird positioning its wings in the manner of the mute swan! But it is precisely the positioning of the wings that so characterizes *C. olor* and to a certain extent *C. atratus*. Indeed, I might almost say that the wing positioning so embodies their life habits that I mistrust the biological conception of the *Cygnus* group in the new Naumann from the very start. Without exaggerating, this is ethologically equivalent to showing an annoyed cat as a model to illustrate an enraged dog!

Before I present my modest, and, unfortunately, very incomplete ethological observations, let us become clear about ethology itself. As is well known, *ethos* means customs and practices in the human sense. For animals, this word actually does not fit at all, for language customs and practices are instilled and learned in us. But a duck brings its language and ritual—as I will later call its manners of interaction—into the world with it, and executes both without ever having even seen or heard a conspecific. Therefore, we speak here only of *instinctive, that is, innate, customs, and practices*. Consequently, we mean by *ethos* something completely different from what it really signifies. Certainly there are in the higher animals also instilled and learned things, but it is better to call them traditions, as Morgan does. For example, mallards hatched and led by a park-bred mother that is accustomed to people behave quite differently from those that grow up out in the hunting areas. This occurs because in the former case, the adult had not warned her children at the sight of people; that is, did not herself show fear. The opposite occurs out in the wild. Therefore, wild ducks manifest by their tameness or shyness not something inherited through instinct, but rather something transmitted by tradition from the adult birds. On the other hand, duck chicks instinctively fear people and large approaching animals when they get close.

[*Editor's Note:* Material has been omitted at this point.]

In what follows, I have omitted what I mentioned about the biology of Anatidae in my "Brautente"[1] *(drinking motion as a sign of greeting, etc.); familiarity with the mentioned treatise is, therefore, absolutely essential for the understanding of these remarks.*

[1]"Beobachtungen bei einem Einbürgerungsversuch mit der Brautente (*Lampronessa sponsa* (L.))" [Observations on an Experiment in Introduction of the Wood Duck], Journ. für Ornith. 1910 **1,** 101–156. Also in: J. Neumann Verlag, Neudamm, 1910.

(592) **SWANS *(Cygninae)*[2]**

Mute Swan *(Cygnus olor)*[3]

The mute swan is the one most available for observation, since it is so frequently kept tame, that is, incapable of flight, in our parks. *It has probably not been changed in any way by captivity.*.At the Berlin Zoological Gardens, I have been able to observe two wild mute swans caught on the Black Sea living among other mute swans raised in Germany. If I had not banded these South Russian individuals, I would no longer be able to distinguish them from their tame conspecifics. They are not any thinner, longer legged or smaller the way wild forms usually differ from their domesticated counterparts. At most, there is a small difference in that, especially in fall and spring, they become somewhat more flight restless and make frequent attempts to escape. This is probably because these two birds were only recently brought to Berlin as adults, whereas our other mute swans feel no desire to leave, since they are long accustomed to this place or were even born here. The Russian swans are just as tame as the normal swans and become no more anxious when caught than their local comrades. Finally, one can detect no trace of the senseless fear usually seen in wild-caught birds under such circumstances. This trust of people, which is quite different from the caution of the whooper swan, probably explains why *C. olor* has become a park bird so readily. It should be noted in passing that mute swans are caught on the Black Sea by running a boat among them during the flight feather molt. Catchers then jump into the water and grab the animals, which are frantically paddling away. None of the birds attempts to escape by diving.

Since the mute swan is so easily accessible and is virtually the quintessential swan, the uninitiated observer tends to transfer the habits and customs of this species onto its (593) congenerics. I hope I shall succeed in what follows in showing that *C. olor,* though it exhibits some similarities with the Australian *C. atratus,* holds a special position among the Cygninae and shows no similarity at all in its behavior with its white relatives.

The *resting position* of the mute swan is shown in Plate 1, Figure 1. The very thick neck distinguishes this bird immediately from other swans, and the relatively cumbersome body is also characteristic. A certain inclination toward aggressive postures is a particular aspect of most of the other body positions of this bird. The young that have just reached adulthood demonstrate this behavior. It seems to us like bragging, and it is completely

[2] In the systematic order and sequence I have quite generally adhered to Vol. XXVII of the British Catalogue *(Salvadori),* but have permitted myself some deviations, because I was concerned with the *biological* cohesiveness of the individual groups.
[3] I have omitted all the authors' names here, since I consider them to be totally irrelevant in all non-systematic treatises, especially in the case of long known bird species. The nomenclature has been taken from the *Hand-List of the Genera and Species of Birds* of Bowdler Sharpe.

absent in whooper swans. The *"body-shaking,"* which we see in Figure 2, occurs in a remarkable violent, noisy manner that seems rehearsed, and it is particularly noticeable in spring when the birds are especially agitated. One may compare this with the shaking of the Bewick Swan *(C. bewicki)* in Plate 2, Figure 6. Similar behavior also regularly occurs in the *wing flapping* that follows bathing, as I have shown in my "Brautente." This is seen in all Anseriformes, especially if water has gotten on their supporting contour flank feathers. The neck is greatly outstretched, the body is raised high, and then two or three jerky wing flaps that are audible for some distance occur. The Black Swan behaves almost exactly the same, while the Whooper Swan group assumes a far less challenging posture, as seen in Plate 2, Figures 7 and 8.

The most striking and by far best known posture in the mute swan is the *wing positioning,* which I have reproduced in Plate 1, Figures 4 and 5. If even in serious works one speaks of "sailing before the wind," then this nonsensical kind of expression cannot be sufficiently condemned, for our bird considers it to be very disturbing if the wind blows up under its raised secondaries from behind. *Here we have one of the nicest examples of an aggressive position.* As soon as the mute swan wants to drive off an opponent, that is, assume a position that instills fear, it raises its elbows along with the secondaries and ultimately draws back the neck somewhat. If the bird recognizes the superiority of the opponent and becomes afraid, then the wing positioning stops and is replaced by the normal body posture. It is clear that this aggressive position is assumed mainly by the mature male, especially before, during, and shortly after the brooding period. In fact, during that time he never really (595) comes out of it. As is well known, the pairs, or rather the families, live strictly separated from other conspecifics at this time. If now a *C. olor* male catches sight of any other swan approaching, then he raises his elbows somewhat and swims toward the intruder. The closer he gets the higher he raises his wings until he reaches the position in Figure 5. Usually the alien swan turns to flee as soon as it sees the lord of the territory approaching; it then paddles quickly away by alternating the legs in the usual manner. The pond owner rushes noisily after it, performing *jerky swimming strokes with both feet together,* so that he repeatedly moves forward like a shot. Indeed, this kind of swimming movement, which is always associated with the aggressive posture, makes a much more arrogant impression than the rapid paddling of the fleeing bird. It should be noted that the speed of both swimming methods is the same. If the alien faces the aggressor, then it naturally assumes the aggressive position. Now both birds swim around each other in narrow circles only a few meters in diameter, with their primaries near the water or even dragging in it. Usually neither really dares to attack the other, and the patience of the observer is severely taxed. As soon as one is about to retreat, the other swims toward it with renewed courage, which naturally causes the retreater to turn around. In the exceptional case of a fight, the animals grab each other's shoulders with their bills and beat at each other with the carpal joints. The mute swan, as is well known, not only assumes this wing position to drive away a conspecific opponent, but also assumes it as soon as a person or a dog approaches. One can immediately recognize, almost better than in any other animal, the surges of anger of our bird in the positions of its elbows. Of the white species, *only C. olor positions its wings in the positions* described above, and *only it paddles in jerks with both feet together*. *C. atratus,* of course, also lifts its elbows, but rarely so strongly or strikingly, and it does not lay its head back (see Plate 2, Figure 2). It always swims with alternate leg movements. In the whooper swan group and the black-necked swan, we find no trace of this wing position, and the corresponding posture in *Coscoroba* looks basically different (Plate 2, Figure 10). Often, the furious *Olor* covers the last stretch to his opponent in flight, and does this in a special manner. The individual wing beats are much more powerful than in a normal takeoff, so that he often rises from the water with a jolt, (596) whereas he usually needs a much longer run. The mute swan can frequently be motivated to this aggressive flight by the most trivial causes: a person passing by, or indeed, even the very sight of any small water bird that is usually ignored can release this movement with no real attack following it. To translate this into human behavior: it appears as if such an impressive warrior is bursting with so much power and braggadocio that he has to release it in one way or another.

If the above-mentioned arrogant behavior reminds us in a certain way of a courting turkey cock, which often maintains its characteristic posture all day long and can even be set off in ardent courtship by the sight of a domestic chicken, then we must clarify the differences in the behavior of the swan and the turkey. In the turkey, the feather erection, tail fanning and wing dragging, which are purely courtship movements, cease if the animal becomes enraged and is about to go on the attack. If we put each turkey cock with some females in the spring, then he will immediately begin to court and, in so doing, always turn toward the hens. He can go on this way for hours until finally a female invites him to copulate by running around him playfully in a characteristic manner. The cock courts more and more vigorously and finally treads the hen lying on the ground before him. If we put a second cock in front of him the courtship posture ceases. The animals go at each other and begin to fight at once. Anyone who has dealt with turkeys knows the peculiar soft rage call, which cannot be reproduced with words. It is emitted *when the feathers are flat against the body* in a quite characteristic position, which is approximately the opposite of that which the animal assumes during courtship. By contrast, in the mute swan, the wing positioning has nothing to do with mating. Expressions of affection of *Olor* occur with very close lying feathers, as I will show more fully later. *The turkey cock wants to impress the female, or rather, stimulate her to mate, with the feather ruffling. The mute swan wants to intimidate the opponent. One has to be on guard, therefore, against transferring the purpose of the threat position of one species to another,* as is often done. This difference in behavior between *Meleagris* and *C. olor* is, of course, based on the entire life habits of the birds; the turkey is polygamous and in the wild the cocks probably perform the courtship strut in the spring at definite sites (597). They are sought out by hens eager to mate. The turkey cock also has a peculiar courtship enticement call, the well-known gobble, which is only heard from the male. I have observed that unmated hens come from far away in the spring when they hear this sound from our pheasant run. Fighting probably does not occur all that often in the turkey, nor in the heath hen or the black grouse. The turkey cock shows no tendency to defend nesting areas, or rather, the female, nest, and young, and for that reason, simply has no aggressive posture to intimidate an opponent. By contrast, in the mute swan it is pretty much the opposite. A courtship dance does not occur in this monogamous bird, and the entire efforts of the male are directed toward purging the nesting area of enemies of all sorts. Among these are, above all, its own conspecifics. One has to have witnessed how a stronger mute swan pair drowns and shakes off the small young of a second pair residing on the same pond to understand the justifiable apprehension that individual pairs have for each other.

The manner of taking off and landing is another characteristic behavior of the mute swan. In it, the crook of the neck always lies closer to the body than in the whooper swan group: compare Figures 7, 8, and 9 on Plate 1 with Figure 9 on Plate 2. When swans are in full flight, the neck is stretched out completely straight, but the crook of the neck immediately appears when the birds slow down and prepare to land. It is remarkable that in all swimming birds the customary neck position of flight also appears when one holds the birds under one's arm in the manner shown in Figure 6, Plate 1 or lifts them by the wings.

Everyone has observed that mute swans at rest or swimming *stretch one leg backwards and to the side with widely-spread toe webs, often holding this position for a long time*. As far as I can recall, I have only otherwise observed this in the black swan. In this way the bird dries its foot and then, usually with a few slinging movements, draws it forward under the lateral abdominal feathers. If the swan is on the water while doing this, then it adroitly draws the dried foot between the feathers without wetting it again. Many other water birds at rest also tuck the foot in the abdominal feathers to warm it, while paddling only with the other one. Yet, as I mentioned above, I have never seen these other birds perform the peculiar preliminary drying action. Most often (598) one sees this stretching of the idled foot in spring and summer sunny weather; toward fall, when the air is usually cooler than the water, it becomes less and less frequent.

To express affection, C. olor assumes the same position that we recognize in the bird toward the right in Figure 10, Plate 1: the plumage is down flat, and the lower neck lies horizontally stretched out on the water. The head and upper neck appear peculiarly thickened. It is pretty much the opposite of the aggressive posture in Figures 4 and 5. If the mates of a pair or otherwise well-acquainted swans approach each other, they assume the position just described or at least suggest it. Each bird apparently immediately recognizes in this the friendly intentions of its associates. *I would like to mention at this point an observation that indicates that the Anseriformes (and perhaps all birds?) recognize one another only by the face rather than body forms.* For example, if a swan is eagerly searching for food and keeps its head and neck underwater for a long time, and a conspecific approaches, then sometimes the latter attacks the unsuspecting one by suddenly driving its bill into the back feathers. In such circumstances, a bird can attack its own mate or a family member, and the attacker looks very chagrined when the head of a close friend suddenly pops out of the water. Then the attacker immediately assumes the affectionate position—as if to excuse itself—and the startled victim calms down.

The *voice* of the mute swan first of all consists of the actual enticement call, a nasal "khr" that is not at all hard to imitate. It is similar to a loud snore. Family members on the water call each other together this way when they cannot see each other. In flight this sound is apparently emitted by all the members of the troop with short interruptions, and it can sound almost as if the noise were caused by wing flapping. The warning call is similar to the enticement call. It sounds a bit sharper, something like "kheerr," and is repeated frequently when the animals are quite uneasy. In addition, the voice is used for showing affection, and, therefore, I prefer to discuss this type of expression further in the following section.

A true courtship dance does not occur in *C. olor,* as already mentioned. Mates are usually chosen toward fall, apparently without externally visible signs. All one can see is that the two birds assume the above discussed affectionate posture every time they meet (599) by smoothing down the body feathers and getting a peculiar thick head. Soon the two animals constantly stick together, and at the end of winter one observes copulations.

The prelude, or perhaps better stated, the agreement and *invitation to mating* is essentially the same for all the swans as well as the geese (*Anser, Branta, Chloephaga*, and relatives), the Egyptian goose *Casarca* group (*Alopochen* and *Casarca*), and the species of *Dendrocygna*—even though small distinctions can be seen in the different species. Therefore, let us here examine mute swan copulation in detail so that we can refer to it again later in the Anseriformes. The mates approach one another in the affectionate position and then pull closer so they appear from a distance to lay their heads together while both softly emit the enticement call. A closer look shows that the heads do not touch—touching is very unpleasant to most birds. Rather they stay a few centimeters apart. Then one begins to plunge its neck under the water as if to feed off the bottom and raise it again rapidly. This is shown by the left bird in Figure 10, Plate 1, which has just pulled its head and neck out of the water as can be seen from the water running off. Both birds frequently repeat this motion many dozen times and it gradually becomes similar to bathing motions, so that ultimately a front to back rocking undulatory motion results in the very long bodies of the swans. During this prelude, the two birds are usually so close together that the neck of one plunges out over the nape of the other, as clearly shown in Figure 11. However, one cannot say that the male always plunges his head over the neck or shoulders of the female. Between these bathing movements, both animals often rub their flank feathers with the side of the head. Each rubs itself, never the other. Gradually, the female becomes flatter and flatter and she finally sinks deep in the water with half outstretched neck, in order to be mounted by the male. In mounting, he holds himself fast with his bill in the feathers of the anterior third of the female's neck. Figure 12 shows the actual mating process, which is always very adroit and lasts perhaps three to five seconds. During this one regularly hears an elongated snorting vocalization, and I can assert with some confidence that this is usually uttered by the female: on the occasion of the mating of a male mute swan (600) with a female trumpeter, I always heard the trumpeting only; the mute swan male remained silent. During treading, the body feathers lie as close to the body as possible, a position which, as in many birds, is correlated with great excitement. The male then drops off to one side, and in the next moment we see them breast to breast. For a short time they push up high against one another with a characteristic snorting, and the water between them is strongly roiled by the vigorous foot paddling. Unfortunately, I have not been able to photograph this peculiar position, which led *Naumann* to the false assumption that swans mate breast to breast. Figure 13 in Plate 1 was taken a few moments later, with the female on the left and the male on the right. He is lifting his elbows, but in a very different way than in the aggressive position. After mating, the birds take a long bath, as do all other swimming birds. Finally, the disarranged plumage is preened.

The mute swan pair has yet another way of showing mutual affection, especially shortly before, during, and after the breeding period. Let us suppose, for example, that the male has driven away a real or supposed opponent. He swims back to the female, who is standing on the bank or dabbling peacefully. The closer he gets to her, the more aggressive his posture becomes. The female likewise raises her wings, and to the uninitiated observer it appears as if there were two enraged mute swans ready to fight. They come within a few centimeters of each other with wings raised high and neck feathers ruffled. But suddenly they stretch their heads straight up as they pull their necks back to the rest position and emit a drawn-out snort preceded by a short initial sound. Then both animals are satisfied and go about their normal business. The observer is always amazed that the female recognizes the male's friendly intentions in spite of his enraged appearance. The whole process affords a marvelous sight, as if both birds enjoy how menacing they can look. I have not yet observed this behavior in any other bird in such a pronounced fashion.

For many years we have kept only mute swan males on a pond at the Zoological Garden. They get along better with each other in general than when both sexes are placed on the pond together. In this case, the most dashing male soon (601) pairs with a female and considers all other males to be intruders in his breeding area that must be fought and driven off. Such a pond tyrant patrols the water almost without pause and allows no swan other than his mate to enter the water. This offers a truly sorrowful sight, with only two swans always on the water and the others lying around pitifully on the bank, finally more or less degenerating. Sometimes the pond tyrant pursues them even on land with ruffled feathers and dragging wings. Woe to the unlucky one he actually reaches. Naturally it never occurs to those in distress to join in defense against their tormenter; instead, they are constantly in great fear of him. Since there is not much value in breeding mute swans—even at the lowest prices, one cannot sell the young in the fall—and since one would prefer to see the water covered with many swans, I got rid of all the females and now keep about half a dozen males together. But if one now thinks that peace and order prevail, then this is a tremendous mistake. The males usually pair off and, to cap it all, the strongest pair does the same thing as a true breeding pair, so that in spring one must finally banish the two worst tyrants. With luck, the next ranking pair will not be quite as bad as their predecessors.

Male pairs are as good as indistinguishable from a true pair in their conduct. The same expressions of affection, aggressive posturing of each towards the other, and even nest building can be observed. Naturally each bird tries to tread the other and often performs the previously described initiation to copulation at length. Ultimately the affair is unsuccessful; I have never seen one male allow himself to be mounted by the other. Just when one prepares to mount, the other turns around and tries to do the same. I have shown in my "Brautente" that this rejection is not always the case: in two males of *L. sponsa* I have seen one bird lie on the water like a female after the usual preliminaries and willingly allow himself to be mounted. Even between female birds, of course, such as pigeons and many ducks, one can frequently observe mating in which one female behaves just like a male. (602)

In conclusion, I would like to add some observations on the *psychic capacities* of the mute swan. Try as one might, one cannot find any indications at all of thought processes, as is probably the case in most birds. In the fence experiment mentioned in my "Brautente," the swan fails miserably. It does not ever deliberately fly over a fence that it can see through even after days, yes even weeks, of experimental duration. Also, in their behavior toward a feared opponent separated from them by a wire mesh fence, swans, especially *C. olor,* make a poor showing compared to many other Anatidae.

If one removes the tyrant in the horrible tyrannies mentioned above and carries him away before the eyes of the anxious oppressed ones, they do not take a deep sigh of relief, go into the water, and frolic with abandon. Far from it! After hours, one of them, plagued by thirst, approaches the water. Finally, since the tyrant does not rush up, it glides into the water, but still remains near the protecting bank. Only after some time—a half day or more may pass—do the others finally follow suit. Evidently the swans had not realized that they were freed from their cruel lord by his removal.

If we observe a mute swan going from land or ice into water, we find that it always lowers itself onto its breast just as soon as the upper neck gets over the water. The animal then laboriously scoots forward until it is completely on the water with enough depth to swim. Thus the swan always tries to swim when it is still not yet quite in the water. The psychic ineptness of our bird is even more striking when it walks across melting ice. If it comes to a puddle only a few millimeters deep, then it lies down in front of it in the previously described manner and swims through the shallow water with the greatest exertion instead of simply *walking* through it. Geese and ducks would not think of trying to swim through water that was not deep enough.

Black Swan *(Chenopsis atrata)*

The black swan has been made the representative of a special genus, *Chenopsis*. Biologically, and especially ethologically, this is certainly unjustified, because in manner it very closely approximates (603) the mute swan. However, it is not at all similar to the whooper swan group. Figure 1 on Plate 2 shows two half-year old black swans intending to take off and the unbelievably long neck of this species is excellently portrayed in the birds standing in front. Also, this Australian bird has an aggressive posture similar to that of the mute swan; however, it rarely raises its wings so high, and the entire effect derives more from the curled elbow feathers. In all this, the neck is never laid back, but is held quite rigidly erect, and the neck feathers stand up very straight so that they stick out nearly at right angles from the skin of the neck. Copulation is initiated and performed in the same way as in *Olor;* only the afterplay is different. The upright posture is entirely absent, as is the raising of the elbows. Both mates swim around each other only with smoothed plumage and with necks held up at a peculiar lateral crooked angle.

[*Editor's Note:* Material has been omitted at this point.]

(604) The brooding process of *C. atratus* is different from the other swans (including whooper swans?) in that the male also enthusiastically participates. In fact, I have observed for several years that the male of a pair in the Berlin Zoological Garden regularly sat on the eggs from about 10 in the morning until 5 in the evening. The brooding mate only left the nest after the other one had already been waiting for a considerable time to relieve it. In *C. olor* and *melanocoryphus,* as is well known, the male often sits by the brooding female, but during the brooding break the mates leave the nest together.

It is striking to me that black swans, in contrast to mute swans, resemble ducks in their strong inclination to fly at dusk. Young birds, especially, become very flight restless even when it is already quite dark. Unfortunately, it seems to be impossible to keep flight-capable black swans on small ponds. Right off they fly far away, and they do not return home. This is perhaps due to the restlessness of most feathered inhabitants of Australia because of the droughts there.

In general, *C. atratus* is quite tolerant of smaller water fowl even during the breeding season. It just about never bothers with the ducks on its ponds, and only troubles with the geese when they attack it or endanger the nest site. In one case, a pair of greylag geese had selected the same small island for brooding as had a pair of black swans, and there were, of course, frequent differences of opinion: for some years the geese won, but later the swans were victorious. Consequently, that pair of swans detested greylag geese, but did not concern themselves in the least with bean, pink-footed, white-fronted, Canada, or other geese. It should be expressly noted that these two swans, even outside the brooding season, tried to chase off all greylag geese that came too close to them. They especially went after the above-mentioned pair of geese, which shows that the swans made an individual as well as a species distinction.

As is well known, even black swans raised in Europe, since they are of Australian origin, tend to breed in the fall. So it often happens that the young hatch during the coldest period of the winter and succumb to freezing shortly after leaving the nest. If one takes the eggs away from the birds to prevent this, they usually soon lay a second or even a third clutch. In the end, (605) this results in a good spring brood. Because of this fall nesting period, when mute and black swans inhabit a pond together, the Australian often has the upper hand over its twice-as-heavy white associate. Since the mute swan tends not to attack toward the fall, and also shows little courage in defense, the black swan often succeeds in routing it. The mute swan usually makes no attempt to compete with the black swan for nest sites. Often this dominance persists until spring, and by that time the much stronger white cousin no longer dares to rise up against the Australian.

In Figure 14, Plate 1, we see a 2½ year-old *hybrid,* whose father was a black swan, and the mother was a mute swan. The Berlin Zoological Garden has two such hybrids, apparently males, which were bred near Berlin by a private fancier. When I first saw these animals, then 1½ years old, they were rather uniformly dark gray with white flecks, colored like a houdan. Now they are uniformly ash gray, with only the head, neck, and middle of the back showing white flecks. The flight feathers are of course white, as in both parents. These birds are just about exactly midway between the parent species in voice, form, size, and behavior. The picture well illustrates that they deviate from the mute swan in their very different tail form, which approaches that of the father, as well as in the longer neck. The bill is uniformly dull red with a black nail. Both the wattle and the dark color at the base of the beak of *C. olor,* as well as the beautiful white band of *C. atratus,* are missing. I have the impression that the birds feel themselves more closely related to the black swans than to the white ones, and they are chased perhaps even more by black swans of the same pond than by their white associates. It would be really interesting to find out whether these two hybrids will be fertile in the coming year. To learn this, I have placed a female black swan and a female white swan at their disposal.

The Whooper Swan Group.

[*Editor's Note:* Material has been omitted at this point.]

(607) All whooper swans in the broader sense exhibit quite different behavior from *C. olor* and *C. atratus,* and this is most strikingly evident in

the fact that they *completely lack all aggressive posturing, wing positioning in particular.* In brief, they are more vigorous, persistent, and adroit in their movements and waste no time acting arrogant or making an impression. Perhaps as inhabitants of northern, barren environments they were unable to become as specialized in a many sided battle for existence as their two relatives from lower latitudes. The whooper swans, except for the long neck, are more like big, busy ducks than swans, as the layman conceives of swans on the model of *C. olor.* Therefore, the whooper swans are customarily called uglier.

[*Editor's Note:* Material has been omitted at this point.]

In great excitement, especially when an opponent is in sight or has been successfully driven off, the whooper swan mates usually raise a loud trumpeting call and set the closed wings up from the side of the body. They then make jerky sideways wing movements. When they close in on an opponent, the attack apparently starts with movements similar to those I have described as being an invitation to copulation in the mute swan. That is, the animals submerge head and neck in the water at short intervals. All whooper swans are very malicious birds; they try to peck at any small duck, especially when it first comes onto the water. Mother ducks leading young are usually soon robbed of their young and in some circumstances even killed themselves. Since whooper swans do not engage in threat or aggressive posturing, they usually soon gain the upper hand over the mute swan: by the time the mute swans have gotten into their posture, the whoopers are already biting and striking. The horrified victims get the worst of it. Whoopers resort much more to biting than all other swans (608).

For a long time we had a swan pair of which the male was *C. olor* and the female *C. buccinator* in the Berlin Garden. Usually they got along together quite well, and I have also observed many copulations. Nevertheless, immediately thereafter, they occasionally came to blows. I observed one instance of this very well, and here is the reason for the conflict. The instincts of the two bird species did not mesh, because immediately after copulation, the male *C. olor* raised up before the female. Now the whooper swan group lacks this afterplay, and immediately after the treading, the female began to flap her wings to shake off the water. The *olor* considered this wing flapping to be an attack, went on defense, and the battle began. I would like to note here that the entire initiation of copulation is the same with all species of swans, so that in this respect, *olor* and *buccinator* understand each other quite well.

It is probably known well enough that the whooper swan group does much better on foot than *C. olor* or *atratus,* and also, because of the long wings, does not need as much of a run to take off as does the mute swan.

I would like to report the experience of a fancier whose credibility is beyond all doubt, especially since he related it to me without any encouragement on my part. It was a very remarkable event. He had on his park ponds a captive wild male whooper swan that had become tame with time and showed an unbelievable attachment to his keeper in the spring. The bird followed the keeper on foot everywhere in the garden. When the master lay down on the lawn, the bird went through all the movements that swans perform in initiating copulation by constantly dipping his head and neck to the ground in short intervals. Finally he *attempted to mount my witness* as if he were a female swan. This demonstration could be repeated at will. The whole thing is not so remarkable to the experienced animal breeder, since many birds that are kept isolated from conspecifics direct their drive for reproduction toward some particular person. This we know, for example, in parakeets. I mention it now because it gives such a splendid *explanation for the tale of Leda and the swan*.

I do not want to end my discussion of the whooper swans without dealing with evidence for the intelligence of *C. cygnus* proposed by Schilling and reproduced as late as the 3rd edition (vol. 3, p. 597) of Brehm's *Tierleben* [Animal Life]. It has been copied over and over again from there. A shot, flightless whooper swan immediately joined up with tame mute swans in time of danger "and in this way knew how to keep safe." What we are dealing with here is clearly the instinct, widespread among animals, to socialize in the time of danger. An individual feels safer in a flock than alone (even people!). One only needs to take a net to a pond filled with water fowl, and the frightened birds immediately forget rank order and animosity and congregate closely on the water. This is done even by young and naive individuals, and I can see no intelligence in this behavior, since it does not represent a processing of personal experience.

[*Editor's Note:* Material has been omitted at this point.]

(613) In conclusion, here are some observations that apply to *all swans*.

All swans show the intention to take off by holding the feathers very smooth and raising the neck steeply erect, as is shown in Plate 2, Figure 1 for the black swan and in Plate 2, Figure 5 for the Bewick's and the whooper swan. In this position, the enticement call is usually given. The head shaking or bill tossing that is very characteristic of many other swimming birds is either not seen at all or only suggested. These larger birds must, of course, always take off against the wind. Therefore, to have the necessary run, they must find the pond site opposite to the wind direction. Even quite inexperienced birds do this correctly when making their first flight. This action depends, therefore, not on experience, but is innate.

The view is widely held that swans cannot dive. I have personally seen *Coscoroba*, black, black-necked, and trumpeter swans disappear under the water surface and emerge again rather far away. I assume that *C. cygnus* and *C. bewicki* behave the same way. In mute swans the downy young occasionally try to get under water, but they do not quite succeed, and grown birds of this species probably never dive.

Swans take their *downy young on their backs* when the latter are tired or cold. Of all the Anseriformes, probably only swans have this characteristic. I have observed this in *C. atratus, olor,* and *melanocoryphus,* and it would be useful in judging the systematic position of *Coscoroba* to know whether or not it also takes its young on its back. I have never seen similar behavior in geese of the *Casarca* group or in swimming ducks.

(615) Geese *(Anserinae.)*

First of all, let us discuss the genus *Anser*. The domestic goose is the one most available for observation, and its ancestor, the greylag goose *(Anser anser)* is so easy to keep in captivity and so commonly bred that it also provides a rich source for ethological investigations. In my experience, the bean, pink-footed, and white-fronted geese behave in precisely the same way: what is said for the greylag goose holds for these closest relatives except for the voice.

It is not my purpose here to go into the walk, flight, and the general way of life, but rather to concern myself with the finer points of living habits and customs of the genus *Anser*. *The fundamental characteristic that pervades the ethology of geese is the highly developed and close family life*. This stands out even more because of the greater versatility and gregariousness of these animals compared to swans. Even the much more developed vocal talent of *Anser* indicates this.

The *vocalizations* of *Anser anser* are naturally exactly the same as in the domestic goose, except that the greylag goose lets itself be heard only when it truly has something to say, whereas the domestic goose has apparently lost the subtle inhibitions so that it sounds off quite aimlessly at the most trivial stimuli. The same is true in a corresponding fashion of wild mallards and domestic ducks, and also for *Bankiva* and domestic fowl. For the wild animals, every outcry represents a genuine danger, since it can awaken the attention of an enemy. It appears, therefore, to be advisable in the wild to use one's voice sparingly; for the domestic animal, this danger is not present. The beckoning call of *Anser anser* is the trumpeting, brassy, nasal "gagagak," or "gigagak," which is emphasized on the first syllable. I have been unable to discern with certainty a consistent difference between the male and female in this call. But the individual deviations are rather great, and one learns to distinguish individuals in a group by their enticement call, especially if one understands nothing of music and pays more attention to noise and timbre than to pure tone. That birds who are acquainted with each other can recognize each other immediately is so well known that I need not mention it here. Nevertheless, it has a surprising effect on the observer when the young, circling high in the air, can recognize their parents among a large number of conspecifics (616) all giving the enticement call. If one moves flightless parent geese to a new pond that is unknown to the young and invisible to them from the air because of tall trees, the young can find that pond by recognizing the calls of their parents. This enticement call is even emitted by scarcely feathered young birds, only it is then much higher and less resonant. Very young animals, of course, peep. I cannot say with certainty whether chicks only a few days old understand the blaring enticement call of their parents, but I do not quite think so. That is to say, if the small young are dispersed by some sort of disturbance, the parents immediately begin a loud calling. The impression is given that the progeny do not quite know that this brassy signal belongs to their parents who otherwise deal with them in quite different softer tones. They only move toward the family in a purposeful manner when they actually see them. Later on, they answer the calling adults and thus are very attentive to the beckoning call. I should note here that geese *acquire much by tradition, and that instinct often plays only a subordinate role. This is certainly an achievement of long and intimate family life.*

The warning, or better, fright call, is a short, nasal "gang" that I could imitate so well, especially while my voice was still changing, that I could drive flocks of geese to despair. Upon hearing it, the animals usually stormed off terrified in hurried flight to the next body of water. This warning is often produced only very quietly, especially when a pair is leading small young and the father notices something suspicious but not yet particularly frightening to him. Upon this barely audible tone, the mother becomes watchful, and the little ones prepare to flee. A quiet sound that one always hears when geese are grazing along contentedly or walking some distance can scarcely be rendered in words, but still might best be written as a very nasal "gangangang" in three to seven parts. It sounds like a conversation with oneself. One could term it the conversational tone. However, it is only emitted, at least by the greylag geese, when they are moving along, and is probably a summons to family members not to lag behind. If there is a prospect of a fairly long journey on foot, or if the animals want to end their rest and go to the meadow, then these tones are emitted somewhat more energetically, and mean, therefore, as much as "Let's go!" This vocalization is much more staccato and considerably amplified when the birds intend to take off; then each tone sounds singularly hard and sharp. So we have here a good example for how *one and the same sound, according to its strength, can express three different degrees of excitement,* (617) *that correspond somewhat to the intensity of the movement about to be undertaken.* Of course, the vocalization is understood in a corresponding fashion by the congenerics, but is probably not emitted with the intention of communication, since even birds that are kept singly vocalize in the same fashion. Yet another vocalization of *Anser anser* belongs here. It is a loud "djirb-djarb" that cannot be confused with other tones. It is often repeated rapidly over and over again. Up to now, I have been able to figure out only the following difference between the meaning of these tones and that of the above described departure signal: after both, departure as a group always occurs, but in the case of "djirb-djarb," the emphasis is "on foot," for then the animals set out on a long *walk*. It is striking that the geese, considering their rather impoverished language, have a special vocalization for departure on foot in contrast to intention for movement in general.

Geese hiss, as is well known, when they are enraged and especially when they fear the opponent. The neck is extended forward and down with the bill opened, and the plumage is ruffled. The ruffling ends in a loud shaking, especially when the animals do not quite dare to proceed to physical violence. This shaking reflex occurs again and again as long as one stands, for example, at the nest of the apprehensive animals. Young geese separated from their parents emit a characteristic monosyllabic elongated *distress* tone that gives even us the impression of being pitiful. I have never heard this vocalization from independent animals. In its place, the beckoning call appears instead.

All members of the genus *Anser,* after attacks on an enemy and especially after successfully driving off an opponent, share a rather odd behavior. The simplest case is this: a pair of geese with or without young approaches a strange conspecific. The gander runs or swims toward the strange bird in a rage, with his neck outstretched. The latter usually heads for the hills. Immediately the attacker turns around and rushes back to his mate. Then both emit a loud *triumphal call:* a blaring followed by a quieter peculiar chattering that can be rendered by a very nasal continuing "gang-gang-gang-gang." While doing this, the animals look as if they would attack each other at any moment. They usually scream their loud calls directly into each other's ears while holding their necks greatly outstretched with their heads only a bit above the ground. Even the tiny, downy young join vocally in this triumphal call and completely assume the posture of the adults. With the (618) delicate "vee-vee-vee" of their little voices, this family gabfest sounds, to be sure, absurd according to our concepts. However, it is an eloquent sign for how intimately parents and children enjoy each other's company, and, being so firmly united, stand in opposition to the rest of the world.

This whole triumphal drama is used in a sort of figurative fashion to ingratiate oneself with another conspecific. If, for example, a bachelor gander, looking for a bride during the winter, approaches a female goose, his earnest intentions are usually best indicated by the fact that he drives off some weaker animal in her presence that is otherwise completely unimportant to him. Then he hurries back toward the female in the described posture and with loud triumphal calls. Initially, the female does not usually join in. But when she does, one can be certain that the marriage is sealed. I have often observed that 1¾ year-old greylag geese, who are thus just reaching sexual maturity, flirt with one another and apparently even mate occasionally. But as long as there is no common triumph call, the affair is not yet quite serious. At this point, I would like to explicitly note that this vocalization, as I have already mentioned, does not merely signify a sexual attraction. On the contrary, young that are still with the adult pair announce their heroic deeds to their parents and siblings in the very same way, and they likewise help to confirm the successes of their father and mother. Even ganders or females well acquainted with each other do the same thing. I would like to conceive of the phenomenon in this manner: the animals mutually express attraction by the fact that they present a common front to the entire outside world. When an amorous gander approaches his intended in this manner, his behavior involves a sort of ingratiation by making others look bad and by flaunting his own heroic deeds. In the discussion of the *Casarca* group, we shall see that these conditions are even much more pronounced in many respects. The vocalization here termed triumph call can also be elicited by another kind of great joy as, for example, a reunion after a long separation. We might assume, therefore, with some certainty, that the father returning to his family after driving off an opponent is really expressing his joy, which the family shares.

These remarks may appear to the uninitiated to be quite anthropomorphic, even facetious. However, they are not meant to be in the least bit comical or figurative. *In gregarious or family-oriented creatures, one of which of course is man, similar habits simply emerge in the most diverse animal groups. For me, the behavior of geese is not a* (619) *sign of human-like intelligence, but rather a sign that many of our forms of interaction, such as making others look bad or glorifying our own group members, are nothing more than social instincts.* These instincts are often, after all, quite inaccessible to logical deliberation and are performed from the beginning by individuals who think nothing at all about them.

Of course, all the characteristics discussed here are much more firmly fixed in animals as family oriented as the geese are than they are in humans, where they develop only by laws and tradition and, therefore, are different in each race. In the Anseres, and probably generally in monogamous birds, negligent fulfillment of parental duty by both parents has always resulted in the death of the children. With that any unstable family dies out completely.

The crucial point of monogamy lies not in the fact that each male produces progeny with only one female and vice versa, but rather that they stick together and both care for the brood. Thus, monogamy proves to be absolutely necessary in all groups whose young are altricial, and therefore, need food brought to them. Furthermore, this advanced form of brood care also occurred where the defensive abilities of the animals, especially the males, came to be used for protection of the young. Consequently, in these cases, we generally find a smaller number of progeny, since their successful rearing is more assured.

Within the goose family there is no manner of rank order or disharmony. By contrast, in many animals, especially apes, the privilege of the stronger and pushier is always highly developed and is often exploited in cruel ways in spite of the strongest group coherence. Probably the reason for the lack of conflict in geese is that, as grazing animals, they do not compete for food. It is this competitiveness, as is well known, that severely disrupts the family peace in many other creatures. And in the generalized sense, jealousy of possessions, even in humans, disrupts the family peace. As a rule, quarrels and chasing between individual troops of captive geese occur only at feeding troughs. It is my view that this is simply because the animals have come too close together while eating, and this closeness is unpleasant for them. Under such conditions, one can well observe the strongly developed rank order of the various families with respect to each other.

[*Editor's Note:* Material has been omitted at this point.]

(621) To study the other customs and habits of *Anser,* we should best observe the events chronologically as they normally happen to the animals in the course of a year.

The actual *courtship* occurs in the winter and naturally involves only as yet unmarried individuals, since the mates, once paired, remain mated for life, barring unusual circumstances. Wild geese become sexually mature no sooner than the end of their second year, and thus usually search for a mate during the second winter of life. Even at this time the adult pairs still associate with the young brooded out in the previous spring. From observations of captive geese, it seems to me that even the young from the previous year, who were unwelcome near the nest during the latest brooding period, might be permitted to join the family unit again in the fall. So mutual acquaintance in geese clearly plays a large role. The attachment of the two-year-old young to the parent family is, however, quite relaxed. Consequently, all single sexually mature and adolescent animals are to a certain degree dependent upon one another. These young animals do not constantly think about defending their families as is so characteristic of geese, and they approach one another easily. I expressly emphasize this, since it is well known that it is exceedingly difficult to introduce a new, strange individual into an already established tribe of geese. If acceptance does not occur by marriage, years may pass before the new animal makes contact.

When the breeding instinct begins to stir in earnest, the ganders begin with an introduction to pairing when they see especially familiar females. These preparations for pairing are reminiscent of those in swans: that is, they dip head and neck under water as if they want to touch bottom. In so doing, they swim in a characteristically haughty posture, carrying the tail and hind body very high. The courting male often carries out this posturing at a considerable distance in front of the goose he has his eye on. Evidently, (622) the female in question knows quite well that the matter concerns her and gradually responds to the proposal with similar but not so precisely executed movements. It is by no means necessary at this point that a union take place. Often the animals swim apart again without having come very close to one another. In time the matter becomes more serious. The gander always stays in the vicinity of his intended, and one can, in the end, even observe matings. Nevertheless, as I have already mentioned, the animals are not yet necessarily really married. The possibility that the goose may yet become involved with several males and the gander with other females is not at all excluded. However, the gander, in making these advances, meets rivals. At this point, jealousy awakens in him, and with it, the desire to fight. If one can finally become convinced that the budding pair of lovers has had its triumphal call in unison, then the pair can be considered firmly married.

Such a honeymoon pair offers anyone who really knows how to study it in detail marvelous material for observation. As is well known, both sexes are colored similarly and are externally as good as indistinguishable except for the somewhat greater size of the male. But remarkably, at this time even the most uninformed observer would notice even from a great distance the difference between the two. It is astonishing how a bird that has absolutely no fancy plumage or form, nor courtship calls, nor even particular courtship postures can do so much for himself as this gander shortly before breeding time. For the pair, then, but particularly the male, the movements are characteristically tensed or, as one might say, haughty. In walking, swimming, or flying, an unnecessary expenditure of energy for such movements is displayed, as seen, by the way, in many birds. In particular, cranes indulge in a high-stepping, stiff-legged march. In the swimming gander, the tail is high, and the neck continually resembles the affectionate position that we have learned to recognize in the mute swan. The mates are rarely more than a meter apart and usually remain very close together. The male attempts to read the female's every move in advance. It is astonishing how he accurately anticipates every turn and acceleration and deceleration; he reacts virtually in synchrony with her movements. While doing this, he also drives off every weaker creature that gets in the way and usually flies rapidly toward the enemy even if it is only a few meters away. After the enemy has been evicted, he also flies (623) the short distance back to his mate. The self-congratulating gander then plops down before the female with wings raised high as she joins him in a triumphal call.

Since the eggs are about to be laid, the female needs a lot of nourishment, mainly grass or at least green plants, and she pays little heed to grain. The gander has no need for this voracious business, and thus has more time to play the cavalier. Now, also, the choice of nest sites takes place. Geese, as all Anseriformes, lack the psychic capabilities of bringing nesting materials from various sources to a particular nesting place. So the pair must find a place to nest where there is already nesting material, such as weathered reeds, rushes, straw, and similar materials.

Copulation in geese, as already mentioned, is initiated in a similar manner to that of swans. The animals move closer to one another, constantly dipping and raising the head and neck in and out of the water. Finally, the female lies flat on the water and is mounted by the male, who grabs her neck feathers slightly behind the head with his bill. During the copulation one hears a loud, drawn-out cry, and I have never been able to determine whether this is emitted by both sexes or whether one of the two produces this vocalization. The male then lets himself slip off to one side, and both birds remain for a short time crop to crop. Then they, especially the male, raise their elbows in a completely characteristic manner, as is apparent in Figure 6, Plate 3, with the pair of Chinese geese. The male is on the right and the just mounted female on the left. Even during this afterplay the animals scream loudly, and this scream is reminiscent of one that is heard when the animals on the nest are particularly agitated.

One can observe consistently in the greylag goose that the male never comes to his brooding female on the nest. He always keeps himself at a considerable distance and only appears with the female when she takes a break from incubating. During, and also shortly before, the brooding period, the geese, mainly the female, have a completely special call, a very loud trumpeting cry. This is usually emitted when she leaves the nest or if she becomes agitated in the vicinity of the nest for some reason. Domestic geese produce this penetrating cry so often that it becomes irritating. In wild geese, of course, one hears it much more rarely. When the goose (624) returns to the eggs after a break, the gander does not accompany her to the immediate vicinity of the nest. If the pair *flies* to the nest site, then he usually passes it and turns back in a wide circle in order to look after her once again from the air. This behavior of the male perhaps has the purpose of not revealing the nest site to enemies. The female moves secretly to the nest while the distant male swims or flies in full view back to his usual station.

I have never seen greylag geese or even hybrids of domestic and greylag geese defend their nest against human intruders. If the brooding female notices a person preparing to approach the nest site, then she first of all crouches in the position shown in Figure 7, Plate 3 of the brooding bean goose. She lies unmoving with her neck outstretched, only to leave the nest hurriedly in a hunched position as soon as the person comes too close. The goose then joins her gander, who does not dare to intervene physically. They observe the intruder from a distance. In the white-fronted, pink-footed, and Canada geese, one is usually attacked by both parents and has trouble keeping the animals away from the nest while candling the eggs to check fertility. These species apparently consider man to be an opponent that can be effectively combatted, while the greylag geese, which have always suffered grievously at the hands of the masters of creation for thousands of years, have, in time, given up every defense against them. Indeed, this appears to be expedient in every way for the perpetuation of the species.

Even before the young are led away for the first time, the male turns up at the nest. I believe I can assert that he comes before they hatch. The young remain under the mother for one or two days after hatching and are then led onto the water by both parents. The female predominantly takes over the actual care of the young by pulling up grass for them and sheltering them under her wings. The male worries more about the safety of the family and defends them with conspicuous courage. In ponds where many geese are kept, it is not possible for a goose to lead young without her mate, for in a short time the young will be killed or scattered by the other geese, or the mother and her offspring will be constantly driven away from the food or from the grazing places. Here one finds the proof for the necessity of the protective male who demands respect. The courage of the parents stands in direct relation to the helplessness of the young, and thus it is that the pair leading the youngest (625) offspring on a pond is almost always dominant. As long as the young cannot quite move rapidly, the parents throw themselves toward every enemy with true desperate courage.

They then stand hissing with outstretched wings, ready to protect the young and to gain time for them to reach the protection of water. Later, when the young have become more mobile, this protective action of the parents becomes less and less necessary. At the first warning call of the male, the young have already scurried together and stand ready to flee. A further alarm call sends them rapidly to the sheltering element, followed by the two adults forming a rear guard. This defensive drive of the parental pair, by the way, is not completely extinguished even when the young are already full grown. For example, if one shoots the young of tamed wild geese when they are about nine months old, the adults even then get very agitated. They search for the unseen enemy who is hurting or killing the young and stay hissing, with spread wings, in front of the dying young to protect them. For such events our birds have a good memory. I have observed that, even before the shooting takes place, the adults apparently notice the preparations and flee and try to protect their youngsters against this sinister enemy. Evidently, they do not quite understand that the person they recognize standing on the bank is exerting this terrible effect from a distance.

In the Berlin Zoological Gardens it often occurs that a small gosling or sheldrake young falls into a dirt hole out of which it cannot climb. Naturally, the disturbed parents do not think of merely pulling the youngster out with their bills. As long as the young bird peeps anxiously, the adults remain by it, but if it ceases to cry, the adults go off grazing with the other young and may move so far away that they can no longer hear the unfortunate youngster even if it begins to cry again. Apparently, they do not notice its absence, and only if the family again comes by chance in to the vicinity of the pit and again hears the calling will the adults again become aware of the lost young. It is known that large numbers of young wild mergansers and wild ducklings perish every year during their trek from nest to water by falling into deep ruts or smooth-walled holes. The mother never retrieves them, even though she could do so easily.

Since the family unit is so strong, one can easily keep flight-capable greylag geese on a park pond as spring arrives. At the age of nine weeks young geese can fly, and (626) a short time later they make round-trip flights over their home ponds during which the adults call continuously so that the young always come back to them. Once the young have taken a good survey on these flights, they make no more mistakes in flying. One must only take care not to alarm the geese on the pond before they have flown rather high over it on their own initiative. Otherwise, they will take off at the warning of their parents and in their agitation forget to note their flight path. They then can easily get lost and often do not return. Once this difficulty is overcome, one can catch other flightless geese on the pond or do other frightening things without danger. For it does not bother the greylag geese at all to circle about in the air for two hours, until the frightening events on the water have ended. If one is compelled to move the waterfowl to another pond somewhat before winter, as is unfortunately the case in the Berlin Gardens, it is only necessary to capture and move the flightless adults. The offspring are guided to the new pond by the screams of the troubled parents.

The great fear of the unknown of wild geese raised by their parents on park ponds is surprising. *They do not fear that which they have had bad experiences with, but rather they are highly distrustful of everything* and only become accustomed with time to new impressions. Only one example here. In the above described resettlement, the adult geese know the new pond and feeding area from previous years. When the young finally land after hours of circling and are led to the feeding place by their parents, they first recoil in horror from the grain trough. Only after some time do they dare to approach this unfamiliar object with extreme caution. This fear of the unknown is probably the basis for the fact that these wild geese will always return to their home pond; they simply do not dare to land anywhere else. Not only in Berlin but also in other park ponds, *it is the rule that wild geese whose parents are flightless do not migrate in the fall.* (The same holds true for cranes, as I learned from Mr. Falz-Fein in South Russia.) To be sure, they become restless, but they always return to the parents after their flights. This is good evidence for me that the fall migration of young greylag geese occurs under parental leadership, and it is also known that young whose parents have been (627) shot wander about aimlessly. They remain in the area until they are finally all killed without having made the attempt to fly off independently. Even in later years the captive birds show considerable flight restlessness in spring and fall, but they do not leave. I am far from wanting to assume from this observation of geese that birds in general migrate under the leadership of adult birds. What is true for the greylag goose may not be true for the song birds, in which long-term family ties do not exist.

The unbelievable adroitness of flightless wild geese in avoiding pursuers by diving and quiet evasion is well known. One may observe marvelously well in captive animals that all of these actions, which look as if they are carefully thought out, are innate, and therefore instinctive. At the warning tones of the adults, our young pinioned park wild geese, which have never had any bad experience before, know immediately how to evade by diving, emerge crouched at the water's edge, and slip into some hidden nook. They do this so cleverly that catching them alive occasionally becomes impossible.

Interesting observations of physical and mental differences can be made by crossing wild and domestic geese to obtain half-breeds or by breeding these further to obtain three-fourths domestic or three-fourths greylag geese. Through domestication, the greylag goose has been considerably changed: it has taken on body weight, and its flying ability has been strongly diminished. This decrease of flight ability is not the result of smaller wings in relation to body size, as in most chicken and duck breeds. Instead, the development of flight musculature has evidently diminished considerably in domestic geese. The lighter breeds can still, nevertheless, fly quite well, but have nowhere near as much agility and endurance as their wild relatives. On the other hand, the domestic geese have much more walking endurance than the wild ones. Now it is noteworthy how these purely physical properties have had a feedback effect on the corresponding psychic abilities. I would like to explain this in somewhat greater detail with an example. In the Berlin Gardens we have two connecting ponds for our swimming birds with a low wire mesh fence separating them. Some meters above this separating fence is a bridge. Now let us assume that on Pond A, (628) a flightless, greylag goose pair is raising young. Then these young, at the age of about three months, occasionally land on Pond B when returning from their circling flights. Parents and children immediately start calling for each other, and the young attempt to swim back under the bridge to their customary Pond A, but, of course, run into the fence. Now it does not take all that long—only a few minutes to a quarter hour—until finally one or the other of them swims back to the middle of the pond, gradually considering, as one can clearly see from its behavior, the possibility of flying over the bridge. Usually the other siblings also swim along because, as soon as one of the family strikes out energetically in a certain direction, the others generally follow. But then they again see the parents under the bridge on the other side, and all the children swim back in haste against the wire mesh. Soon they again discover the impossibility of penetrating it and turn around again. Maybe on the third try they raise up from the middle of the pond and fly over the fence and the bridge to the parents. After this has occurred a few times, they completely give up the attempts to swim through that separating barrier that they can see through so well. Now, as soon as they want to return from Pond B to their parents, they move purposefully to the far side of the pond in order to get the necessary clearance and take off. The matter is quite different in domestic geese, and in this the domesticated form of the Canada goose *(Branta canadensis)* is exactly the same. If these birds fly onto Pond B and want to return to the family on Pond A, they swim with mounting anxiety back and forth quite literally for days. Only under special accidental circumstances will they fly over the fence and bridge. Even by experience they learn very slowly. One Canada gander that flies often from Pond A to Pond B with several greylag goose companions frequently fails to return with them. I have very often witnessed that the poor fellow cannot make his way back across the fence. I have then often shooed him out from under the bridge by throwing stones to force him to the middle of the pond, under the assumption that he would finally get the idea to use his wings. This thought apparently also dawns on him, as I could very clearly discern by his cries and his head movements revealing the intention of taking off (629). But after only a few seconds he discards the idea and again swims in a beeline to the fence. If one drives him to the open water again, he often attempts to travel over the side bank to the other pond by foot, which, because of another barrier up on the land, does not work. If one forces him into flight by suddenly running toward him, then, after a flash of inspiration, he finally flies over the bridge, since, in his fleeing, he finds himself in the air. When he takes off from the bank he does not yet have this intention (at this point he is much too dominated with fear of the approaching person); instead he takes off with a lot of exertion only with the intention of landing on the water again. In spite of frequent repetitions of this experiment, the animal still has not learned to fly over on his own. After all this, one must assume that the *decision* to fly is especially hard for domestic geese.

Hybrids between greylag and domestic geese inherit very little of the ingenuity of their wild relatives, and, if they are not dragged along with the flock into the air, wander back and forth along the fence for days, crying piteously. Even three-fourths blooded wild geese, which are quite inferior to the purebred greylag goose in their flight endurance are, in the experiment just mentioned, still considerably inferior to the wild form. Even seven-eighths greylag geese are easily distinguishable from the full blooded birds. Naturally, the experiment often becomes quite complicated in some cases, because half, three-fourths, seven-eighths, and full-blooded birds often form a group and greatly influence each other.

On the other hand, I have noticed that a female domestic goose living with our wild geese displays much more ability to find negotiable holes in a fence, and also remembers these escapes much better. For example, if one divides the bank from a grassy area with a transparent fence, the geese naturally try to reach the grass to graze by running up and down along the fence. If one part of the fence does not hug the ground, or if there is a hole near the ground, the domestic goose will soon find it. She always looks *downward* for a way through, while the wild geese search with raised heads and thus do not notice the opening. Even when the animals finally reach the desired grazing ground through such a hole, it is unbelievably difficult for them to find their way back. With repetition (630) they still do not search methodically in any way for the opening, that is, they do not go directly toward it, but rather only with time develop a feeling about approximately where the passage was located. Here they search for a quarter hour or more until they finally get through the opening mostly by chance. Of course, I am referring here only to flightless wild geese; the others naturally soon fly over the fence.

This "labyrinth experiment" on flying and crawling makes wild and domestic geese appear talented in different ways corresponding to their physical predispositions. It is consequently very hard to draw from this general conclusions about their intelligence.

If hybrids of domestic and greylag geese seem closer to the domestic than to the wild form in flight performance and ingenuity, this also holds true for their external appearance. In addition, I have always had the remarkable experience that even up to three-fourths greylag geese prefer to mate with domestic geese and their hybrids than with pure-bred wild geese. I can give no reason for this.

In expressing the *intention to take off*, geese use special head movements as well as the above mentioned vocalizations. The animals understand this sign language superbly. In this head signal, the bill is jerked from side to side in a way similar to what the geese do when they shake off adhering water. This movement becomes more and more vigorous until finally a peculiar flinging of the bill results. Immediately after this the birds usually set out.

As is well known, all birds, and especially the larger ones, always lift off against the wind. In order to take off, therefore, the geese must first find that side of the pond that is away from the wind. If the animals want to fly in an east wind, then they must first go to the west side of the pond. Even when they are wading in the water on the east side, a good observer can tell by looking that they intend to swim across the pond to the west side in order to get the required run into the wind. A very small sideways bill movement, which the uninitiated takes for pure chance, reveals the desire to fly in this animal or that. Then, with a few soft tones, the whole family swims quietly in a tight group to the west side. Here they suddenly turn about face, and the individual family members set the right distance from one another. Now even the most uninformed person can see the intention to take off in all the geese (631). Immediately thereafter they rise up. Here again, I expressly stress that even single *individuals* give these flight signs, as I usually call these intention movements.

The *flight* of the wild goose is much too well known for me to describe here. I would just like to point out a few peculiarities that one can hardly observe in open hunting areas because of the shyness of the birds. When greylag geese really want to work themselves out in flight, they often cut remarkable capers. Not only do they suddenly begin to rush along at furious speed, but also they throw themselves on their sides in a peculiar way so that one thinks they might tumble over or fall down. It is not at all rare that the long axis of the body is turned by 90 degrees so that the back is oriented exactly sideways, with one wing tip up and the other down. Other swimming birds, mainly *Casarcas*, frequently take exercise in such flight play, which occurs especially when the animals cruise about for amusement after a long rest. They show an actual feeling of cabin fever in this way.

During these exercises, it often occurs that *peregrine falcons* that normally spend the winter near the Zoological Garden *take aim at our wild geese*—a truly splendid sight! On clear beautiful winter days, I have, by making the geese fly up, given many an enthusiastic ornithologist the pleasure of observing this mad chase. As the geese are peacefully circling one can spot immediately, by their behavior, when the falcon is approaching. They warn each other and try as fast as possible to reach the water, often zooming nearly straight down. However, they usually do not succeed because of the high trees lining the pond bank. So then they seek salvation in rapid escape. No matter how fast the goose flies, the falcon is soon right on its tail. It trails the goose about a meter behind, and in a few seconds, the animals are out of sight, even in clear weather. In all this, I have never observed that the falcon really dared to attack the goose, in spite of the fact that both male and female peregrines take part and each storms after one goose. If this chase has taken place toward midday, the scattered geese often do not return until rather late in the afternoon. Naturally, I cannot observe how long the falcon actually chased the goose.

I have already mentioned that the family unit is the determining factor for the entire intimate life of geese. The consequence (632) of this is not only a great attachment of both spouses to one another, but also the fact that, in general, *all well acquainted individuals stick together loyally* so that friendly relations are not obscured even by jealousy. Thus, for a long time, we had a so-called triangle relationship consisting of two ganders and a goose who were, all three, very much united with one another. Each of the two males peacefully tolerated the other mating with their common female. This loyal alliance of the two ganders had as a consequence that in a brief time they rose to the position of fearsome pond tyrants. For while the goose was brooding, the two males proceeded together to destroy all other nests since a single gander alone could not stand up to their combined forces. Finally, I was compelled to remove one of the males.

In the life of spouses with one another, actual copulation plays a relatively small role. It is also infrequent, and the testicles of the male stay rather small, in comparison with those of duck species, as is well known. I already mentioned that one can observe matings at courting time without real marriages resulting. On the other hand, the triumph call, the external sign of mutual agreement and a common front toward the external world, is much more important for the establishment of a life-long union from which brood care proceeds with its complex family life. The minor significance of sexual intercourse for living together may explain why one finds, often enough, that two ganders join together and are inseparable for years, at least in captive geese. It is then not at all possible to pair one of them with a female. He simply has no interest in her, even if one locks him up with her for a long time. He just calls for his friend. Also, if there is an established group of geese on a pond, even when there is a majority of males, one cannot introduce individual females. Even over a long period, the strange females will not be accepted. Sometimes they lead quite a lonely existence for years. I cannot tell with certainty whether ganders tread each other or not. I myself have not seen it; however, a knowledgeable expert has assured me that it happens.

Occasionally, one has an opportunity with captive geese to make nice observations about the *extent* (633) *to which they accept strange young*. As already mentioned, the gander does not arrive at the nest until the young are about to hatch or have just hatched. One can get an unmated goose to hatch out young by exchanging her own infertile eggs for fertile eggs from another goose. In that case, some gander or other, acquainted with her but not married to her, shows up and faithfully leads and defends the young. The leading instinct of the males is rather easily released at the appropriate season and does not need to be stimulated by previous brooding as in the female. In general, one has the feeling that, for the gander, the main joy in all of marriage is the leading and defending of the children.

Rather often I have had to try to introduce goslings hatched in the incubator to a pair leading very small young. In doing this, one runs into many kinds of difficulties that are, however, quite indicative of the entire psychic and instinctive behavior of our bird. In the case of young ducks, if one opens the incubator where they have just hatched and become dry, they then crouch down motionless at first only to shoot away as fast as lightning when one is about to grab them. In the process, they often jump on the ground and quickly hide under nearby objects, so that one has a terrible time catching the little things. It is completely different with little geese. Without displaying fear they look at one quietly and do not object to being picked up. If one handles them for even a short time, one cannot then get rid of them so easily. They peep pitifully if one goes away, and they steadfastly follow after one. I have found that, very few hours after being removed from the incubator, such a little thing is satisfied if it can settle down under the chair on which I am sitting. Now if one takes such a goose chick to a goose family with young of the same age, then the matter generally develops as follows. The approaching person is observed with mistrust by father and mother, and both attempt to get to the water with their young as quickly as possible. If one approaches them so quickly that their young have no time to flee, then the adults, of course, put up a furious defense. One then quickly slips the orphan among them and hastily departs. In their great excitement, the parents at first consider the newcomer to be their own child and want to defend it from the person from the very moment when they hear and see it in his hand. But the bad ending comes next. *It does not occur* (634) *to the gosling to see the adults as conspecifics*. It runs off peeping and, if by chance a person passes, it joins him. It simply considers people to be its parents. Now, this aberrant behavior of the chick naturally draws the attention of the goose pair also, and, according to their mood, they start biting at the little creature, or they completely forget about it and go off with their own children. Then, of course, it is a hopeless case, and the only thing left to do is go back home with the little gosling. Such a little thing is just too trusting of man—a behavior that can become quite unpleasant for it in a zoological garden.

On the other hand, if one takes the precaution of making sure that artificially incubated geese do not become acquainted with a person, that is, if one does not handle them and sticks them quickly into one's pocket when they are about one day old, then the probability of succeeding to associate them with a goose family is greater. I have always found that such chicks like to join up with the other little geese, and consequently they run after their age mates, usually from the very beginning. The parents usually notice nothing at first because of their fear at the whole disturbance. But if one continues to observe them from a distance, then one finds that they quickly pick out the strange young one. They are in an awkward position, and the instinct for combatting an intruder and the leading instinct seem to be in conflict. Often they go directly at the little orphan so that one must fear the worst, but in the end they do not do anything to it after all. At the last moment, they bite down only gently so that nothing serious happens to the youngster. If the little fellow is sturdy, and if it seeks contact with the other chicks, then finally it is tolerated and is accepted into the family after about a day. But if it hangs back in the first hours or if it separates itself from the other goslings, then the parents make no further efforts to encourage it to come along. The following case is quite illustrative of the weakening of instincts or of the discriminative ability in domestic animals. I wanted to put a little orphan in a family where the father was a pure greylag gander and the mother was a greylag-domestic cross. The event occurred essentially as just described. But it soon became evident that the mother could scarcely distinguish the newcomer among her children if at all, whereas the father, after the family had settled down, very quickly noticed the little alien, and at first showed a great desire to pounce on it. (635) It was quite some time before he was no longer disturbed by the presence of the foreign gosling that, as far as we could see, looked just like the other chicks.

Goose pairs, in contrast to wild duck mothers, who do not easily accept chicks of other species, readily lead conspecific or congeneric young that they have hatched from eggs. So, for example, a pink-footed goose, *A. brachyrhynchus*, that had a white-fronted goose, *A. albifrons*, as a mate, hatched out a greylag goose egg and the substituted child is now led with great sacrifice (December, 1910) after seven months, even though it is much larger than its step-parents. One might well wonder whether this bird will later seek attachments among greylag, pink-footed or white-fronted geese when it reaches sexual maturity. In the meantime, of course, it is only concerned with its two foster parents.

It often occurs that the single animals on a pond do not always group according to species. Instead, often quite remarkable friendships are formed that can become very intimate so that the animals can scarcely be separated again. This also occurs with *Chloephaga* and *Casarca*. Thus, I had the experience that a Canada gander *(Branta canadensis)* attached himself most closely to a bean goose *(A. arvensis* [=*A. fabalis*]*)* and constantly courted it, emitting his triumph call. He trumpeted to it at every opportunity despite the fact that an unmated female Canada goose lived on the same body of water. I investigated the matter more closely and found that the bean goose was a male (!), and I did not succeed in pairing the Canada to his own species even by confinement. Such cases frequently occur and are evidence for the intimacy of the bond formed by mutual habituation even in animals of different species and of the same sex. Hence the extraordinarily loyal devotion of mates for whom sexual intercourse is quite immaterial, and of the entire families well into the next spring.

CHLOEPHAGA (CHENONETTINAE)

[*Editor's Note:* Material has been omitted at this point.]

(636) In other life habits, these South American [sheld] geese are distinguished, as is well known, by the fact that they do not go into the water as much as other geese, as is indicated by their long legs and short toes. In Figure 5 of Plate 3 we see a lesser Magellanic gander in his rage posture. Standing erect, with ruffled forehead feathers, he emits his fine,

high whistle. Shortly before he actually goes on the attack, his snow-white joints (perhaps threat coloration!) appear with strongly developed striking knobs. *Chloephaga*, therefore, does not approach an opponent the way *Anser* does with neck horizontally outstretched and head close to the ground. The reason for this is probably that these South Americans do not resort so much to biting; the much weaker bill seems highly unsuited to it. The birds run erect toward the opponent and pummel it with the carpal joints, and it is for this very reason that the horny calluses on the wrists are so strongly developed. On such occasions the female restricts herself to sounding her creaking call at short intervals, and even when the enemy is triumphantly chased off, we hear only the above described whistle and creaky voice in the triumph call of the two partners. In calling, they stand erect side by side and at most move their head and neck up and down.

Even in *Chloephaga*, the gander tries to show off a bit in a dashing way. He runs suddenly at some harmless creature (637) and returns triumphantly to his mate after he has driven it off. But here we find the first indications of a habit that is quite remarkably developed in the *Casarca* species; namely, *the inciting by the females*. They spur their husbands on to attack conspecifics or other geese that happen to be around them by making a sham attack on the strangers, only to then come near their husbands with loud creaky calls and continuous threatening gestures toward the strange geese. The male then very frequently attends to the actual banishment.

To express affection, the same movements are used, and they also serve to initiate or invite pairing. The only difference from the head and neck movements of *Anser* is, however, that they usually are given on land and less often while the animals are swimming. The two mates thus bow before each other by moving the head and neck down to the ground and then up again for a while. Meanwhile, they hold their closed wings somewhat to the side, so that the shining colored black feathers, the tail, and usually also the white wing elbows and the speculum formed by the large wing coverts can be seen. This posture corresponds to that assumed by the female during treading. But it is by no means certain that pairing necessarily follows these movements; the animals often show only mutual attraction this way, as one best sees when two strange individuals gradually come together and prepare to become mates. The advance in all this can, by the way, be initiated by the female, as I have only recently been able to observe.

Copulation itself occurs on land or in very shallow water. Since it is quite certain that *Chloephaga* is descended from more aquatic ancestors, the dipping or diving motions, which are actually designed for pairing on the water, must also have been transmitted. The same motions are seen in the other relatives and also in the swans. But it has a quite peculiar effect when a ritual is carried out on land that was originally designed for water. As is well known, domestic geese also carry out the usual dipping introductory movements on firm ground when they have no water and must mate on land.

[*Editor's Note:* Material has been omitted at this point.]

(644) *The particularly striking characteristic of the genus Casarca is the unremitting inciting by the females.* We have, of course, previously met with the first indications of this habit in *Chloephaga*, and we also find it in many surface feeding ducks. But for the *Casarcas*, it is truly the basic characteristic of their whole being. As soon as a conspecific, or congeneric in the broad sense, approaches a pair, the female makes a kind of sham attack on the stranger either by running at that bird in a rage or at least showing the intention of doing so. She stretches her neck out and holds her head and bill right down close to the ground the way geese do, giving the rage call throughout. If the opponent is weaker, it frequently withdraws at her approach. A stronger one, however, or at least one that is above the *Casarca* pair in the social order at the home pond, generally pays no attention to the impudent behavior of the female. She now runs back to her mate scolding all the way, and repeatedly makes threatening gestures back at the stranger, especially if the latter gives no suggestion of surrendering the field. The male stands by with head raised high, just about as we see in Plate 4, Figure 7. He expresses his agitation eloquently with quite definite calls. As if in a rage, the furious female sails around her husband with lowered head and frightful screaming, pointing again and again at the enemy. Finally the husband sets off toward the stranger and sends him on his way, if possible. One only needs to remove the females from the otherwise quite unsociable *Casarca* pairs, and total peace reigns. The paired female can scarcely see another swimming bird pass in the neighborhood without immediately inciting her mate at it. Occasionally she gives him a quite sharp jab in the breast feathers when he gives no indication of setting off after the hated fellow pond dweller. The behavior of the *Casarca* wives appears to the observer virtually insane at times: they become so extremely agitated at every opportunity. If it were up to them, (645) their husbands would have to be in constant battle with the most superior enemies. One now discovers that the females in fact prefer those males that excel in particular strength and desire for battle. This is especially true of *Casarca* females that are not yet firmly mated; they virtually play their suitors off against one another. It is not actually quite correct, however, to speak of suitors here, *because usually the male does not do the courting. Instead, the female searches out a male* and incites him to fight the conspecifics. If the chosen male consistently loses, then she will take up with a stronger one and carry on the same way with him.

[*Editor's Note:* Material has been omitted at this point.]

Unfortunately, I have never been able to keep free-flying common shelducks *(Tadorna tadorna)*. As soon as their wing feathers are grown, they circle over the trees, but have great difficulty landing in our small park waters. They try again and again, but they are unaccustomed to this problem in their flat seacoast habitat. (657) They probably are not able to descend steeply between high trees as the forest-dwelling wood ducks and mandarins can do so excellently. Finally they give up their attempts and disappear; therefore, I cannot say much about *Tadorna*. The *voice* of the two sexes is as different as it can be, as suggested by the greatly developed tracheal bulla of the male. The male has a soft whistling call, and the female produces a creaking and quacking noise. The enticement call of the female is a very loud, uniform, nasal quacking: "taht-taht-taht-taht." During inciting, and generally as an expression of anger, she emits a rough "roe-wow," as she just about throws her head and neck toward the opponent in her anger.

The previously described head dipping is seen only sporadically as an *initiation of mating* in *Tadorna*. Instead, the entire bird disappears under the water surface; the animals go through the same performance that we are used to seeing in the initiation of bathing behavior in the Anseriformes. Then, after the male has climbed on the female's back under water and grasped the hind feathers of her head with his bill, both birds emerge simultaneously. Let me note that I have also seen this initiation to mating in *Casarca tadornoides*, although it is not the rule there. Even ducks tend to perform matings during bathing initiations; one often sees a female emerge from playfully diving and fluttering and suddenly lie down flat on the water in front of her mate to be mounted. One even sees this in animals in which a different mating ritual otherwise exists.

Let it be expressly noted that the invitation to or *initiation of pairing*, which is often carried out in very different ways for the various groups, is *apparently absolutely necessary for the achievement of copulation*, as long as we are not considering the rapes that occur frequently in most ducks and almost all the time in *Cairina*. If, for example, a goose or swan pair is regularly disturbed during mating by stronger conspecifics (and the latter do not easily give this up), the suppressed pair never uses the moment when they are out of sight of the pond tyrants to perform their copulation quickly. Instead, they always begin their involved preparations anew. The preparations usually last just long enough to attract the attention of the stronger opponents. It apparently never dawns on birds in this situation that speed is necessary in the described circumstances, and that they must quickly use the time when they do not happen to be observed.

(658) I would like to mention briefly that quite distant species of Anatidae well understand the significance of the initiation to mating of each other and get involved in them. Thus, an unmated *Casarca variegata* female approached two male mute swans that considered themselves to be a pair, and she joined vigorously in their neck dipping.

Among the Anatidae, as is also known with many other animals, *the sight of another species member copulating fills them with rage*. This is especially true of geese and swans. A swan is angered if it merely sees the inititative neck dipping. As soon as a pond ruler notices this in two other swans even at a great distance, he storms angrily over to chase them off. I observed that a gander always flew angrily and hurriedly between two members of a goose pair when he heard the call that is emitted during treading. This is not a case of actual jealousy, for it does not even occur to the swan or goose male to court the strange female. I probably do not need to mention specifically that the first-ranking pair of a pond is not disturbed during mating by the others. Anatidae females are evidently not particularly agitated by the sight of the intimacies of others. On the other hand, female golden pheasants, as reported by an experienced breeder, always peck at another female that has just copulated. He termed this very accurately a "*Lex-Heinze*-attitude," [*moral indignation toward prostitution; refers to a proposed law*] and *it is very remarkable that this being angered over the sexual intercourse of others is by no means related to, for example, the human intellect, but rather expresses a natural feeling that is also widespread elsewhere in the animal world.*

Of course, I have often wondered *how one can explain the origin of the earlier described dipping movements that Anatidae perform when they are eager to mate.* In geese and *Casarcas* especially, one frequently notices that the animals really do bring things up from the bottom; with every dipping of the head, branches covered with algae and similar objects are brought up and then dropped. Only when the dipping motions become more violent does the fetching up of objects cease. Now if the Anseriformes under discussion are leading young, one sees them bring food up from the bottom in the same manner. The young then assemble around their parents and eagerly accept the water plants presented before them. One can (659) therefore well imagine that the origin of the head dipping was concerned with an offering of food, even to the mate. This original meaning later was lost. It is precisely those actions that are associated with brood care that are often used by both animals and man as expressions of tenderness toward the opposite sex. I remind readers that most mammals use licking, which originally represented the cleaning of the young by the mother, as an expression of affection. This also applies to the human in love referring to the object of his affection as "baby," "little one," or other diminutive expressions and cooing with the other in baby talk.

Another phylogenetic manner of origin has occurred to me for the ritualized invitation to mate where the neck dipping is included. In the rape attacks, as carried out by the male *Cairina* against all females and by the drakes of most surface-feeding ducks against strange females, the females often try to escape by diving. In *Tadorna*, and somewhat in *C. tadornoides*, the initiation to mate is a rather accurate imitation of these rape chases. The dipping of the head and neck perhaps represents the last reminder of this most primitive circumstance of forced copulation by the male.

I learned to my astonishment *of a beautiful example of how a very utilitarian act can develop into a form of communication*. I had a somewhat nervous goose put into a stall with her downy young. When I approached the animals a short time later, the female, standing in one place, began to trample the way geese do when they stand in shallow water and try to stir up the mud with their feet. They especially perform these movements when they lead young, which immediately assemble around the parents to search for stirred up bits of food. For the goose in the pen, there could be no possibility that she wanted to provide food for the young; the movements served here purely as a means of enticement. Since the children were already close by, there was no purpose in calling them loudly with the enticement tones to come even closer. Once I recognized this enticement trampling, I later observed it frequently in other geese. It especially serves to collect the young very close around the parents.

I would like to cite one more *ritual that has an immediately evident origin*. It is *the expression* (660) *of affection* in a completely different bird group: the *herons and storks*. For example, if one approaches a tame *Ciconia ciconia* or *boyciana*, then it expresses its agitation with the well-known clattering. At first, from this instrumental music, we can draw no conclusions as to how the bird is disposed toward us, but immediately afterwards, the stork either rushes at us in a rage, or it uses the tip of its bill to grab a twig or stem and with it digs the earth under itself as in nest building. Then we know that it has friendly intentions toward us. This expression of friendliness is very understandable; it means originally, "Come, let us build a nest together!" This form of communication is practiced in every season, and, therefore, it bears hardly any relation any more to the real collection of nesting material.

[*Editor's Note:* Material has been omitted at this point.]

I would not like to leave this group without saying a few words about their *psychic behavior*. One can probably say that *Casarcas*, and especially the ruddy shelduck, are birds that learn well by experience. They comprehend relatively early the impenetrability of a wire mesh fence, recognize individual persons, as I discussed earlier, and in general show an active interest in their environment. Since they are physically adroit, they can exploit a locality in many ways. They like to fly up on roofs and other prominent points, sometimes even alighting on the narrow rails of the pond fences, and they visit the lawns, etc. Egyptian geese are usually less ingenious; one can repeatedly capture even flight-capable individuals in front of a wire fence without having them try to fly over it.

As I have already mentioned about psychic behavior in my "Brautente," the behavior toward a fence is no incontestable criterion of intelligence. Thus geese, *Casarca, Chloephaga,* and also Egyptian geese learn very rapidly (662) (swans much more slowly!) that even a superior opponent on the other side of one of these transparent fences is quite harmless, and they tear after it with great heroic courage. The two hostile parties can aggravate each other in this manner for a quarter hour at a time, but as soon as one pair chances to fly over the fence or otherwise enters the space of the other pair that is lower in social rank on the pond, the latter immediately stop bragging and flee with great haste. But they immediately repeat the same performance as soon as they again see a fence between them and their opponents. In the above sentence, I intentionally said "*chances* to fly over," because the obvious idea to cross over the barrier, whether by flying for fully-winged individuals or by crawling through a distant hole that the pinioned birds have known about for a long time, never occurs to the animals. One will claim here that the strong emotion of the birds is the cause of this behavior, and this is indeed the most obvious explanation. But one must keep in mind that strong emotional turmoil of the animals does not cause them to forget the impenetrability of the fence—otherwise the weaker would not tear out after the otherwise feared stronger opponents. However, their intelligence is not sufficient for the much more complex process; that is, to decide to go over or under the fence, whereby the memory of the successful route would have to emerge. They are much too bound to their immediate sensory impressions to do so.

[*Editor's Note:* Material has been omitted at this point.]

(675) **Anatinae**

(With the exception of *Dendrocygna, Alopochen, Tadorna* and *Casarca*)

The surface feeding ducks proper, which we want to compare next, have so many traits in common that biologically they must be distinguished from the genus *Dendrocygna* and the *Casarca* group in the broader sense, with which they have been grouped in the *Anatinae* by Salvadori. However, I also include here the genera *Lampronessa* and *Aex,* which probably do not have much to do with the *Plectropterinae*.

The *wood duck, Lampronessa sponsa,* can be omitted here, since I have already described it in detail in a monograph elsewhere.

But its East Asian relative, the *mandarin duck, Aex galericulata,* shows so many peculiarities that we will have to consider it in some detail. The male, as is well known, is one of the most striking birds that exists, not only in the color, but above all in the form of the feathers. The innermost secondary feather is especially striking: its inner vane is spread into a large light-brown fan that always astonishes the observer. Let it first of all be noted that *this so remarkable feather is completely invisible when the bird flies.* In flight, the bird slips the fan under the shoulder feathers, and one is quite startled by the changed appearance when the animal takes off. One recognizes the mandarin drake in flight best by the large yellow-white patch that runs along each side of the head. Even without seeing that, one can easily distinguish flying wood ducks from mandarins with some practice, when one can see no (676) colors. The wing beat of *Aex* is—except in late fall and winter in very fat individuals—as good as completely silent, and the whole manner of locomotion does not give the characteristically steady impression that we are used to in most ducks. The animals always look as if they feared bumping into something at any moment. Also, the neck appears to be shorter and not stretched out as straight as in the related *Lampronessa.*

One always seems to recognize the crook in the neck quite clearly, but perhaps the crook is only suggested by the crest, which is, after all, quite well-developed even in the female. The flight looks slower than it is, at least I have never observed that mandarin ducks fall behind their North American cousins when they fly together. The *enticement call* of the mandarin female is much shorter than that of the wood duck and sounds somewhat like "hu-ett" or "veck." The flirting call is a "kett" or "ke," with which the animals are very generous. The alarm call is similar to the enticement call, but somewhat sharper. When the birds are searching for a nesting site, thus especially when they are sitting around in the trees, one hears a soft "gegegegeg" from the female that is similar to the corresponding call of the wood duck female, but yet quite distinguishable from it. The voice of the drake shows very little variation and consists only of a short whistling "weeeb." It is altered somewhat according to mood, and in the presence of certain stimuli, one can quite clearly hear a grunting tone when one is close enough. In searching for a nesting site, the "weeeb" is uttered more softly and is repeated over and over.

Along with the highest development of nuptial plumage of all the Anatidae, *Aex galericulata* excels to an equal extent in coquettish behavior. I use this term intentionally, since this trait is innate even in humans and does not proceed from deliberation. Male mandarins display at virtually every opportunity from fall until May, and the females can never desist from provoking these performances. Only when the mates are off by themselves is it rare to see that the drake does a special strut before his female. But as soon as a third individual joins them, especially another male, then the prancing and parading knows no end. Even when several pairs are assembled, a large part of the dusk and daylight hours are spent in this showing off. We then see the drakes lay their heads far back with their splendid crests somewhat erected, so that it almost touches the bright reddish-gold fan feathers that are raised upright with their surfaces parallel to each other. In the process, the beautiful silver-white edged primaries do not lie close against each other as they otherwise do when the wings are closed. Instead, they are somewhat spread and cover the base of the tail toward each side like (677) eight silver stripes. In this position, the birds appear peculiarly short or, to use an equestrian term, reined in tight. Then they jerk the head in a lightning-fast nod, usually emitting the already described "weeb" and point the beak just as suddenly back over the back towards the inner vane of one of the two raised fan feathers. One then also hears the short whistle. If the drakes come too close to each other, they rush at one another. But a fight hardly ever occurs; and instead, the one attacked flees rapidly, but only to immediately turn around again and continue in the courtship drama. As with the mallard drakes, the mandarins also consider it here—to express myself for once in purely human terms—unrefined to swim hurriedly while fleeing. Even when a drake has gotten only a meter or two away from his group, he usually *flies* back to his comrades and, immediately after landing with a jolt, freezes in his imposing courtship display. The bird thus shows here the same arrogant swaggering that I have already discussed in the amorous ganders and exerts this effect by the instantaneous transition from most vigorous motion to complete calm. During the whole drama the females make movements similar to those of the males and accompany them each time with a loud "kett," lifting their elbows somewhat and often making threatening motions toward a strange male, inciting their mate on him. If one observes such a group of mandarins from afar, then one might believe that the birds were all crazy, since their constant bragging, whizzing back and forth and sounding off seem to us completely purposeless and senseless. This impression is strengthened even more when one of the males suddenly takes off out of the water only to land by his comrades just a very short distance away. One thinks one has a group of colored giant water striders *(Hydrometra)* and locusts. Whether the females really prefer the most beautiful and actively courting males I cannot say, but it is nevertheless possible. I have resolved to clip off the fan and the crest sometime from a drake that plays a prominent role among his companions, in order to see if he still displays or shows any change of demeanor, or whether he then plays a less prominent role among his colleagues. But this experiment is not very easy to carry out since our Berlin mandarins are all flight-capable. They are thus almost impossible to catch, except when kept in a flight cage, which would lead us to false conclusions.

Wood ducks and mandarins stick together to a certain degree, but this association is not generally very close. (678) If an excess of mandarin drakes is present, they often go with a wood duck pair, since they are interested in the female. Because I would like to raise hybrids of these species, I have in some cases like this shot the wood duck drake to place a North American duck in the East Asian's unchallenged possession. Unfortunately, I have never had a breeding success from this.

At the actual breeding time, the mandarin drakes are very inclined to rape, somewhat as *A. boscas* does. They then harass not only strange conspecific females, but attempt above all to rape quite specific individuals of the Australian wild duck and of the pygmy duck (a pygmy form of the domestic duck). However, I have never seen them chase wood duck females for this purpose. I also sacrificed the mate of an *A. superciliosa* female that was being severely beleaguered by mandarin drakes. She finally surrendered to one of her assailants and became his mate, but unfortunately, even here, the resulting clutch was infertile. The fertility of *Aex* with other duck forms thus appears to be very slight, in contrast with that of most other *Anatidae* (even in comparison with *Lampronessa*!). Perhaps this may also be in part attributed to the fact that the relatively small and short drakes cannot quite tread the larger *Anas* females.

Mandarins are to a much greater degree twilight and night animals than are wood ducks and probably more so than most other ducks in general. Their large eyes of course attest to this. Since the animals spend a large part of their day hidden under the bushes, they usually enliven a pond only a very little. But once they are in motion, then all their acts take place more quickly and hastily than in their American relatives. In conclusion, let it be mentioned that the *Aex* females unfortunately often destroy a foreign clutch when they find burrows that are already occupied. Wood ducks usually do not do this. In their habits of taking off, copulating, searching for nests, etc., *Aex* resembles completely the *Lampronessa* that was described in detail.

The *mallard (Anas boscas)* and the domestic duck descended from it is of course considered without further qualification to be the archetypical duck. It is therefore advisable to make it the basis for the study of the following forms, especially since even the wild forms can be observed very easily. It occurs completely free in many parks and is unusually tame there. The *voice of the drake* is a quiet, rasping "rab." I do not know if it really sounds different when it serves as an enticement or an alarm call. When the ducks are angry, especially when two males are going at each other, this call is shorter and frequently repeated; (679) then the whole thing sounds like "rabrabrabrab." This "rab," somewhat altered, rather quiet and extended, serves as a call of enticement to the nest when the drake, while searching, believes that he has found a suitable nesting place. This call, likewise audible only at a short distance and accompanied by corresponding head and bill movements, serves as a summons to move along or to fly off. In addition, the drake has at his disposal a loud high whistle, but one only hears it during the courtship ritual or also frequently immediately after copulation. The *voice of the female* is the well-known duck quack: full, loud and very blaring, especially in a confined space. From fall until the actual breeding time—and after that, no longer!—one hears the enticement call, the resounding nasal "quackquackquackquackquack" especially from the unmated females, and this, to be sure, most often in the evening twilight when other conspecifics are passing over. It never seems to be emitted by flying birds. One also hears from the duck, and this throughout the year, an extended "quack," often singly but sometimes also repeated, which represents the warning call or the expression of fear. For this reason, it is most frequently heard when the duck is leading young or is being chased in spring by strange drakes. If she wants to take off, she shows this with a quiet [Editor's note: A printing error in the original caused at least one line to be left out. The following 14 words were reconstructed with the help of K. Heinroth and A. Koehler . . . "queg queg queg queg," and if another "quack" is emitted more loudly, . . .] then the duck is looking for something or intends to walk or swim away. It therefore is an expression of more or less strong unrest. If she is leading young and does not spot anything suspicious, then one hears a very soft quacking that therefore corresponds to the clucking of hens. Like the *Casarcas,* the *Boscas* female also has a special vocalization to incite her male against an annoying conspecific. With a very peculiar head and bill movement, she scolds over her shoulder "queggeggeggegqueggeggeg" to which, in great agitation, a strongly emphasized "quag" is added. If we go from sound to *sign language,* then we find that our birds reveal the intention to march off or swim away by a down to up bill movement, which becomes even stronger and jerkier before taking off in flight. It seems to me that even quite small young understand this sign language of the mother since they often run rapidly in front of the adult duck when the described head movements occur.

If mallards or domestic ducks want to mate they move their heads up and down with the bill held horizontally, not unlike the movement they make when they intend to take off in flight. But instead of the neck becoming longer and longer and the bill tossing getting faster and faster as in intending to take off, the movements of the female gradually proceed to her lying down flat on the water. (680) Immediately thereafter the male mounts. When the copulation is over, he swims rapidly in circles around his mate with his neck greatly outstretched, as we see in the domestic duck pair in Figure 8 of Plate 5, where the female is about to bathe and the dappled male carries out his afterplay.

It is not difficult to *correlate sign language to intention movements* in *Anas boscas.* The sign language is not intentional in the individual animals, but is very well understood by the others. The jerky upward head and neck movements that precede taking off are nothing more than suggested jumps off the water, and the head nodding from up to down in the preparations for mating initiates squatting in the female. If a female intends to incite her husband against an opponent in the described manner, then she threatens the opponent by turning towards her mate, and, by so doing, seeking his protection.

In my "Brautente," I have described the winter courtship and the resulting friendships and engagements of *A. boscas* in such detail that I will not go into it further now.[1] Likewise I refer to the so *strongly developed inclination of the mallard drake to rape alien females at breeding time* that I described there. The knowledge of all these habits described in "Brautente" is absolutely necessary for the understanding of most duck species. The only thing I omitted there is the inciting, which is of less concern in *Lampronessa,* but which is done at every opportunity by the female mallards during fall and winter when another drake appears. In the discussion of *Cairina,* I have of course already mentioned how a repeatedly widowed *A. superciliosa* female had incited all of her husbands against a Muscovy drake that she especially hated, and they attempted to fight with this overwhelming opponent. There is the widely-held assumption that in doing this, the female is sexually interested in the strange male so that she more or less flirts with him. I do not consider this assumption to be valid, since in this case our female immediately disappeared when she was without a mate, and only came back to the Muscovy when she was again paired. In the meantime, she would have nothing to do with him.

[*Editor's Note:* Material has been omitted at this point.]

(687) The actual courtship display in the genus *Dafila* [pintails] is quite similar to that of the mallards; only the flirting with outstretched neck of the female seems to be absent, and instead of the "rab" of the males, a pretty "krick" or "brib" occurs. This is accompanied by head raising and

[1]Plate 5, Figure 7 shows such a winter display of domestic ducks. The wild-form drake raises himself upright in the water and, while he almost touches his bill to his breast, he emits the high whistle. His three white consexuals give the rasping "rab" with their tails slightly raised and heads straight up.

tail erection. Also, the high whistle, during which this bird, like *A. boscas,* stands erect and nearly touches the breast with its bill, is much softer. In addition, the drakes assiduously wait on the females somewhat in the manner of *L. sponsa,* with dainty postures and a very tender, soft "krickkrick" that resembles that of the common teal. They also emit a soft whistling enticement call. The female has a very rough and sometimes surprisingly unpleasant call that sounds like something between the quack and the creaking call of the wigeon. It is somewhat spooky, especially at night. She also has a soft, inciting call "rarrerrer" with accompanying sideways head movements, but in general, these birds are very quiet. They reveal the intention to take off and to copulate with movements that are similar to but less clear than those we have found in *A. boscas.* The ethological similarity between mallards and pintails is enough to suggest a close relationship, and this is brilliantly confirmed by the successful breeding of hybrids between these two species over many generations.

In the Berlin Gardens, where we have raised many free-flying pairs of pintails, one can often observe their soundless, adroit flight. In addition to this, I have found that *the pintail ducks dabble on the bottom, probably the most enthusiastically of all of the surface feeding ducks; certainly their long neck is correlated with this.*

Dafila acuta does not appear to possess good psychic capabilities. The animals simply cannot grasp the separating property of the fence, and the females usually show themselves to be incredibly inept in the choice of nesting sites; they manage to set up their nests on footpaths. If I might conclude from the few cases I have observed, they seek to place their clutches at least several meters higher than the water surface; this would suggest an innate fear of flooding.

I would like to say a few words here about how ducks that breed in the open look for a nesting place. During the early morning hours especially, the pairs walk around the bank margins. At places that appear suitable, they poke about with their bills under the bushes (688) in the grass and leaves lying there. Even the drake participates in this, and if he stays somewhere for a rather long time, the female approaches him in order to inspect his choice. The animals proceed in this way for several days, until finally a nesting depression that the duck has made can be seen. She then lays one egg into it every day. From the last or next to last egg on, she broods. However, as I mentioned in my "Brautente," at each laying, she stays sitting on the eggs for some time, especially when she is laying the last eggs. In the Zoological Gardens, it often happens that the animals place their nests right in heavily used paths, and since the nest searching and laying occur in early morning hours, during which there are very few or no garden visitors, the establishment of the clutch in such places is relatively undisturbed. But as soon as the duck begins to sit on the eggs for longer periods, it very quickly happens that people unwittingly approach the nesting place, and then, of course, the bird hurriedly dashes off every time. These disturbances occur more and more frequently, or, to say it better, on days when constant streams of people flow by her nest within only a few centimeters of the eggs, the female finally gives up her brooding entirely. Since I am aware of this, I of course remove such clutches to protect them from ruin even before they are complete and have them brooded out by a hen. Many songbirds also act like these ducks; as is well known, nests are often started in places where brooding and feeding the young is impossible. But in the early morning hours during which the animals searched for their nooks, disturbances just had never occurred there, and, for that reason, the place seemed to the pair to be safe. So the birds do not give a thought to the fact that they must also consider the other hours of the day during which they are not out searching for a suitable undisturbed site. Our female ducks, many of which had observed for years the immense human traffic that occurs almost all day in spring and summer, nevertheless are again and again fooled by the quiet morning hours and build their nests next to the most heavily beaten tracks!

[*Editor's Note:* Material has been omitted at this point.]

(694) *To sum up, if we consider the vocalizations of the male surface-feeding and diving ducks,* then we find that most of the drakes actually have only a courtship sound that, in a number of species, is also emitted, sometimes altered a bit, in other conditions of great agitation. The courtship call can, at that time, have a different meaning. But frequently, the vocalization in question is used only in courtship; thus the males of teal, garganey, red-crested, rosybill, common pochard, and bufflehead, for example, are completely silent except in matters of love. The females, on the other hand, have enticement calls, warning calls, and sometimes other different calls as well. The most striking exceptions are the drakes of *Anas boscas, superciliosa, and poecilorhyncha;* in these we hear a soft "rab," which is perhaps somewhat differentiated under various conditions, that corresponds to the loud "quack" of the female. This "rab" is in addition to the high whistling that is produced only in courtship in these species and is always associated with quite specific movements. In mandarin drakes and especially in the wood duck drakes, the high whistling voice is also used with some alteration for the expression of various emotions, as I have of course shown earlier in detail in my "Brautente." Among the other surface-feeding ducks, we hear in *Mareca* the whistling of the males throughout the year even in cases where courtship is not necessarily involved. *It is nevertheless quite striking that the males in a systematic bird group with many species have become otherwise silent in order to produce a unique courtship sound.* This has occured especially in cases in which we find no great variety in the females, such as in all diving ducks.

The Significance of the Wing Speculum

In the ducks considered above, the *Casarca* group, and partly also in the *Chloephaga* group, we find a *wing speculum* that is quite magnificently developed. It usually (695) occurs almost to the same extent in both sexes, and the question naturally arises as to what may be the meaning of this structure, which is rather rare in the world of birds. The speculum is usually invisible when the wings are in the rest position, during walking or swimming; it is instead concealed entirely beneath the contour flank feathers. This shiny coloration of the secondaries or their coverts has, in my observation, nothing at all to do with courtship. This shiny coloration does, after all, occur in many species that have absolutely no courtship through the display of nuptial color. In the diving duck group, the speculum lacks the metallic sheen, there replaced by white coloration, which frequently also extends to the primaries. In the geese in the narrower sense and in the swans it is absent, and in *Sarcidiornis, Plectropterus,* and *Cairina,* the entire upper side of the bird more or less shines, so that consequently the speculum does not appear to especially stand out from the other coloration.

It strikes us that true geese, which totally lack a speculum, frequently show a bright carpal joint in flight. But they always show a shiny coloration at the end of their bodies because of the white tail coverts or brightly colored parts of the tail feathers. We are no doubt justified in conceiving of this white as a beacon for the conspecific flying behind as is precisely the case after all for many mammals—the deer, for example. The swans, with one exception, are entirely white or at least mostly so, and even in the black swan, the white flight feathers are visible in flight; so even it too would still be visible to its conspecifics even on a rather dark night. Anatidae that have a metallic or white wing speculum do not possess the beacon of the geese and swans. We must assume, therefore, that the speculum has taken its place: *the speculum in flight therefore constitutes a means of directional aid and enticement for the bird following behind.* In the diving ducks, which fly with extremely rapid wingbeats, a shiny metallic reflection would probably be ineffective; it is there replaced by white that remains altogether quite visible even during the uncommonly rapid motions of the wing tips. The shimmering speculum coloration is limited with a very few exceptions to the secondaries, which, of course, are not flapped nearly as much in flight, so that their image is not blurred by the movement. Only in the case of *Nettium brasiliense* does the shimmering color extend considerably onto the primary part of the wing, but the wing itself is strikingly rounded compared to that of related forms. I unfortunately do not know the flight pattern of this species; I presume, however, that it moves with a slower wingbeat than is customary for the other relatives. Then, therefore, the shimmering color would remain well visible even on the peripheral areas. (696) The shiny white contrasting coloration of the small wing feathers of *Chloephaga* and the *Casarca* group can no doubt be interpreted more as a threat color, since there the carpal joints constitute the main weapon. Moreover, these birds also seem to recognize each other in flight by these white patches.

Of the *Dendrocygna* species I am more familiar with, *autumnalis* and *discolor* have a broad, white speculum; *arborea* a clearly silver-grey one; and in the cases of *arcuata, fulva,* and *eytoni,* one sees in flight a shiny band going around the rear end of the body produced by the flank feathers and upper tail coverts. *D. viduata* has its contrasting white head, and in the case of *D. javanica, which has no bright colors whatsoever, an acoustic enticement device takes the place of an optical one:* the inner vane of the outermost primary in this species bears a very peculiar tongue-like projection that causes a whistling with the wingbeats. I wish other observers would check these assertions of mine for the remaining Anseriformes that have not yet been available to me for study!

I will show by the following experience *that the shimmering colors in fact have a certain attraction for our birds.* Once, by mistake, a number of wild-form domestic ducks and mallards were confined in a pen without water. When I inspected the animals, which by then had become very thirsty, I noticed that the ducks were huddled together in one corner because of fear of me, and *they were touching the shiny head and neck feathers of the drakes with their bills and performing drinking movements:* the glistening surface therefore to a certain extent gave them the illusion of water. Let it be noted, by the way, that thirsty ducks, like birds in general, attempt to drink from all shiny reflective things. For example, they attempt again and again to get water from a varnished floor, and it is of course also well known that grebes, as well as water beetles, frequently take glass roofs for ponds: evidence for the fact that these birds and insects perceive water by sight rather than by smell. Now, however, for our ducks, water plays a very vital role in their lives, and it is clear that for a swimming bird, the illusion of this element on the feathers of a conspecific must be the most attractive means of enticement imaginable. This is, of course, only my supposition, but perhaps it will stimulate many a reader to contemplation and contradiction. Through the observation of two rather distantly related species of fruit-eating pigeons, *Carpophaga rubicera* and *Ptilopus insolitus,* which both have a shiny red structure on top of the base of the beak that looks exactly like the berries that one finds (697) in their gizzards, I surmise that certain colors and structures that at first seem to us to be no less than impractical perhaps imitate vital things for those birds, and then are used for communication among them. I have already expressed myself on this matter in my "Ornithologische Ergebnisse der ersten Deutschen Südsee-Expedition [Ornithological Results of the First German South Sea Expedition]" in the *Journal für Ornithologie,* 1902: 413-414, and would like to term this phenomenon *enticement mimicry* [*Lockmimicry*].

[*Editor's Note:* Material has been omitted at this point.]

(698) Relationships Between the Size of the Testicles and Polygamy, Nuptial Plumage, etc.

If one shoots a mallard drake in April and opens the body cavity, the immense size of the testes is (699) astonishing, and one immediately understands the hunger for copulation with which the bird throws himself onto any female that he catches sight of at this time. But it is considerably more difficult to understand why his own wife needs to entreat him to copulate for such a long time, sometimes even without success. The anatomical findings in many other Anatinae are similar to those in *A. boscas,* at least as far as I have been able to collect data in this respect; the same is true in *Cairina moschata.* In all these species, we see a more or less great sexual desire in the male before the brooding period correlating with the growth of the gonads. On the other hand, we see the marriage-like fall and winter matings that have nothing to do with reproduction in these very species and in no others. One can therefore well imagine that we are dealing here with a remnant of that same sensuality that is very great in the spring, and that consequently is entirely dormant only during the molt. Unfortunately, this assumption can in no way be justified anatomically; outside the breeding period, approximately until the development of the urge to rape, ovary and testes are just as regressed as in other birds in the corresponding seasons. As is well known, in those seasons, one cannot even distinguish the tiny gonads of the mature animals from those of young, sexually immature individuals. It is difficult to understand how these ducks can find enjoyment in mating while they have completely inactive gonads that have scarcely one hundredth of the weight of the fully capable ones.

In geese in the broadest sense, in swans, and in the *Casarca* group, I have never found very large testicles; they do not mate as often as ducks before the brooding period, and they always mate only with their own female after introductory preliminaries. Occasional fall and winter matings occur, and also half-year-old female swans sometimes with full abandon allow themselves to be mounted, but that is rather the exception. In all these forms, the male takes part in the brood care; he is therefore not distracted from his duties as a family father by gigantic swelling of the testes. I would like to regard the behavior of the Anatinae to be the more primitive: the male attempts to pass on his traits as often and as generously as possible. The duck stays true to her mate, since her opportunity for passing on traits is not increased by the association with many drakes. On the contrary, she derives advantages from living together with her mate since he defends her, at least to a certain extent, from ardent rivals, and he sees to her security especially when she must be almost constantly out searching for food in order to make eggs (the clutch produced in eleven days almost approaches the weight of the mother bird!). (700) Selection for male brood care came about only in exceptional cases *(Mareca sibilatrix).*

The defense of eggs and young by the male evidently became more important only in the larger forms, which were therefore more capable of defense, and therefore we find it especially in swans, geese, and *Casarcas.* At the same time, because of better brood care, a decrease in clutch size could occur without jeopardizing the existence of the species. Instead of an average of eleven eggs, we usually find only five or six eggs in these larger birds.

Now, does the presence of the nuptial plumage have a correlation with the size of the testicles? In fact, we find the decorative colors and forms in the diving and surface-feeding ducks, in which the male performs no brood care, while these decorations are absent in geese and swans. One might therefore conceive of a connection between great reproductive capability and nuptial *plumage.*[1] In individual cases, however, this is often not the case because one cannot discover a difference in the sexual desire between the female colored males of *A. superciliosa* and *A. poecilorhyncha* and the magnificent mallard males. But as I have already mentioned, it seems evident that tropical ducks, for some reason unknown to us, can produce no nuptial plumage, or at least not of a sort that alternates with a dull eclipse plumage. It is nevertheless certain that the most striking forms, such as *Lampronessa* and *Aex,* do not have the largest testes. I would therefore prefer that this question remain open. But this much seems certain to me: *strictly monogamous males; that is, brood-tending males, are always quite similar to or at least differ only very little in feather color and form from the females,* and the voices of the two sexes are very often also alike. *The reverse of this statement, however, is not legitimate.* I question whether one can consider the sexual dimorphism of the *Chloephaga* species here; probably in these cases we are dealing not with nuptial plumage proper; thus in *Chl. inornata* and *magellanica,* the black foot color of the male is more primitive than the yellow of the female. In addition, a gander, even when it is only a few weeks old, already has the black and white vestments that it keeps from then on. Perhaps we are dealing here with a scare coloration for nest defense or something similar. (701)

In order to warn against false generalizations, I would like to point out that peacocks and pheasants *(Pavo, Phasianus, Chrysolophus)* have quite small testicles even at mating time. These gallinaceous birds copulate rarely, and the nuptial plumage, as well as the incessant and so striking courtship dance, suggest to us a sexual capability that seems to be much greater than it actually is. The domestic rooster has immensely developed testes, which might be a phenomenon of domestication; I know nothing of the behavior of the wild form. Unfortunately, there are no investigations of those gallinaceous species in which the males tend the brood, are monogamous, and almost always lack decorative plumage, such as *Perdix, Caccabis, Numida* and others. Also, in the world of small birds around us, for most forms, we know neither the size of the testes in reproductive state nor the behavior of the males toward strange females. The gigantic development of the spring testicles of *Passer domesticus* is known, to be sure, but apparently no one has yet gone to the trouble of observing in detail whether male sparrows ever display before their mates, or whether they always only pursue strange females as drakes do. I myself have never observed them copulate after a courtship display, and in the married pair,

[1]Today, a direct anatomical relationship probably no longer exists, as is evident from the fact that mallards and wigeons that are castrated in the spring assume their eclipse plumage toward summer and their nuptial plumage toward fall exactly as their reproductively capable companions do. (H. Poll: "Zur Lehre von den sekundären Sexualcharakteren [On the Theory of Secondary Sex Characteristics]." *Sitzungsber. d. Ges. Naturf. Freunde,* Berlin, 1909, p. 344.

the female always appears to me to play the inviting role. I have been able to learn even less about how things stand with the related *P. montanus,* in which the sexes are identically colored. I believe that such questions are more important for our understanding of nature now than the new description of a 57th sparrow species!

My biological comments throw much light on the systematic relationships of the forms discussed here. Unfortunately, however, some quite important groups are lacking here which are as yet absent on our park ponds. But one can indeed hope that at least some of these in part very interesting genera will in time reach us alive. The arctic sea diving ducks, with few exceptions, form the only group that does so poorly in captivity (i.e. in fresh water) that we will most likely have to do without ethological observations on them in the long run. *I think that in voice, manner of interaction, and similar things, we often have a very good basis for the degree of relationship of species, genera, and subfamilies:* such things, which are not exactly essential for the survival of the forms in question, certainly have been better able to remain unchanged in many cases than external and internal features, which are important for our system of identification, but are caught up in eternal flux in the struggle with the external world.

I am convinced that the *Anseriformes* would be good subjects for working on the systematic relationships of a (702) bird family in detail. It is above all necessary to compare the downy young of all the forms and to collect the tracheal bullas. In doing this, one ought to consider certain species where both sexes have the same vocal talent to see if this organ is also found in the females, as is in fact the case with *Dendrocygna.* Even the bill and leg color certainly gives many a good hint of the lesser or greater differentiation of this or that anatid; the very fact that almost all downy young have black bills and feet is food for thought.

If I have succeeded in convincing the reader with my ethological studies that there is still, in the world of animals and birds around us, an unbelievable amount to observe that is absolutely necessary for an understanding of their way of life, their coloration, and their form, then I have accomplished my purpose. In systematics, that is, to the extent that it is concerned with the description of species, and in faunistics, much has been achieved in the last decades, and nice biological observations have been collected, but they must be made and evaluated from guiding perspectives. Let us guard against giving ornithology the reputation of dreary accumulation of things and uncoordinated specialization!

In conclusion, a word to the psychologists. We know from many excellent papers about the manifold and compounded instincts with which bees and ants achieve their highly organized community life. Whoever wants to fathom the psyche of warm-blooded animals, and especially the birds, will see that even here most things are innate and are therefore characteristic of all individuals of the species in question, even if mental associations play a substantial role. In this work, I have drawn special attention to forms of interaction, which, as far as social birds are concerned, are often quite astonishingly human, especially when the family (that is, a father, mother, and children) constitutes such a long lasting and closeknit group, as in the geese. *The Sauropsidian group has developed here emotional states, customs, and motives that are quite similar to those in us people, and that we customarily consider to be meritorious, moral, and to have developed from understanding.* The study of ethology of higher animals—unfortunately still a very poorly tilled field—will bring us ever closer to the realization that in our behavior towards family and strangers, in courtship and similar things, we are dealing with much more primitive and purely innate processes than we generally believe.

[*Editor's Note:* Plates 1-5 follow.]

Explanation of Plate 1

Figures 1-13. Mute swans, *Cygnus olor*

Figure 1. At rest: short thick neck.

Figure 2. Shaking: neck outstretched, bill tip directed straight up.

Figure 3. Flapping its wings (tip of left wing has been amputated): note the straight neck.

Figures 4 and 5. Aggressive posture.

Figure 6. Mute swan carried under arm: note that almost the same neck position results as in Figures 7-9.

Figures 7 and 8. Mute swans taking off.

Figure 9. Mute swan landing.

Figure 10. Initiation of mating. To the right: expression of affection.

Figure 11. Initiation of mating. One bird plunges its head into the water over the body of the other.

Figure 12. Copulation.

Figure 13. Completion of the copulatory afterplay (male on the right).

Figure 14. 2½ year old hybrid of a male black swan and a female mute swan.

phot. Heinroth.

Lichtdruck von Albert Frisch, Berlin W.

Explanation of Plate 2

Figure 1. Nine-month old black swans, the two in front preparing to take off.

Figure 2. Black swan pair with young. Note the wing attitude of the male in the left front.

Figure 3. Trumpeter swan *(C. buccinator)* in rest position. Compare the quite different neck shape of *C. olor*, Plate 1, Figure 1.

Figure 4. Dwarf swans *(C. bewicki)* asleep: as in all anatids, the bill is always stuck under the shoulder feathers on the opposite side of the body from the supporting leg.

Figure 5. In the front, singing swan; left rear, dwarf swan *(C. cygnus)*, preparing to fly.

Figure 6. Dwarf swan shaking.

Figures 7 and 8. Trumpeter swan wingflapping.

Figure 9. Trumpeter swans taking off. Compare these last four pictures with the mute swans in Plate 1, Figures 2, 3, 7, 8, and 9.

Figure 10. Coscoroba: wing position of arousal.

1 2 3 4 5 6 7 8 9 10

phot. Heinroth.

Lichtdruck von Albert Frisch, Berlin W.

Explanation of Plate 3

Figure 1. Black necked swans: female on the nest in defensive position, male screaming in excitement.

Figure 2. Male black necked swan attacking.

Figures 3 and 4. Black necked swan pair emitting the triumph call.

Figure 5. Male Andean goose *(Chloephaga inornata)* in attack stance.

Figure 6. Swan goose, breeding stock of *Cygnopsis cygnoides*, after mating. Male, right; female that has just been tread, left.

Figure 7. Female bean goose *(Anser fabalis)* slipping away from the enemy while brooding.

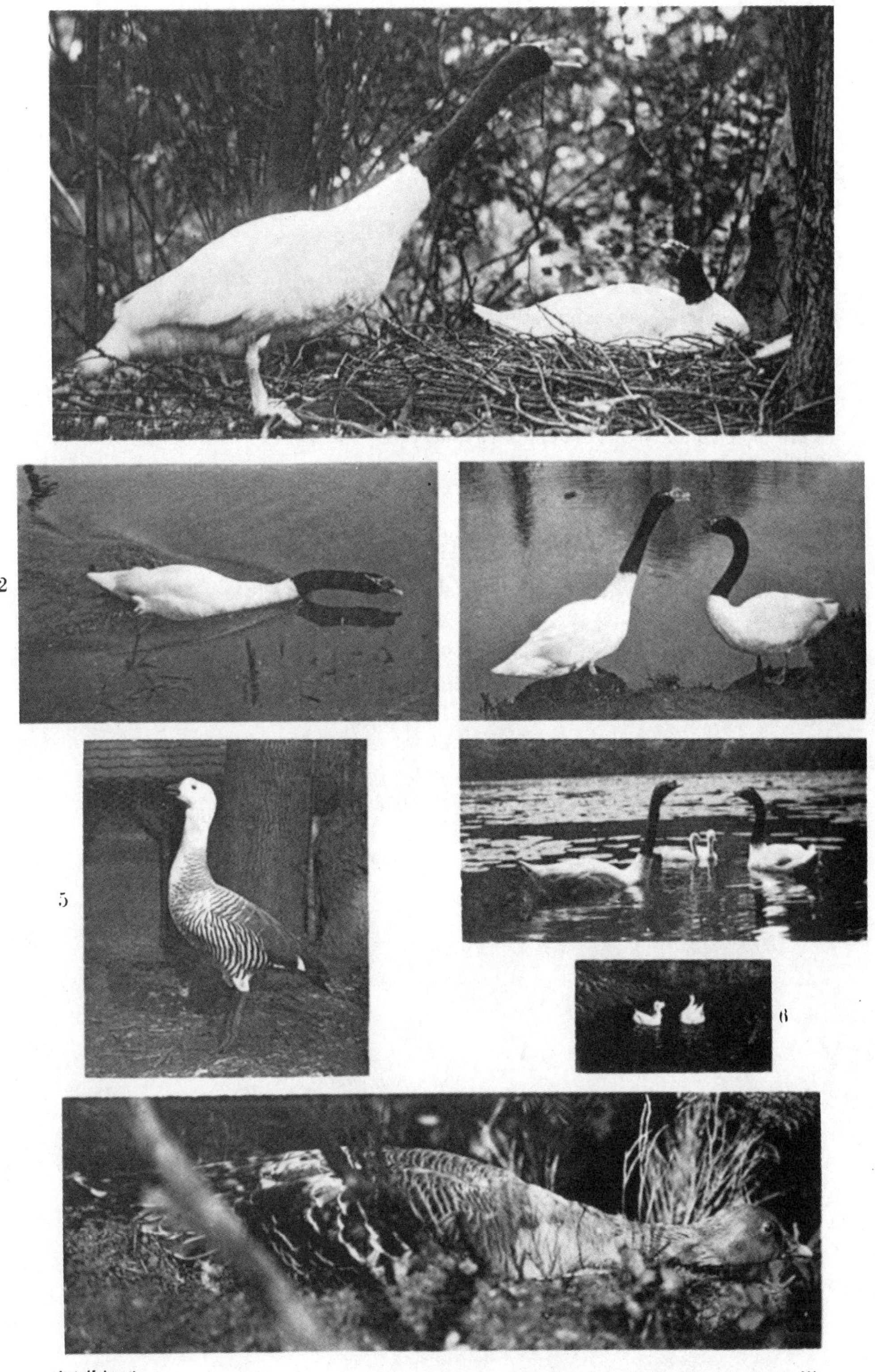

1 2 3 4 5 6 7

phot. Heinroth. Lichtdruck von Albert Frisch, Berlin W.

Explanation of Plate 4

Figure 1. Pair of Egyptian geese *(Alopochen aegyptiacus)*. Male left, female right.

Figure 2. Egyptian geese copulating.

Figure 3. Egyptian geese after copulation. The female that has just been tread is to the right; the male, seen directly from the front, is left of her with his elbows upraised.

Figure 4. Hybrids (male, male) from a male ruddy shelduck and a female Egyptian goose.

Figure 5. Hybrids (male, female) of a common shelduck and an Egyptian goose.

Figure 6. A pair of ruddy shelducks *(Casarca casarca)*, the male to the left. (The same sleeping posture as seen in Plate 2, Figure 4.)

Figure 7. Male ruddy shelduck in arousal posture.

Figure 8. Australian shelduck *(C. tadornoides)*. Male right, female left. (Wing Knobs partially protruding from the wing coverts.)

Figure 9. Hybrids of an Australian shelduck and a common shelduck, female in front.

Figure 10. Pair of common shelducks *(Tadorna tadorna)*.

Figure 11 and 12. Muscovies *(Cairina moschata)*, domesticated form. Both pictures show the manner of movements of the head and neck in arousal.

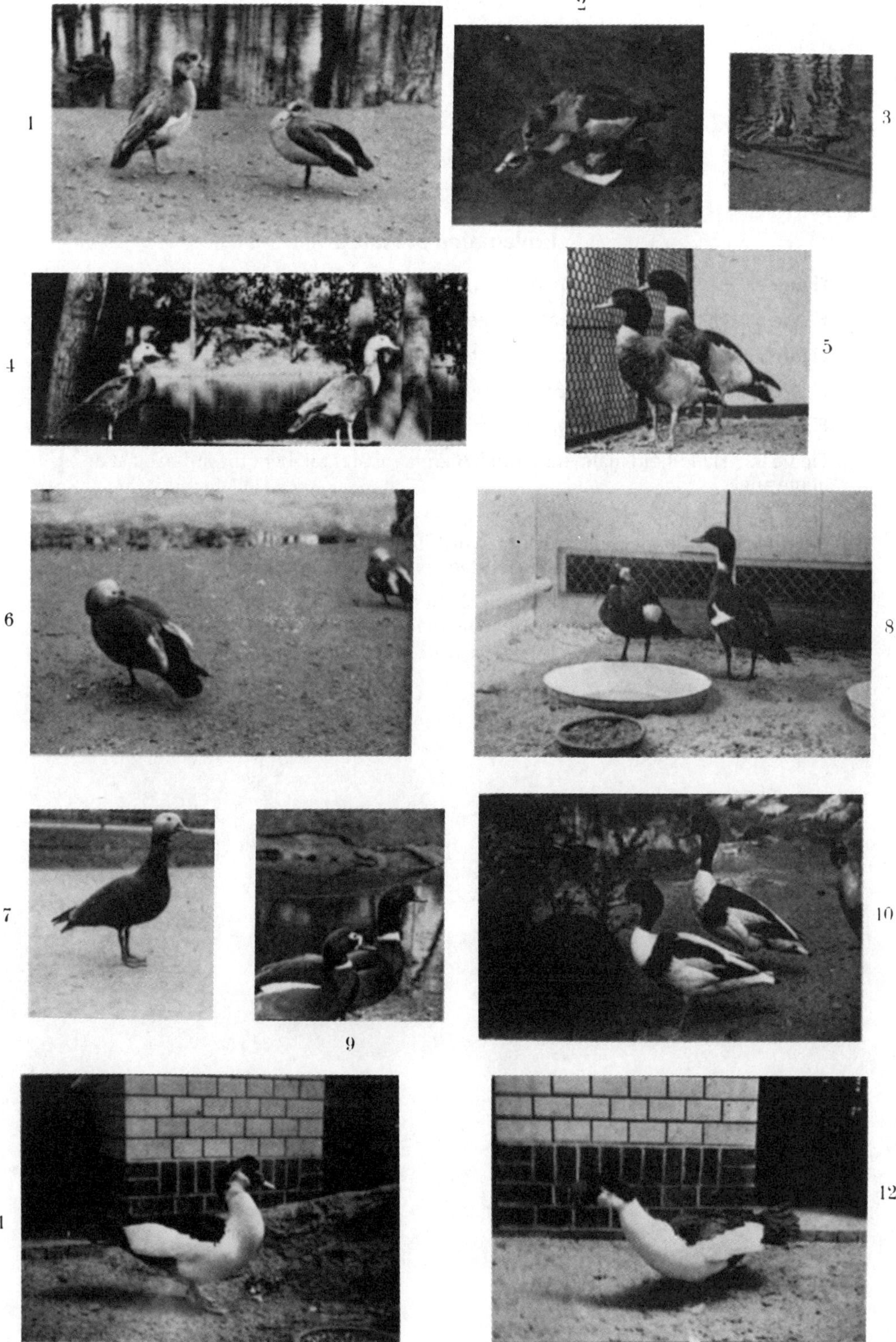

phot. Heinroth.

Lichtdruck von Albert Frisch, Berlin W.

Explanation of Plate 5

Figure 1. *Dendrocygna viduata,* 2 days old.

Figure 2. *Dendrocygna viduata,* 6 weeks old.

Figure 3. *Dendrocygna viduata,* adult.

Figure 4. Male musk duck, *Biziura lobata,* aroused.

Figure 5. Male musk duck, walking.

Figure 6. Hand-held male musk duck *Biziura lobata:* tail bent toward back at an acute angle.

Figure 7. Courtship dance of a flock of domestic ducks courting: the wild-type drake with erect body is emitting the whistling sound; the three white drakes are raising the rump plumage with the curled feathers.

Figure 8. Copulatory afterplay of the domestic duck: the male, with his neck outstretched, is circling the white female that has just been tread and is preparing to bathe.

1

2

3

4

5

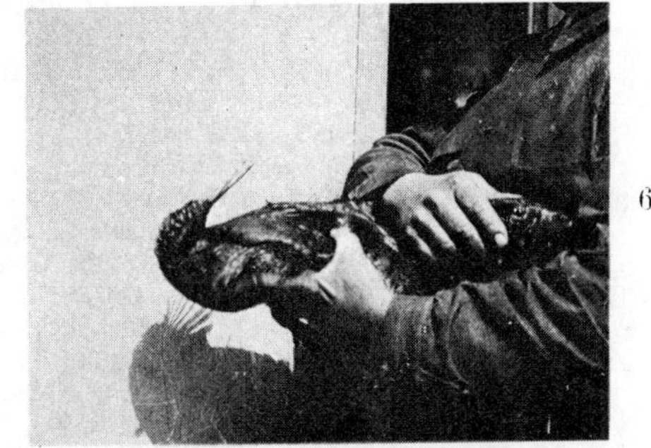

6

7

8

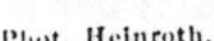

Phot. Heinroth.

Lichtdruck von Albert Frisch, Berlin W.

15

Copyright © 1914 by the Zoological Society of London

Reprinted from pages 491-517, 522-528, and 560-562 of *Zool. Soc. London Proc.*
35:491-562 (1914)

The Courtship-habits * of the Great Crested Grebe (*Podiceps cristatus*); with an addition to the Theory of Sexual Selection. By Julian S. Huxley, B.A., Professor of Biology in the Rice Institute, Houston, Texas †.

[Received March 14, 1914: Read April 21, 1914.]

(Plates I. & II.‡)

Index.

---

* It was not until this paper was in print that I realized that the word *Courtship* is perhaps misleading as applied to the incidents here recorded. While Courtship should, strictly speaking, denote only ***ante-nuptial*** behaviour, it may readily be extended to include any behaviour by which an organism of one sex seeks to "win over" one of the opposite sex. It will be seen that the behaviour of the Grebe cannot be included under this. "Love-habits" would be a better term in some ways; for the present, however, it is sufficient to point out the inadequacy of the present biological terminology.

† Communicated by the Secretary.

‡ For explanation of the Plates see p. 561.

PART I.

1. Introduction.

In these days the camera almost monopolizes the time and attention of those who take an interest in the life of birds. It has rendered splendid service, but I believe that it has almost exhausted its first field. At the present moment both zoology and photography would profit if naturalists for a little time would drop the camera in favour of the field-glass and the note-book. For the many who do not care about using a telescope, the prismatic binocular has more than doubled the possibilities of field-observation; and when full advantage shall have been taken of those possibilities, not only will science be the richer for a multitude of facts, but then, and only then, will the photographer, now hard-pressed for new subjects, suddenly find a number of fresh avenues opened up to him.

This second paper on the courtship-habits of British birds, like the first, will, I hope, help to show what wealth of interesting things still lie hidden in and about the breeding places of familiar birds. A good glass, a note-book, some patience, and a spare fortnight in the spring—with these I not only managed to discover many unknown facts about the Crested Grebe, but also had one of the pleasantest of holidays. "Go thou and do likewise."

I shall first give a connected account of my own and others' observations, followed by a discussion; and in a second part or appendix I shall give in detail some of the material worked up in the first part, as well as some notes on various points not connected with the main subject of the paper.

2. Appearance.

Structure first, function afterwards: I must describe something of the bird's appearance before attempting to give an account of its habits, though I shall try to be as brief as possible, since any standard work of descriptive ornithology will give full details of the plumage and taxonomic characters. The Great Crested Grebe, then, is of course a water-bird, and essentially a diving-bird. Its tail is remarkable in being reduced to a few tiny feathers, and its legs are set as far back as possible, so as to have the position of a ship's propellers. Its body is long and approaches the cylindrical; the neck is very long and flexible, the head flat, the beak sharp, long, and powerful. In colour, the Great Crested Grebe has back and flanks of much the same smoky mottled brown as its small cousin, the Dabchick; the underparts, however, including the chin, throat, and front of the neck, are of an exquisitely pure white (furnishing the "Grebe" of commerce). The back of the neck is very dark brown.

The chief ornament of the bird, the crest from which it takes its name, is reserved for the head. In these pages I shall use the word *crest* to denote *all the erectile feathers of the head taken together*. The crest, as thus defined, consists of two parts—the *ear-tufts* (or *ears*, as for brevity's sake they may be called) and the *ruff*.

Both are composed of special narrow, elongated feathers, stiff, and formed of comparatively few barbs. Those constituting the ears are black, all of about the same length, and spring in two tufts from the top of the head, above the tympanum. The ruff is bigger and more elaborate: it consists of a broad band of feathers springing from the sides of the face and head, their free ends pointing downwards and backwards on either side of the neck. If we take the part of the head behind the eye, we find that at first the feathers are of the ordinary length, then slightly elongated (the beginning of the ruff), and then longer and longer till we get to the hinder border of the ruff. Corresponding to the increase of length there is a change in colour. The proximal (upper) part of the ruff is white, then we get to vivid chestnut, and this deepens gradually to glossy black (see any good picture of the Crested Grebe).

Both ruff and ears are extremely erectile; and as the birds make great play with them during all the actions of courtship, the various positions into which they can be put must be described.

Let us begin with the ears. These, when depressed or shut, stretch straight out backwards, continuing the line of the flat head's crown. When shut forcibly, the feathers of which they are made are close together and all parallel.

Often, however, they are not thus "at attention," but "standing at ease," to use a military metaphor: then the tufts as a whole point in the same direction, but their component feathers diverge and bristle-out a bit. This seems to be the usual and most restful condition.

Further, the tufts may be erected: and they may be erected in two ways—either *laterally* or *vertically*. When erected laterally, they stick out horizontally at right angles to the head, so that from the sides they can scarcely be seen, as they are end-on to the eyes. When erected vertically, they seem, when viewed from the side, to be sticking straight upwards; but when they are seen from in front, it is found that they diverge from each other at a considerable angle (Pls. I. & II. figs. 4, 5, 11). During erection, the individual feathers always diverge fanwise very considerably. Thus there are four conditions of the ear-tufts to be distinguished.

They may be:—(a) *Depressed.*
(1) *Shut tight.*
(2) *At rest* (relaxed).

(b) *Erected.*
(3) *Vertically.*
(4) *Laterally.*

The ruff is more complex in its attitudes, as in its structure. During depression it, too, may be either shut tight or lying easy. When really shut, it bears from the side a curious resemblance to the gill-covers of some eel-like fish: its rounded hinder border lies along the side of the cylindrical neck, whose outlines its own scarcely overlap, either dorsally or ventrally (fig. 1). When relaxed (at rest), this resemblance disappears, for the feathers all diverge slightly, and the smooth appearance of the surface is lost.

When the ruff is erected, the feathers composing it may be made to diverge in a single plane only, the original (longitudinal-vertical) plane of the "gill-cover," or they may diverge outwards as well, making an angle with the side of the head (movement in the transverse, as well as the sagittal plane). I do not think that they are ever moved in the transverse plane alone. As a result of these movements, three chief forms can be taken on by the ruff. First there is the *curtain form*, in which motion in the vertical plane alone takes place, the ventral edge being brought forward till it makes an acute angle with the line of the chin (fig. 3), the two halves thus hanging like curtains on either side of the head. Then there is the *pear-shaped* condition (figs. 4, 11), where there is a considerable amount of forward and a moderate amount of transverse motion. The ruff in this state has its vertical height greater than its breadth (fig. 5). Owing to the transverse bristling of the feathers, the two halves of the ruff almost blend into a single whole; they can scarcely be distinguished either from the front or, still less, from behind, whereas in the curtain form they are very distinct. Finally, there is the *elliptical form*, when, added to the same amount of longitudinal motion, the greatest possible amount of transverse bristling has taken place. The ruff is now actually broader than it is high (fig. 9), and the blending of the two halves is practically complete. There are, of course, intermediate states. Instead of "full pear-shaped," you may have "half pear-shaped"; and between pear-shaped and elliptical there comes the *circular.*

The three I have named, however, are those which the bird usually adopts.

The ruff, therefore, may be:—

(a) *Depressed.*
 (1) *Shut tight* ("gill-cover").
 (2) *At rest* (relaxed).

(b) *Erected.*
 (3) *Curtain-like* (motion of feathers in one plane).
 (4) *Pear-shaped* } (motion of feathers in two planes).
 (5) *Elliptical* }

By a combination of particular positions of ruff, ears, and neck, and sometimes wings and body too, the birds can assume a number of characteristic and often-recurring *attitudes*, which are the raw materials, so to speak, of all the elaborate *habits of courtship*.

Before giving any more definitions, I will now give an outline of the Grebe's annual history, and then go on to describe some of the actual happenings that I saw, in order to give an idea of the problems to be solved. Then I shall try to define and classify the various courtship-habits, and discuss the general bearing of the facts.

3. Annual History.

This is somewhat as follows *. About the first week of February they leave the sea-coast and fly back in bands to the inland waters where they breed. They live in flocks for about three weeks, and then start pairing-up. Pairing-up lasts altogether about a fortnight, bringing us to mid-March. From this time on to the end of summer, the unit is neither the flock nor the individual, but the family, represented at first by the pair. About the beginning of April nest-building begins, and by the end of the month every nest will have eggs. The family parties live together through the summer, though apparently the cock leaves the hen to look after the young when they are half-grown (Pycraft, '11). There is usually no second brood unless the first is destroyed. At the end of September they gather into flocks again, and live thus for well over a month, finally leaving for the sea-coast in the second week of November.

The period of pairing-up itself I have unfortunately not been able to observe; the keeper tells me that there is much flying and chasing about. The part played by the "courtship" in the actual pairing-up is thus left uncertain. From analogy with other birds and with ourselves we should expect that the chasing was the expression of felt but unreasoned likes and dislikes, and that the courtship-actions were only gone through *after* the two birds had become fairly well-disposed towards each other. The courtship-

* The dates refer to the movements of the birds at Tring Reservoir, and have been given me by the head-keeper there.

actions, I am told, are at any rate to be seen immediately after pairing-up.

4. Some Descriptions.

(*a*) Let us start with the commonest of all the scenes of courtship—the one which had first attracted and puzzled me years ago, and led me to choose the Grebe as a bird to watch.

As the birds ride on the water, very little of their under-surface is usually visible; but now and then a twinkle of white is seen. This may be merely a bird rolling half over to preen its belly; but if it proceed from two birds close together, this form of courtship is almost sure to be in progress. In such a case, the glass reveals that the two birds are always a pair, cock and hen; they are facing each other, their beaks perhaps a foot, perhaps a mere couple of inches apart, their necks held up perfectly straight and elongated to a truly surprising extent. It is this holding up of the neck that shows some of the white of throat and breast. Their ears are erected vertically and their ruffs are full pear-shaped. The few little feathers that do duty for a tail are cocked up as far as they will go—that is to say, about half an inch (fig. 11).

In this attitude the birds proceed to go through a curious set ritual.

Let us describe a particular case. A pair of birds, cock and hen, that had been fishing not far apart, suddenly approached each other, raising their necks and ruffs as they did so, till by the time they had got face to face they were in the attitude I have just described. Then they both began shaking their heads at each other in a peculiar and formal-looking manner. Each bird began by waggling its head violently from side to side, some four or five times in quick succession, like a man nodding emphatic dissent. Then the quick side-to-side motion gave place to a slow one, and the beak and head were swung slowly across and back, with a seemingly vague and enquiring action, as if the bird were searching the horizon for it knew not what. The head was moved back and forth perhaps a couple of times, and then the violent shaking began again. This alternation of shaking and slow side-to-side swinging was repeated over and over again by each bird: strangely enough, the pair kept no time with each other—the violent shakings of the two neither coincided nor alternated, but each shook and swung without any apparent reference to the other's rhythm.

After six or seven repetitions of the performance another action came in. After the slow swing and before the waggling (or sometimes, I think, taking the place of the slow swinging), but not every time, the bird bent its neck right back and down as if to preen its wings, put its beak under some of the wing-feathers near the tail, raised them an inch or so, let them fall, and brought its head swiftly back into position for another of the violent shakings. This action had obviously something to do with preening, but had an extraordinary look, as of a stereotyped and meaningless relic. The birds seemed to be performing some routine-action absent-mindedly and by mere force of association, as one may sometimes see a man wind up his watch in the day-time, just because he has been changing his waistcoat.

Finally, after each bird had given about a dozen or fifteen violent shakes, with a corresponding number of slow swings and liftings of wings in between, they veered up into the wind almost simultaneously, lowered their crests, brought their necks down, and, in a word, became normal once more both in appearance and behaviour.

One must have names for things if one is going to discuss them, as any philosopher will tell you. So I propose to call a whole performance such as that just described *a bout of shaking*; each little time of violent waggling I shall call simply *a shake*, and shall measure the length of a bout by the number of shakes in it (counting the shakes of both birds added together). Finally, the curious actions resembling preening I shall call *habit-preening*, because I believe them to have (or rather, to have had) something to do with real preening, but, in the performance as gone through to-day, to have become a mere habit, vestigial so far as its original function is concerned.

(*b*) That is one little scene: now take another.

A solitary bird, which proved to be a hen, came flying over from one reservoir to another. She alighted near one shore and began swimming slowly across towards the other, meanwhile alternating between two attitudes.

First of all she arched her neck right forward till the bill, which pointed slightly downwards, was just above the water. The ears meanwhile were scarcely erected; the ruff was thrown forward in curtain-form, and thus, since the head had been brought so far forwards and downwards, actually swept the water on either side (fig. 3). So she progressed, looking from side to side, and now and then giving a short barking call. After four or five of these calls, which represented perhaps 20 or 30 seconds of time, she put down her ruff and raised her neck nearly straight up to enlarge her circle of vision.

After some seconds of looking about her in this position, she relapsed again into the first attitude—"with neck outstretched, you fancy how." This was repeated eight or nine times, till at last a cock, thirty or forty yards away, appeared to notice the calling bird. He pricked up his neck, looked towards her for a short time, and then dived. At this she changed her whole demeanour. Up went her wings: back between them, with erected ruff and ears, went her head. A glance at fig. 7 will show her attitude. The wings were brought up, half-spread on either side of the body, with their anterior border pointing downwards. They were almost in the transverse plane, but sloped slightly backwards from the water.

In this position the beautiful white bar formed by the marginal wing-coverts, along the anterior margin of the wing, and the broad white blaze formed by the secondaries, which are quite invisible when the wings are closed, shone out vividly. The gap between the wings was filled by the head; this from the front somewhat resembled an old-fashioned picture of the sun, with the ruff rayed out considerably all round, and the ears were erected laterally so as to fit on to the top of the ruff on either side. Below the head shone the white of the puffed-out breast. The bird's whole appearance was wonderfully striking, and as unlike as possible to that of its everyday self.

All this took but an instant; directly the cock had dived she was in this attitude. As she waited for his re-appearance she turned eagerly from side to side, swinging nearly to the right-about and back again as if not to miss him. Eventually he came up, three or four feet on the far side, and facing away from her in the most amazing attitude. I could scarcely believe my eyes. He seemed to grow out of the water. First his head, the ruff nearly circular, the beak pointing down along the neck in a stiff and peculiar manner; then the neck, quite straight and vertical; then the body, straight and vertical too; until finally the whole bird, save for a few inches, was standing erect in the water, and reminding me of nothing so much as the hypnotized phantom of a rather slender Penguin.

As I say, it grew out of the water, and as it grew it gradually revolved on its long axis until at its fullest height it came to face the hen. Though all this was done with an unhurried and uniform motion, yet of course it took very little time. Then from his stiff, erect position he sank slowly on to the surface; the hen meanwhile put down her wings and raised her neck; and the pair settled down to a bout of the head-shaking. Their attitudes and actions were practically the same as those of the pair described above, but the bout only lasted about half as long. It was ended by the two birds ceasing to shake and gradually drifting apart. Finally they put down their crests and went off together to preen themselves and fish.

These actions, too, must now be named. I propose to call the attitude of the hen as she searched and called for her mate the *Dundreary-attitude*, for the two halves of the ruff in curtain-form give the bird, especially when seen from the front, a considerable resemblance to that famous personage of the drama. The attitude later assumed by the hen, with head back and wings arched, shall be the *Cat-attitude*, for the round ruff gives the bird the look of a very contented and somewhat fat cat. The cock's combined dive and emergence I shall call the *Ghost-dive*. The whole ceremony I have called the Discovery Ceremony (see p. 512).

(*c*) Now for the highest development of the courtship-actions that I have seen. The incident I am going to describe took place in the middle of the hour-and-a-half's watching the results of which are recorded on p. 549. A pair of birds had been preening themselves and fishing, with occasional languid bouts of head-shaking. After a dive they came up not far apart and swam together with outstretched necks, which, as they neared each other, they gradually raised, beginning to shake their heads a little at the same time. The raising and the shaking progressed simultaneously, till when the birds were face to face they were in the regular "shaking" attitude and waggling their heads with a vengeance. The bout of shaking thus begun was the longest I ever saw: between them the birds shook their heads no less than 84 times, and with as much vigour at the eighty-fourth as at the first shake. There was rather a curious difference between the cock and the hen. At first neither of the birds did any of the wing-lifting, the strange parody of preening described above, and there named habit-preening. After the fifteenth shake, the hen began to give an occasional wing-lift, and these became more and more frequent on her part, until after about the sixtieth shake she was turning round and putting her beak under her wing-feathers between nearly every shake. The cock, on the other hand, did not begin this habit-preening until after the fortieth shake, and even after that only repeated the trick at rare intervals.

At the close of the bout, the pair swung parallel, but did not bring their necks down. Nor did they lower their ruffs; on the contrary, they put them up still further, from the pear-shaped form customary for shaking to the extreme elliptical, bringing down their ears meanwhile from the vertical to the lateral position, so that the whole crest now appeared like a large chestnut-and-black Elizabethan ruff. This change in the crests made me think something exciting was going to happen. Sure enough, the hen soon dived. The cock waited in the same attitude, motionless, for perhaps a quarter of a minute. Then he, too, dived. Another quarter of a minute passed. Then the hen appeared again, and a second or two later, some twenty-five yards away, the cock came up as well.

They were in a crouching position, with necks bent forward, ruffs still elliptical, and both were holding in their beaks a bunch of dark ribbony weed, which they must have pulled from the bottom. The hen looked about her eagerly when she first came up; when the cock appeared she put her head down still further and swam straight towards him at a good pace. He caught sight of her almost immediately too, and likewise lowering his head, made off to meet her. They did not slacken speed at all, and I wondered what would happen when they met. My wonder was justified: when about a yard apart they both sprang up from the water into an almost erect position, looking somewhat like the "ghostly Penguin" already described. *Sprang* is perhaps too strong a word; there was no actual leap, but a very quick rising-up of the birds. The whole process, however, was much quicker and more vigorous than the slow "growing out of the water" of the ghost-dive. In addition, the head was here not bent down along the neck, but held slightly back, the beak horizontal, still holding the weed. Carrying on with the impetus of their motion, the two birds came actually to touch each other with their breasts. From the common fulcrum thus formed bodies and necks alike sloped slightly back—the birds would have fallen forwards had each not thus supported the other. Only the very tip of the body was in the water, and there I could see a great splashing, showing that the legs were hard at work. The appearance either bird presented to its mate had changed altogether in an instant of time. Before, they had been black and dark mottled brown : they saw each other now all brilliant white, with chestnut and black surrounding the face in a circle.

In this position they stayed for a few seconds rocking gently from side to side upon the point of their breasts; it was an ecstatic motion, as if they were swaying to the music of a dance. Then, still rocking and still in contact, they settled very gradually down on to the surface of the water; so gradually did they sink that I should think their legs must have been continuously working against their weight. All this time, too, they had been shaking their heads violently at frequent intervals, and after coming down from the erect attitude they ended the performance by what was simply an ordinary bout of rather excited shaking; the only unusual thing about it was that the birds at the beginning were still, I think, actually touching each other. The weed by this time had all disappeared : what had happened to it was very hard to make out, but I believe that some of it was thrown away, and some of it eaten by the birds while settling down from the Penguin position.

In their final bout they shook about twenty times, getting less excited towards the end ; they eventually drifted apart, put their crests down, and almost at once began to pick food off the surface of the water.

Let us call the diving for water-weed and the appearing again with it in the bill the *weed-trick*; and the rapid swimming together, with the subsequent figure erect breast-to-breast, let us call the *Penguin-dance*, for here once more the general resemblance to Penguins (exceptionally graceful ones, let us admit) forced itself upon the mind.

(*d*) One last scene before we pass from mere description to the heavier task of analysis.

Sitting on the bank one day, looking out over a broad belt of low flags and rushes which here took the place of the usual *Arundo*, I saw a Grebe come swimming steadily along parallel to the bank, bending its head forward a little with each stroke, as is the bird's way in all but very leisurely swimming. I happened to look further on in the direction in which it was going, and there, twenty or thirty yards ahead of it, I saw what I took to be a dead Grebe floating on the water. The body was rather humped up; the neck was extended perfectly straight in the line of the body, flat upon the surface of the water; the ruff and ears were depressed (fig. 8). So convinced was I that this was a dead bird that I at once began revolving plans for wading in and fetching it out directly the other bird should have passed it by. Meanwhile, I wanted to see whether the living would show any interest in the dead, and was therefore much interested to see the swimming bird swim up to the tail-end of the corpse and then a little way alongside of it, bending its head down a bit as if to examine the body. Then it came back to the tail-end, and then, to my extreme bewilderment, proceeded to scramble out of the water on to the said tail-end; there it stood for some seconds, in the customary and very ungraceful out-of-water attitude—the body nearly upright, leaning slightly forward, the neck arched back and down, with a snaky Cormorant-look about it, the ruff and ears depressed. Then it proceeded to waddle awkwardly along the body to the head end, slipping off thence into the water and gracefulness once more. Hardly had it done this when the supposed corpse lifted its head and neck, gave a sort of jump, and it, too, was swimming in the water by the other's side. It was now seen that the "corpse" had been resting its body on a half-made nest whose top was scarcely above the water, and it was this which had given it the curious hunched-up look. The two swam about together for a bit, but soon parted company without evincing any further particular interest in each other.

Both these birds had crests of very much the average size, so that it was hard to tell their sex; but I think that the "corpse" was a hen, the other bird a cock.

The meaning of this action (which I only saw this one time) remained extremely problematical to me while I was at the Reservoirs. The mystery will, however, be solved in the next section, and so let us anticipate and call the attitude of the "corpse" the *passive*, that of the bird that climbed on the "corpse's" back the *active pairing attitude*.

5. The Relations of the Sexes in the Great Crested Grebe.

(i.) The Act of Pairing.

As I say, it was especially the proceeding last described which puzzled me; and it was not till I had got home and looked up the literature, that I found a welcome paper by Selous ('01) * which exactly dovetailed into my own observations. I had been mainly concerned with the behaviour of the birds on the open water and during incubation; he had paid special attention to nest-building and pairing His observations solve the mystery that has so far surrounded the Grebes' actual pairing; by them it is now established that the attitude which so puzzled me is adopted always, and only, for the purpose of coition, and that coition takes place solely on the nest. I should

* A short summary of this paper will be found on p. 529.

perhaps have said "on *a* nest"; for the birds may build several incomplete nests or platforms before one, finally chosen to be the true nest, is finished and laid in.

From Selous's observations, the actions and ceremonies connected with coition are quite elaborate—almost of the same order of elaboration as the courtship-ceremonies, though the two rituals are completely independent and appear to have developed along quite different lines.

We have already got to know the *passive* and the *active pairing attitudes.* To complete the description of the mere attitudes, it remains to add that, before sinking down into the passive pairing attitude (which Selous calls "lying along the water"), the birds usually assume a curious fixed and rigid pose. I will quote Selous's words:—". . . curling his neck over and down, with the bill pointing at the ground [weeds], perhaps six inches above it, he stood thus, fixed and rigid, for some moments (as though making a point) before sinking down and lying all along. There was no mistaking the entirely sexual character of this strange performance, the peculiar fixed rigidity full of import and expression."

We must now see how the *attitudes* are combined in the *actions* themselves.

In the first place, we have the *active pairing attitude* and the actions associated with it. These actions in pairing have been already once described, and they seem to show little variation. The bird leaps up, and comes down almost upright near the other's tail. Copulation is then attempted. Selous found it hard to decide if such an attempt was successful or not. When it seemed successful, the birds apparently uttered a louder cry than usual, and afterwards their behaviour had a satisfied look. Once, however, when the birds seemed thus satisfied, he adds: "The time occupied was extremely short, and one would hardly have thought from the position of the two birds that actual pairing had been possible." In other cases he could be fairly sure that the attempt was not successful. Whether successful or not, the act always ends in the peculiar way already described: the active bird waddles forwards along the other's body, and walks somehow over its head into the water, upon which the passive bird raises its neck, leaves the nest or platform, and swims away in normal position.

In the second place, we find that the *passive attitude* ("lying along") may take place either on the nest or platform itself, or else on the open water (but then apparently never far from the nest); the act of pairing itself, however, is possible only when the "passive" bird is lying on some firm support. In the second place, both cock and hen go into this attitude (the precise attitude which the lower bird assumes during the act of coition) indiscriminately: Selous's records give an approximately equal number of times for the two sexes.

However, before trying to draw any general conclusions, let us take a particular case—that described by Selous on pp. 180–181 of his paper :—

About an hour earlier in the day there had been an attempt at pairing. Then, after a period of rest on the open water, the birds swam together towards the nest (which had been built the day before). When just outside the bed of reeds in which the nest was situated, the hen went into the passive attitude, on the open water. The cock came up to her, swam a few yards past her, went twice back to her and away again, then went right into the weeds and himself lay along the water in the passive attitude. While he was doing this (or immediately afterwards) the hen swam to the nest, leapt on to it, and sank down in the passive attitude once more. Upon this the cock came up to the nest, jumped on to the hen's back, and they apparently paired successfully, both birds meanwhile uttering a special shrill screaming cry.

Here are various points to be noticed. The *joint approach* of the birds to the neighbourhood of the nest is invariable when they have previously been some distance away. When one bird is sitting, or when both are already close to the nest, as when building is in progress, the case is of course different (p. 533); but in the period between nest-building and incubation they seem never to approach the nest singly.

The passive attitude on the open water close to the weeds and nest may or may not be assumed. In the three cases where this happened and Selous is absolutely sure of his facts, the bird that assumed this position was the female, and was also the leader in the procession towards the weeds. (We want to know more about this. It seems probable, from other considerations, that it is a mere coincidence for the leader to have been always the hen; but, this being granted, it is quite likely that the leading bird would be the more eager, and so would hasten to put itself into the attitude which apparently expresses readiness to pair.) In other cases the birds swam straight to the nest, and one of them ascended it and then went into the passive position.

Next, the way in which the cock swam about close to the hen while she was in the passive attitude, but still on the open water, "as though about to pair" (I quote Selous), is interesting. There must be a strong association established between the sight of the passive attitude and the desire to pair, so that the active bird shows its thoughts, so to speak, even when pairing is impossible (as when the passive bird is on the open water).

When the passive bird has gone into position on the nest, it is very nearly always the case that the active bird comes up to the passive one and examines it or swims about a bit, whether an attempt to pair is afterwards made (Selous, '01, pp. 180, 345) or not (*loc. cit.* pp. 165, 344, etc.). Sometimes the second bird is not eager, and refuses to come near at all (*e. g.*, *loc. cit.* pp. 172, 456). At other times (*loc. cit.* p. 341, and perhaps p. 181) an attempt to pair is made, and the active bird jumps up apparently at once, without any delay. In the case observed by me the active bird seemed very definitely to examine the passive one, poking its beak down close to it; but the examination was very short, the attempt at pairing following immediately.

When the active bird is moderately eager but not quite eager enough to attempt to pair, it may swim up to the passive bird a number of times, each time make as if to spring up, and then decide not to, but swim away again.

The assumption of the passive attitude by one bird is generally, so far as I can see, used as an invitation to the other bird to pair: perhaps I should express myself rather differently, and say that it always denotes readiness to pair, and is generally used as a *primary excitant*—*i. e.*, it is the first sign given by either of the birds of readiness to pair. This is well brought out by incidents such as this:—Both birds are building the nest; suddenly the hen jumps up on to the nest and goes into the passive attitude, every now and then raising her neck and looking round at the cock (Selous, *op. cit.*). At other times it may be only a *secondary excitant*—a mere symbol. This may happen when one bird is sitting and the other approaches the nest; the sitting bird may then assume the passive position at each approach of the other. Here the *approach* is the primary stimulus, and the assumption of the passive attitude is called forth by it, and not by internal causes. In this second case a less degree of "sexual feeling" is presumably needed to induce the passive attitude than in the first case.

But we are going too fast. We must not omit to notice the curious action of the cock in himself copying the hen's passive attitude. This action—one bird going into the passive attitude, the other coming up and examining it, and then going off and assuming the same attitude—appears only to occur when the first bird goes into position on the open water (and not on the nest), and even then not always. It seems, however, to happen in the majority of cases (though we are perforce generalizing from very few instances). It looks as if it were a signal to the first bird that the second was ready and willing to proceed further in the matter; for the birds after this may proceed together to the nest, where the first (*loc. cit.* p. 180) or the second (*ibid.* p. 456) bird ascends the nest and assumes the passive attitude once more. In other cases, however (*ibid.* pp. 179, 454), the affair ended with the second bird's assumption of the attitude. Here it looks as if a ritual ceremony was developing out of a useful action (see below).

As regards the actual act of pairing (or its attempt, which for our present purpose comes to the same thing), *the two sexes seem here also to play interchangeable roles.* In 1900 Selous saw three attempts to pair, one apparently successful, two unsuccessful: in all three cases the active bird was the larger of the pair. In 1901 he saw two attempts, both of which he thinks were successful: here the active bird was in both cases the smaller of the pair.

Now, if we could be sure that the 1901 pair was the same that was there in 1900, all would be well: but we cannot be sure. There was a marked difference in the pairing-behaviour of the 1900 and 1901 pairs—a difference that cannot be referred back to the fact that in 1900 the birds were building a true nest and were incubating, while in 1901 they had only got to the length of building a pairing-platform. In 1900 the smaller bird (that we have so far presumed to be the hen) was more forward in invitation, while the active pairing-position was adopted by the larger bird alone; in 1901 the case was exactly reversed. It would be, in my opinion, more remarkable that such a change of character should take place in two birds in the space of one year than that the same water should be occupied by two different pairs—albeit but a single one—in two successive years. Mr. Selous, however, writes to me that for various reasons (*e. g.*, the site of the nest, etc.) he is practically convinced that the birds were the same in both years. However, whether the pair was the same pair or not, in both years there was a marked difference in size both of body and crest between the two birds of the pair, and, if all the books are not wrong, this should be quite enough to distinguish the sexes. Sometimes, it is true, the two birds of a pair are almost exactly alike; but nowhere do I find it stated that the hen is ever larger or has a better crest than the cock. It is the part of the professional ornithologist to find out if this is ever so; till then, we must be content to say that it is extremely probable that either cock or hen can play the "active" part in copulation—what we should usually call the male part. This can be more easily imagined in birds than in almost any other animals in which copulation takes place, but even in a bird is remarkable enough. Definite attitudes of the two participating organisms have been evolved to facilitate the passage of genital products in a definite direction: and here, hey presto! although the genital products continue to pass in the same direction, yet the attitudes, developed only in relation with and accessory to this direction, are at will reversed.

This facultative reversal of pairing-position would certainly be remarkable; but even for the moment supposing that it does *not* occur in our Grebe, it would merely appear as the as yet unattained end of a process of sex-equalization which in this species has already run a considerable course. This process consists in a gradual transference of all the secondary sexual characters of the male to the female, and *vice versa*. In its general aspect it will be discussed later; here it will be sufficient to consider it in relation to the pairing actions alone.

Let us see what is without doubt common to both sexes in the Crested Grebe to-day. First of all, we find that either cock or hen may lead the way towards the pairing-platform. Secondly, either cock or hen may assume the passive pairing-position (the position that one would naturally call *female*) on the open water. Thirdly, either cock or hen may assume this position on the nest or pairing-platform. This is important, for the pairing-platform is never ascended except for the purpose of pairing, or for this position, which we may call the beginning of, or the invitation to, pairing, and the nest only ascended for these two purposes and for incubation. Fourthly, when one bird is in the passive position, the other, be it cock or hen, may come up to it, examine it, and make as if to leap up on to it, just as it often does before an actual attempt at pairing is made. The natural end of this sequence would be that, fifthly, either cock or hen might not only make as if to ascend into the active position, but actually do so. If the text-books are right in their descriptions of the sexes in the species, then we can say that this end has been reached, and that, as far as pairing-positions go, the sexes are interchangeable. If the text-books are wrong, then our evidence is simply insufficient. Here it can only be shown that, however incredible this reversal may appear, yet it is quite certain that in the Great Crested Grebe all the preliminary steps towards it have been already taken.

Further, Selous (*loc. cit.*) places on record some remarkable facts which show that reversal of pairing-attitude does take place in tame Pigeons. Here he several times saw, immediately after the act of pairing, the "male" bird crouch, and the "female" then get into the normal male attitude. The act of pairing was then gone through a second time, but with the attitudes of the birds reversed. See also Selous '02.

We have therefore evidence that the full reversal can take place, and now only want to be certain that it has taken place in this species. In any case we can say that characters (in this case *attitudes* and *actions* only) of the female have been transferred to the male, as well as characters of the male to the female.

We must now go on to consider a very different question, which is also well brought out in the pairing-habits of the Great Crested Grebe: I mean the gradual change of a useful action into a symbol and then into a ritual: or, in other words, the change by which the same act which first subserved a definite purpose directly comes later to subserve it only indirectly (symbolically), and then not at all. The action in question here is the passive pairing-attitude, and the Grebe is interesting as showing all three stages of the process at one time—the passive attitude employed sometimes directly, sometimes symbolically, and sometimes ritually. Speaking phylogenetically, we have the following steps:—

(1) The ascent on to a nest or platform, and the assumption of the passive attitude, are necessary if pairing is to take place, and the passive bird must get into position before the active bird can even begin its part in the coition act.

(2) The ascent and the attitude are used by the passive bird as

an incentive to the active bird, as a sign of readiness to pair. The active bird may or may not respond.

(3) As a symbol, the *attitude* is obviously more important than the actual ascent on to the nest, since the attitude is used only in pairing, while the birds may ascend the nest for various purposes; and, in addition, the assumption of the attitude comes after the ascent, and is thus in time more immediately associated with the act of pairing. Thus the attitude by itself comes to be used on the open water (though always close to the nest) as a sign of readiness to pair. We may say that readiness to pair is indicated precociously—it is pushed back a step. Such processes of pushing back are very common in early ontogeny; embryologists then say that the time of appearance of the character is *cænogenetic* (even though the character itself, as here, may be palingenetic). The phylogenetic change has here been precisely similar; the only difference is that the displacement affects a mature instead of a very early period of life.

(4) The attitude being now sometimes a mere symbol can be, and is, employed by either the active or the passive bird. In fact, when one bird employs it thus symbolically, the other usually responds by immediately repeating this symbolic use.

(5) From useful symbolism to mere ritual is the last step—one that has taken place often enough in various human affairs. It appears that these actions and attitudes, once symbolic of certain states of mind and leading up to certain definite ends, lose their active symbolism and become ends in themselves. When I say that they lose their active symbolism, I mean that they are now not so much associated with readiness to pair as with the vague idea of pairing in general. Thus associated with pleasurable and exciting emotions, they may become the channels through which these emotions can express themselves, and so change from purposeful stimuli to further action into merely pleasurable self-exhausting processes (see below). It is at least hard to see how to explain such happenings as that described on p. 534, (*c*) 6, where first one bird and then the other goes into the passive position on the open water, after which there is simply a resumption of feeding or preening.

Another general point worth noticing is this:—In the case of this Grebe the male has even less possibility of enforcing his desires than the majority of birds. In a few birds the male is not so helpless. The ordinary Barndoor Cock, for instance, is often rather forcible in his methods. In the Wild Duck (*Anas boschas* L.) the drakes often kill the ducks by continued treading *. Somewhat similar forcible pairing is recorded of the Mute Swan (*Cygnus olor*). In such species it is by no means necessary for the race that the act of pairing should be particularly pleasant to the female. In most birds, however, the female has the upper hand: she can always prevent the cock

* Huxley, Biol. Centralbl. 1912.

from pairing with her, by simply running or flying away (*cf.* the Redshank, Huxley, '12_1). In our Grebe we are a step further still: not only must the female (or the passive bird, if we want to be precise; but this is, for the present, complicating the issue unnecessarily) be willing to pair, but she must also take the first steps--must ascend a nest or platform and assume a special position—before the cock can think of pairing. Here, therefore, supposing that the functions of the sexes had not been almost equally distributed, it would have been necessary for the hen to have had a strong impulse towards pairing; it might be that she was impelled directly by a violent physiological stimulus, or more indirectly by association, through the act being extremely pleasurable.

The phylogenetic course of events is hard to disentangle; we might suppose it to have been somewhat as follows:—

(1) Owing to the need of a firm support for pairing, it became necessary, as above set forth, for the female to take the initiative in the act of pairing, by assuming a special position.

(2) The male had thus no means of expressing his readiness to pair [whereas in most monogamous birds it is the male, as one would expect, who takes the initiative: *cf.* the Warblers (Howard, '13), the Redshank (Huxley, '12_1), etc.].

(3) Meanwhile, quite independently, a process, or tendency—call it what you will—had shown itself, by which the characters of one sex might be or tended to be transferred to the other, and *vice versa.*

(4) This was seized upon by Selection (we cannot as yet speak less metaphorically) and employed to supply the present want; the pairing attitude of the female was transferred to the male to give him, too, a means of expressing his readiness to pair—to enable him, should he wish it, as well as the hen, to take the first step towards the performance of the act of copulation by the pair.

(5) As so often occurs, the process did not stop precisely at the desired spot (we still speak in metaphors, for brevity's sake); with the female pairing-attitude was transferred the female pairing-instinct, and so came about the complete or nearly complete facultative reversal of the pairing habits.

This naturally does not pretend to be more than a possible scheme; but it is worth while setting out such a scheme, merely to show how this "reversal of the sexes" could have come about.

(ii.) Courtship.

I have started with the subject of coition, because the first thing I want to make clear about the courtship-actions is their total lack of connection with the act of pairing itself—a notable fact, in which the Grebe differs radically, of course, from many other birds, especially those in which the sexes differ in appearance, *e.g.* the Bustard or the Peacock, but also some in which the sexes look alike, *e.g.* the Redshank.

In relation to this, no doubt, is the fact that pairing only takes place on the nest, and that the nest is hidden away among the reeds, while the courtship actions are, I believe, always gone through out on the open water. This, in itself, would not be conclusive evidence of total separation of the two sets of actions, for the performance out in the open might be followed directly by a return to the reeds and subsequent pairing. But there are two further facts which make it conclusive. In the first place, one of the reservoirs at Tring is completely bare of reeds, and consequently of Grebes' nests too. It is, however, the richest in fish, and numbers of Grebes fly over to it from the other reservoirs every day, and at all hours of the day, to feed. Now, in spite of the absence of reeds, and so of nests, and so of the possibility of pairing, the birds interrupt their fishing, or sleeping, or preening, to go through the ritual of courtship just as often on this reservoir as on any of the others. That is point number one.

Point number two goes still further.

I frequently kept individual pairs under observation for a considerable length of time, and then, if I watched long enough, always found that one set of courtship-activities would in point of fact be followed by a pretty long interval of resting or fishing, and that then this time spent in every-day affairs would be again succeeded by another series of courtship-actions—a proof that these actions are what we may call *self-exhausting* and not *excitatory*. The best record, because the longest, was on this same reedless reservoir. I had one pair under observation for an hour and forty minutes (section 10, record 11). During that time they had six simple bouts of shaking, and also two prodigious long bouts, followed each time by the diving for weed and then the strange Penguin-dance. And between all these elaborate displays of sexual emotion, no sign (or possibility) of pairing—nothing but swimming, resting, preening, and feeding.

I was thus—much against my preconceived ideas—driven to think of all the complicated postures and evolutions of courtship in the Grebes as being merely *an expression of emotion*.

The particular form of expression used is no doubt determined—predetermined—by the arrangement and innervation of certain structures which the birds possess: but the impulse to use the muscles and nerves is an emotional one—during courtship there must be in the mind of the bird an excitement, a definite feeling of emotion. Let us, to satisfy the physiologists, try to put it in terms of nerve-currents. One member of a pair is continually seeing its mate at its side. This, in its present physiological condition, stimulates certain tracts of its brain, charging them up and up until they are in a state of considerable tension (mental accompaniment:—state of diffused emotional excitement). Finally, the tension reaches the critical point, and a discharge follows. This discharge flows down hereditarily-determined paths, and actuates the muscles concerned in courtship (mental accompaniment:—violent and special emotion, quickly dissipating itself with a sense of "something accomplished, something done.")

This merely indicates the possible material mechanism; of the actual, we know next to nothing. However, by comparing the actions of the birds with our own in circumstances as similar as possible, we can deduce the bird's emotions with much more probability of accuracy than we can possibly have about their nervous processes: that is to say, we can interpret the facts psychologically better than we can physiologically. I shall therefore (without begging any questions whatever) interpret processes of cause and effect in terms of mind whenever it suits my purpose so to do—which, as I just said, will be more often than not.

Let us take the parallel from human affairs. Far be it from me to go into the matter with a heavy hand; let us merely look at a few familiar facts in an unfamiliar biological light. The "courtship-actions" of man are mostly predetermined by heredity: any young couple that you like to take will be pretty certain to "express their emotion" by holding each other's hands, by putting their arms round each other's waists, or by kissing each other; and of this last action kissing on the mouth is the "highest development." Let us merely notice that these actions are not perhaps exactly parallel with what we find in the Grebe—that they are altogether more fluid, less fixed, and that they are sometimes less self-exhausting and more excitatory in character: on the whole, however, they are not very different. Moreover, in their case we know a great deal about the accompanying emotions, either from our own experience or from what others tell us. To take only the most specialized form of human courtship-actions, the kiss; although we know that it may act as an excitant (*cf.* Dante's famous lines on Paola and Francesca) yet the accompanying emotion is in itself quite special, different from all others, and the emotional process is usually something *an und für sich*, expressing itself in the action, and exhausting itself in the process with a feeling of inevitability. In the memory, however, it leaves its trace, and as it were desires to repeat itself, but only when the emotional tension shall again have risen (think of Plato's epigram to Agathon: or the lovers in Richard Feverel; or Romeo and Juliet). That will suffice to show what I mean by a self-exhausting expression of emotion. Such a process would be one that to the doer of it feels at the time almost inevitable, though he can only do it at certain moments. At other times, determined by his general mental state (*cf.* section 10, record 1), the action, however pleasant to recollection, is not "spontaneously" possible, and if performed is forced or at least not fully pleasurable. When normally executed, the action is accompanied by violent and pleasurable emotion, which usually dies down, or changes, into a quite different feeling, one of satisfaction, meanwhile leaving its mark in the memory. Its recollection then acts as a partial stimulus, so that next time it is a little more easily performed.

This will, in the first place, show how difficult, and almost inevitably futile, it is to try and deal with the emotional essence of things by the methods of "ordinary biology"; I think, however, that it will serve to explain what I mean by a self-exhausting expression of emotion, and will give the point of view from which to look at the facts of the Grebe's courtship: let us now go on to examine the facts themselves more systematically. I will take the different forms of courtship-action one by one, describe their usual occurrence and their relation to other actions, and then mention the most important variations or exceptions that I have seen.

The various attitudes already described are combined into definite *actions* or *ceremonies*.

(*a*) The simplest form of courtship-action is the *bout of shaking*, of which I have described a typical example. As already seen, shaking may take place either before or after other courtship-actions, but in perhaps the majority of cases it is not thus a link in a chain of processes, but a single self-originating and self-exhausting process. It varies a certain amount in intensity and in length, and also in the amount of habit-preening that takes place. Of this there may be none, or, towards the end of a bout, there may sometimes be more preens than shakes. The bouts seem, to the casual onlooker, to start themselves—in reality, I think, each bird excites the other. One gently shakes its head under the force of rising emotional tension; the other bird had not quite got to that stage, but the sight of its mate shaking acts as a stimulus, and it too pricks up its head a little and gives a shake. This reacts on the first bird, and so the excitement is mutually increased and the process fulfils itself—a very good example of "crowd-psychology," and also a good example of an epigenetic process.

There is one well-marked variation of this form of courtship which seems to denote a higher level of excitement; it is especially common when a third bird has intruded into the domestic harmony of the pair and has been driven off (section 5, iv.).

Here the beaks are pointed somewhat downwards, the neck brought a little forward instead of vertical, the whole head brought forward and curved over, and the ruff erected more than usual (fig. 6). This attitude is almost always confined to the beginning of a bout, the birds sooner or later relapsing into the ordinary position.

The bout of shaking is not only the commonest form of courtship-action, but it also forms part of all the other more elaborate forms. It always ends the series of actions, and often begins them as well. It is as it were the foundation on which they are built, and was probably (if I may express a mere opinion) the earliest to appear in phylogeny.

The other ceremonies of courtship are all formed by the

combination, in various arrangements, of the shaking-bout, the Dundreary, the Cat, the Ghost-dive, the weed-trick, and the Penguin-dance; in addition, they may be slightly modified by jealousy.

They can be divided into two groups: (1) those in which the Cat-position plays a prominent part, and (2) those into which weed-carrying enters. Let us consider them in this order.

The Cat-position forms a part of two quite distinct ceremonies, which, simply for the sake of ready reference, I shall call the *Ceremony of Discovery* and *the Display*. The first of these is gone through, as far as I can make out, when the two birds of a pair find and rejoin each other after being separated for some time. The second always occurs in the middle of a bout of shaking; on such occasions I presume that the shaking has not been "self-exhausting," but that the emotional excitement that accompanies it has reached a slightly higher level than usual, with the result that it overflows into a new nervous channel, and so expresses itself in this new way.

(b) *The Ceremony of Discovery*.—A typical case has already been described (p. 497). I should interpret the facts thus:—The two birds of a pair have become separated—perhaps they have gone off fishing in different directions, or one has been on the nest and the other has not stayed near by. They wish to rejoin each other. To this end the bird that is searching puts itself into a special attitude, which is probably adapted for uttering the special cry only heard on such occasions, and cruises about, alternating its signal-calls with moments of looking about it. On hearing the call, several neighbouring birds will usually prick up their necks and look about them; but I believe that it is usually only the searching bird's true mate who takes any further interest (this would doubtless depend on the emotional state of the neighbouring birds). Once this discovery of the missing mate has been made, a special ceremony takes place to celebrate the event. This ceremony is a peculiar one, and is practically confined to these occasions of discovery; very possibly the memory of the ceremony and its excitement adds to the eagerness felt by one bird of a pair to rejoin the other. The ceremony itself usually consists in this—the bird has been discovered dives, upon which the searcher puts itself into what I have called the Cat-attitude, a bizarre but beautiful position obviously recalling the elaborate displays of many other birds. In this attitude the searcher waits, almost always in a state of great excitement, as shown by its turning itself hither and thither, from side to side. It is stimulated to this excitement by the diving bird: first of all, as the dive is very shallow, the diver's approach is marked by a swift ripple of the surface; and then, when the diver at length appears, it is in a shape as unlike that of everyday life as is the "Cat-position" of the searcher—albeit the two are at opposite poles of the Grebe's capabilities. Sometimes the diver emerges when only a few feet from the searcher. This is merely to reconnoitre his position; head and neck alone appear, the crest not erected, and are swiftly withdrawn again. The final appearance takes place almost always beyond the searcher, and the bird emerges with its back to the other, facing it only as, revolving on its axis, it settles down. The performance always ends with a bout of shaking.

Although in my "typical" cases the searcher has always been the female, yet the male may also search for his mate in the same way. I have watched an obvious cock in the regular Dundreary (search) attitude for a long time; only on this occasion no mate responded to the call. Further watching is necessary to see whether it is merely an accident that my searchers have usually been hens, or whether my observations represent reality. (It may possibly be connected with the fact that the hens seem to spend more time on the nest than the cocks; but this is mere conjecture.)

(c) *The Display Ceremony* is quite different. The birds are already together, and the display is simply a form of excitement similar to the bouts of shaking. In typical cases the pair will be indulging in a bout of shaking; suddenly one of them flies off a few yards and puts itself into the full Cat-position, showing its circular ruff and white-striped wings to its mate. There is no diving, however, and, after some seconds' display, the birds swim together and there is another bout of shaking; after this they simply swim off, or separate, or feed. Either cock or hen may go into the Cat-attitude. In one case the first bout of shaking had been preceded by a "flirtation" (p. 521) on the part of the cock.

To show how one ceremony may blend into another, I adduce the following instance:—There was a regular Discovery Ceremony, the hen calling to the cock, but with this difference, that they swam together, shook, and the hen flapped off and went into the Cat-position, and that only then did the cock remember, so to speak, to do the ghost-dive (p. 498). Another mixed "ceremony," this time more closely related to pure display, is related on p. 547.

(*d*) Finally comes the ceremony of the *Penguin-dance.* I have little to add to the description already given (p. 499). Twice, curiously enough, a single pair was seen to perform the dance twice in a morning. This might imply that some special physiological state, probably of high excitement, was necessary for the act, for I only saw it on two other occasions, and Selous only saw it once.

The performance can only be gone through when both are equally excited; for instance, once (p. 547) after a bout of shaking the cock dived and fetched weed from the bottom. The hen, however, was not stimulated to do so too, and when he came up he found no answering stimulus, and so dropped the weed as he swam towards his mate.

There is no reason for supposing even this elaborate ceremony to have any direct relation whatever with coition. It is a form of excitement and enjoyment, seemingly as thrilling to the birds as it is to the watcher, but, like all the other courtship-actions, self-exhausting.

Very interesting "incomplete stages in development" of this ceremony were seen in one pair of birds, ranging from simple diving to the complete ceremony (see below).

(e) *Other Courtship Ceremonies.*

(1) *Back-to-back Ceremony.*

There are considerable individual variations in the courtship-activities of the Grebe, and I have seen occurrences which may well be interpreted as rudiments of new ceremonies. In one pair, for instance (section 10, record 12), the birds almost always went into a formal back to back, or rather tail to tail, attitude after each bout of shaking.

(2) *Diving Ceremony.*

Other actions seem to stand in some relation to the more highly-developed ceremonies. For instance (pp. 545, 552), simple diving, either by one or both birds, without any fetching of weed from the bottom, is introduced as part of the courtship between the two bouts of shaking.

(3) *Weed-trick Ceremony.*

In still other cases, the weed-trick is gone through and is followed immediately by a bout of shaking. In both pairs in which this was seen, one bird alone brought weed, although in one pair (not in the other) both had previously dived. I am inclined to believe (but more observations are needed) that two distinct ceremonies are here involved: first, the fetching and offering of weed by one bird, usually the cock, to its mate; and, secondly, a typically "mutual" ceremony involving simultaneous diving of both birds of a pair: and on to this latter the "penguin dance" has been grafted.

The offering of weed is strongly reminiscent of occurrences in the sex-differentiated, non-mutual courtships of other birds, such as the Warblers, where the cock often carries leaves or twigs in his mouth during sexual ecstasy (Howard). It seems to me probable that, since diving is necessary for weed-fetching, the one has come to be associated with the other, and the two ceremonies have come to be mixed up: in the extreme case on one side there is no mutual ceremony—only an offering of weed by one bird to the other; at the other extreme we have the complete penguin dance as described on p. 499, where both cock and hen bring up weed; and as intermediates we have mutual diving (p. 552) and mutual diving where the cock alone brings up weed (p. 552, end).

Taken all in all, the courtship is chiefly *mutual* and *self-exhausting*: the *excitatory*, *secondary-sexual* forms of courtship such as weed-offering or pure display serve not as excitants to coition, as in most birds, but as excitants to some further act of courtship; and this is always a mutual and also a self-exhausting one. The excitants to coition are of a very special nature, and are symbols, rather than mere general excitants.

Habit-Preening. (See p. 497.)

This is very frequent, occurring in about half the bouts of shaking seen. The more excitement, the less preening, seems to be the rule; long bouts may sometimes degenerate into practically undiluted preening, the head simply being brought more or less up, but not shaken, between the "preens." It is always the hind end of the wings, I believe, which is raised and let fall by the beak.

In some way there must be a strong association between preening and head-shaking in the Grebe, for solitary birds who were really preening themselves I have several times seen raise their heads, slightly bristle their crests, and give a rudimentary shake. Why or how the association has taken place is more difficult to say. I certainly believe that the action I call habit-preening has been derived from true preening, and has been ceremonialized in the process of becoming part of a courtship-action. For the present we must leave it at that.

"*Habit-Shaking.*"

That for some reason there is a very real association between shaking and preening is shown by the following facts. When actively engaged in real preening of themselves, the birds are often seen to lift their heads, give a rudimentary shake or two (without erecting ears or ruff) and then go back to business. This is generally seen when the bird is engaged in preening its hinder parts. We have observed it in autumn as well as in spring, and so it presumably takes place the year round; there is thus obviously a real association between the preening and the shaking, and the shaking is not a mere release of simpler sexual energy.

This is exactly the converse of what I have called "habit-preening," and may therefore appropriately be styled "habit-shaking."

There is thus a single association with a two-fold result. How it can possibly have arisen, or what purpose it can serve, remains to me at present an absolute mystery. I leave it as a puzzle to future bird-watchers and comparative psychologists.

Fighting between Cocks.

I saw very little of this beyond mere hostile expression (p. 521). Once, however, I saw two birds actually grappling: one was struggling half-submerged, while the other was more or less on top of it, and had hold of the feathers of the back of its opponent's head. After some considerable splashing and struggling, they separated and swam apart.

In birds which pair up early and remain "married" for the season, like the Grebe, one would, of course, not expect to find any of the regular combats seen in other species. It would be interesting to see whether there is more fighting in February, during the actual process of pairing-up.

The question must now be put—"What for?" What is the good of all these divings and posturings, these actions of courtship, these "expressions of emotion"? To what end are colours and structures developed solely to be used in them, and what return is got for the time and energy spent in carrying them out? They are common to both sexes, and so have nothing to do with any form of true sexual selection; they are self-exhausting processes, not leading up to or connected with coition, and so cannot be sexual excitants in the ordinary sense of the term.

It must be, however, that they fulfil some function; and I believe I know what this function is. I believe that the courtship ceremonies serve to keep the two birds of a pair together, and to keep them constant to each other.

The Great Crested Grebe is a species in which the two sexes play nearly equal parts in all activities concerned with the family. The cock shares equally in nest-building, nearly equally in incubation and early care of the young (it is only later that the young pass into the care of a single parent, probably the female, see Pycraft, '11). Thus, from the point of view of the species, it is obviously of importance that there should be a form of "marriage"—a constancy, at least for the season—between the members of a pair. The same result—marriage—is observable in such a species as man; but in man the main cause is a division of labour between male and female, whereas in the Grebe the sexes have been made as similar as possible. It would seem that the Grebes' family affairs had simply required more labour to be spent on them, and that Evolution had happened to go along the simple path of increasing the quantity of labour, by bringing the male in to do female's work, instead of improving the quality by adopting the principle of specialization.

Birds have obviously got to a pitch where their psychological states play an important part in their lives. Thus, if a method is to be devised for keeping two birds together, provision will have to be made for an interplay of consciousness or emotion between them. It would be biologically enough if they could both quite blindly, and separately, attend to the common object—nest, eggs, or young; but with brains like theirs there is bound to be a considerable amount of mental action and reaction between them. All birds express their feelings partly by voice, and very largely by motions of neck, wings, and tail; and not only this, but the expression can be, and is, employed as a form of language. This being so, we have here a basis on which can be reared various emotional methods of keeping birds of a pair together. As always, selection of accidental variations has led to very diverse results; so that we see this "emotional companionship" playing a part in many apparently very different actions of birds. Herring-Gulls sit or stand close beside each other for hours together, occasionally rousing themselves to a joint ceremony of shaking their necks. As the Snipe drums overhead, there is often a call from the marsh below. Many birds when paired are always calling to each other, and probably singing birds sing partly to their mate; Dabchicks have a special spring note, usually given as a duet. As a very simple case, I have seen a pair of Blue Tits very recently paired up who, although feeding, were perpetually calling to each other and at frequent intervals coming close up side by side; it was perfectly obvious that they simply took pleasure in each other's presence, like the engaged couple that they were.

We have thus the following train of reasoning. Many birds must be kept in pairs during the breeding-season. This may be partly effected by the instincts of the separate birds—the instinct to build a nest, to sit on eggs, to feed young; and partly by instincts which only can find play when the two birds are together. These latter are often very emotional, and the courtship habits of the Grebe afford a very specialized example of this emotional bond between members of a pair.

If my contention is correct, it is clear that many actions and structures solely used in courtship are of use to the species, and not only to one sex of the species; these therefore must be maintained by Natural as opposed to Sexual Selection.

[*Editor's Note:* Material has been omitted at this point.]

6. Discussion.

There are various considerable difficulties concerned with the courtship-structures and actions of the Great Crested Grebe. In the first place, it is clear, from what has been said, that in this bird there is no sexual selection in the ordinary sense of the word; the crest and the courtship-actions are almost identically developed in cock and hen alike.

On the other hand, the crest is only fully developed in the breeding-season, thus resembling true secondary sexual characters; and, as I have pointed out (Huxley, '12_2) it is used only in courtship, so that if not "secondary," it is at least "sexual." Further, the crest is smaller (though but slightly) in the female than in the male, a fact which it is, at first sight, simplest to explain by assuming that the crest was acquired by the cock as a secondary sexual character, and has now been almost completely transferred to the hen (*cf.* similar transference, complete or incomplete, in Lycænid and other butterflies (Weismann), Reindeer, mammæ of mammals, colours of many birds). We will revert to this point.

The courtship *actions*, however, can scarcely be explained by transference. The Penguin-dance, for instance, can never have been anything but a joint ceremony, equally shared by both sexes. Furthermore, even in the Dabchick. although it (and it alone in the subfamily) lacks all courtship-structures on the head, there is a *joint* courtship-action—the two birds come face to face, stretch up their necks, and emit the well-known cry. This being so, it is fairly clear that the ancestral courtship-actions of the Grebes were not in the nature of a display by one sex, but were joint actions of the pair. There is nothing especially remarkable in this. The display-courtships are, on the whole, more striking, and so have been more frequently described; but (to draw on my own limited experience) Razorbills and Herring-Gulls have very well-marked joint courtship-actions, although the actions are associated with no special structures whatever, and Selous has described other such actions in Swans, Divers, Guillemots, Fulmars, and other species.

I should put forward the theory that the courtship-habits of birds are based upon at least two totally different foundations: in the first place the actions gone through by males alone, apparently as the direct result of sexual eagerness (*solitary actions*), and, in the second place, the actions gone through by male and female together, and perhaps often (though by no means always) connected or associated with nest-building (*combined actions*). Primitively in neither case would there be any special structure or colour associated with the action. For solitary actions this is well seen in the dowdy Warblers, so fully described by Eliot Howard; here the cocks resemble the hens, but go through elaborate droopings of wings and fannings of tail, with bristlings of feathers on throat and crown. Later, Sexual Selection has stepped in, and naturally enough has taken what was already given, and added to it. The same instinctively-displayed parts—wings and tail, throat and crown—are the parts which are especially singled out for the development, first of special colours (Finches, Woodpeckers), then of special colours and structures combined (Turkey, Argus Pheasant, Blackcock, etc.). In combined actions a similar process has been gone through. In the Herring-Gull and Razorbill we have the instinctive actions pure and simple—a direct outcome of nervous excitement. Then, again, something has stepped in and used what was thus provided, and we get combined actions displaying colour (coloured mouths of Fulmar Petrels, Selous), and finally colour and structure, as in the Grebe. The members of the Heron tribe in general, and the Egrets in particular, have also ornamental structures common to both sexes; it would be very interesting to know the course of courtship in these birds. Pycraft ('13) figures a mutual display executed by the Kagu.

The question now arises, How have such colours and structures arisen? By Sexual Selection followed by transference, or by some other process? Such other process can easily be imagined, and I feel confident that it has played a considerable part. We may call it *Double* or, better, *Mutual Sexual Selection* (*Mutual Selection* for short). Where combined courtship-actions exist, and a variation in the direction of bright colour or strange structure occurred, it would make the actions more exciting and enjoyable, and those birds which showed the new variation best would pair up first and peg out their "territories" for nesting before the others could get mates. The level would tend to be raised generation by generation. Mutual Selection is in a way a blend between Sexual and Natural Selection. The structures and actions arising under it have their immediate origin in the preferences of individual birds, not in anything outside the species, and in their immediate function they are entirely confined to the courtship. On the other hand, the mutual courtship itself, the activities of both birds taken together, may be of use to the species as a whole, in keeping the sexes together when necessary. Then the indirect function of all the shaking-bouts and displays of the Grebe is a function of use to the species, and besides the direct origin there is added an indirect origin under the pressure of Natural Selection.

Mutual Selection has a certain similarity with assortative mating, but is by no means the same thing. Like true Sexual Selection, it encourages an ever higher level in the development of a character, once variation has given it a basis to start from. In the Grebe the line of variation encouraged by Mutual Selection has been the tendency to produce ruffs and tufts of feathers on the head, and to go through actions involving, besides the use of these structures, diving and sporting with water-weed.

The question in the Grebe is complicated, as noted above, by the slightly less developed crest of the hen; this, however, might easily be accounted for by differences in the metabolism of cock and hen. The Discovery and (especially) the Display Ceremonies are also rather stumbling-blocks in the way of an explanation by Mutual Selection; they seem so very like the Displays of solitary courtship. However, even here the second bird plays a part, which in the Discovery Ceremony is at least as important as that of the displaying birds.

What is quite clear, however, is that, even supposing (what to me personally appears very doubtful) that ordinary Sexual Selection has "produced" the structures and the cat-position (we must know more about the habits of other species of the genus to decide this), yet it has gone hand-in-hand with a process of Mutual Sexual Selection as regards the majority of the actions. These actions (like the display of the Peacock, but unlike that of the Warblers) are much too elaborate and much too specialized to be considered as the immediate outcome of any form of physiological excitement. They obviously have a long and complicated evolution behind them, and, as they can only be performed by the two birds together, there is nothing to account for them as they now stand but some such process as I have just sketched under the name of Mutual Selection

Then there comes the question of the facultative reversal of the act of pairing (or, possibly, only of preliminary pairing-attitudes). The other cases noted by Selous (Pigeon and Moorhen) differ in that the male crouched to the female directly after the act of pairing, who at once proceeded to play the male's part. In the Grebe there was always a long interval before the "reversal of instinct" took place.

In all, however, it is very difficult to see how to account for it, except on the assumption that there has been a reciprocal "transference" of pairing-instincts. This transference may be apparent or real. It is apparent if we believe that the units for such sexual characters are equally present in the germ-plasm of both sexes, and that the characters themselves do not appear in the other sex (or only appear as rudiments) as a result of the great primary sex-difference.

If the transference is real, then one must assume that the zygotic constitution of the two sexes is different in regard to secondary as well as primary differences, but that there is a constant tendency—depending on some as yet unknown process—to transfer such characters to the opposite sex. (Hybridization experiments, where the female of a species can transmit to her male hybrid offspring the secondary sexual characters of her own species, indicate that the first method is the true one.) How else than in one of these two ways can we explain transference in both directions? This is seen, for example, in man, where a male organ, the moustache, appears rudimentarily in the female, and female organs, the mammæ, appear rudimentarily in the male: in abnormal cases, besides, the transference may be complete, the organs being completely developed in the wrong sex. Such moustached women and men with breasts again support the idea that the transference is not a real transference, but consists in the removal of an inhibition only.

(I would not trouble to mention the theory that these appearances of characters of one sex in the other are due to descent from a hermaphrodite ancestor, were it not actually the case that Metchnikoff has advanced it. It is enough to point out that if this were so, the primitive mammal must have been a hermaphrodite.)

To us it makes little odds whether there is inhibition alone or transference followed by inhibition. In both cases the character will be in antagonism with the inhibitor: supposing that there is no longer any need to inhibit a character of one sex in the other, then on Darwinian and Weismannian principles the inhibiting "force" will atrophy, and the character, remaining as strong as ever, will appear equally in both sexes.

Apply this to the present case. Birds are for the most part

so constructed that impregnation would take place equally well whether the sexes are in normal or reversed position: that is to say, there is no necessity for keeping to the customary position—and accordingly "reciprocal transference" of the pairing attitudes (whether the transference be apparent or real) may, and quite probably will, take place. If so, then in one of our Grebes the instincts and reflexes for the pairing-actions proper to its sex co-exist side by side with those for the pairing-actions proper to the other sex. It is also obvious, first, that both cannot be gratified simultaneously; and, secondly, that these two very different sets of actions must be associated with two very different sets of emotional states. The bird may "feel female" or it may "feel male," and according to its feelings, so will it tend to act. But, as we saw before, in discussing the pairing-attitudes, it appears that, owing to the difficulty of coition in the Grebe, the "female" (passive) pairing-attitude has become a mere symbol of readiness to pair. Thus Natural Selection has come in to assist the slow process of transference (at any rate, so far as pairing-attitudes are concerned), and since whatever involves them will probably involve coition itself as well, we have an additional reason for believing that actual reversal of pairing does take place, as Selous supposes, in the Grebe.

At any rate, there can be no doubt about the reversal in the Pigeon and Moorhen. The sudden reversal that here takes place is rather different, but may be explained somewhat as follows:—Here, too, both active and passive instincts are now represented in either sex. A bird is in a state of sexual excitement; this excitement releases itself in the performance of, say, the male part in the act of pairing. The excitement is not always completely exhausted by the act, and, if so, the act is repeated (just as the shaking-bouts of the Grebes are continued for a longer or shorter time, according to the degree of what we may call courtship-excitement). But supposing that general sexual excitement arouses both the male and female emotional states, then the performance of the act once in the male attitude will only exhaust the feeling of "male excitement," leaving the "female feeling" still a-tingle. The result will be, first, an inducement to repeat the act and, secondly, an inducement to repeat it with attitudes reversed.

Thus such immediate reversal is more or less an accident of heredity, while the Grebe's reversal is an accident aided by the usefulness of the transferred actions, which thus bring the accident within the sway of Natural Selection.

This treatment of the question is of necessity sadly speculative, but it is our duty at least to try to construct a coherent mechanism of theory to explain the isolated facts of observation.

Finally, a word as to terminology. I have already pointed out (Huxley, '12_2) that the phrase *secondary sexual* cannot be applied to the Grebes' ruff and ears or to their courtship-actions, because this term always implies a difference between the two sexes, and yet the crest of the Grebe has a sort of secondary sexual look about it—unreflectingly, one would at once write it down as such. This is due to our incomplete classification. We begin by separating out *sexual* characters from all others—these being characters that are different in the two sexes. We divide them into *primary*, *accessory*, and *secondary*. The mammæ of mammals (with the exception of man) have nothing to do with courtship or mating, yet they are usually included under the same heading as the tail of the Peacock, while the Grebes' courtship-structures would be left out in the cold.

Besides the mere criterion of difference in the two sexes, we must have some other criterion—a criterion of use.

It is naturally impossible to draw up any completed classification that will satisfy every case. To do so would be beyond the powers even of a Herbert Spencer—and not of much use when done.

It is enough to point out, first, that our group of Secondary Sexual Characters is a bit of an *omnium gatherum*. Some of them, as the mimicry of the female *Papilio*, or the brown colour of the female Pheasant, are protective, of use to the individual and to her offspring. Others, such as the mammæ of mammals, are of use only to the offspring; others, like the sexual differences in the beak of the Huia, where male and female hunt in couples, one splitting open the wood, the other picking out the hidden grubs, have arisen by a division of labour, and are of use to the couple as a couple. One might go on, but it would be unprofitable.

In the second place, we must recognize as a fact that the existence of individuals of separate sexes with wills of their own has led to the development of what we call courtship—simply a process in which a series of actions is carried out as the outcome of an emotional state based on sexual excitement. All courtship is based on sexual excitement, and characters connected with courtship merit a separate name of their own. This name lies ready to hand in Poulton's term *epigamic*; we must, however, remember that the literal meaning of the term must not be pressed, for in many cases the courtship ceremonies do not lead, directly or indirectly, to the act of pairing. Let us rather turn it the other way about, and, defining an *epigamic character* as one that is used in courtship, go on to define *courtship* as a series of actions based immediately or remotely upon sexual excitement, and, to make ourselves clear, we must add that sexual excitement is not merely sexual desire, but that whole emotional state into which a member of one sex may be thrown by a member of the other. The necessity for the distinction is obvious, if we think of the conditions in Man. Sexual excitement, of course, includes mere sexual desire, and also includes the fighting of males among each other as a result of sexual desire.

If we want a tabular statement, we can draw up something like the following *:—

(A) *Characters different in the two sexes.*
(*Sexual characters*).

(1) *Primary.* Of the gametes and gonads.

(2) *Accessory.* Concerned with the union of the gametes.
[Copulatory organs, pairing attitudes, sexual desire.]

(3) *Secondary.* All others.

(*a*) Developed through Natural Selection.
[Huia beak; mammæ; marsupium; incubation by ♀ alone in birds, &c.]

(*b*) Capable of being developed through Sexual Selection.
[Horns of deer; tail of Peacock, &c.]

(B) *Characters similar in the two sexes.*

(1) Capable of being developed through Mutual Selection.
[Grebes' courtship and crest; Herring-Gull's courtship, &c.]

(2) All other characters.

[(3b), (B)(1): *Epigamic.* (Courtship characters, i. e. *all* characters concerned with the *relations of the sexes*, excepting those connected immediately with coition.)]

It might perhaps be better, as has been suggested to me, to restrict the term *Secondary Sexual* to 3*b*, and employ *Sex-limited* where I have employed *secondary sexual*. For one thing, however, this would conflict with Darwinian use; also, I am at present more concerned to show the necessity for new thinking than for new terminology, which will be more suitable in a more general and definitive paper.

I will conclude by hoping that anyone who has the opportunity will observe the habits of the Crested Grebe during the time of pairing-up in early spring; the full courtship of the Dabchick would also be of very great interest. In the near future, I hope to publish a more general paper upon Mutual Selection, so that any notes sent to me on this subject will be gratefully received and acknowledged.

[*Editor's Note:* Material has been omitted at this point.]

List of Literature.

'13. Howard, H. Eliot.—The British Warblers. Parts 1–8. London, 1907–13.

'12$_1$. Huxley, J. S.—A First Account of the Courtship of the Redshank (*Totanus calidris* Linn.). P. Z. S. 1912, pp. 647–655.

'12$_2$. „ „ The Great Crested Grebe and the idea of Secondary Sexual Characters. Science, n. s. vol. xxxvi. pp. 601–602.

'11. Pycraft, W. P.—Habits of the Great Crested Grebe. Field, vol. cxviii. Oct. 7, 1911, pp. 823–824.

'13. „ „ The Courtship of Animals. Hutchinson & Co. London, 1913.

'01. Selous, E.—An Observational Diary of the Habits—mostly domestic—of the Great Crested Grebe (*Podicipes cristatus*). Zoologist, May 1901, pp. 161–183, Sept. 1901, pp. 339–350, Dec. 1901, pp. 454–462, Apr. 1902, pp. 133–144.

'02. „ (Reversed pairing in Moorhens.) Zoologist, May 1902, pp. 196, 197.

'13. „ (Display of Swans.) Zoologist, Aug. 1913, pp. 294–313.

Scientific Names of Birds mentioned in the Text.

To save constant reference to birds by both English and scientific names, and to help foreign readers, I append this list.

Blue Tit.	*Parus cœruleus* L.
Bustard.	*Otis tarda* L.
Coot.	*Fulica atra* L.
Dabchick.	*Podiceps fluviatilis* (Tunst.).
Egret.	*Ardea, Egretta, Herodias, Garzetta.*
Fulmar Petrel.	*Fulmarus glacialis* (L.).
Guillemot.	*Uria troile* (L.).
Heron.	*Ardea.*
Herring-Gull.	*Larus argentatus* Gmel.
Kagu.	*Rhinochetus jubatus.*
Mallard.	*Anas boschas* L., ♂.
Moorhen.	*Gallinula chloropus* (L.).
Peacock.	*Pavo cristatus* L.
Prairie Hen.	*Tympanuchus americanus.*
Razorbill.	*Alca torda* L.
Redshank.	*Totanus calidris* (L.).
Shoveller.	*Anas clypeata* (L.).
Snipe.	*Gallinago cœlestis* (Frenz.).
Swan (Whooper).	*Cygnus musicus* Bechst.
Tufted Duck.	*Fuligula cristata* (Leach).
Warblers.	Sylviidæ.
Wild Duck.	*Anas boschas* L.

EXPLANATION OF THE PLATES.

[The figures were drawn from my notes and rough sketches by Miss Woodward, to whom I am much indebted for the interest and care she has shown. Taken as a whole, they give a far more graphic and accurate idea of the birds' general appearance and behaviour than any other illustrations of which I know.]

All figures refer to *Podiceps cristatus*.

PLATE I.

Fig. 1. Head and neck, showing ruff and ears relaxed.
2. Resting attitude. Note the position of the head, and the curve of breast and rump. In most figures these are erroneously represented.
3. Search (Dundreary) attitude. Note the ears relaxed, the crest spread longitudinally (sometimes it may touch the water).
4. Head and neck in Shaking-attitude (ears erected vertically, ruff pear-shaped).
5. Shaking-attitude from behind. Note the curious shape of the lower part of the neck.
6. A pair in the Forward (excited) Shaking-attitude. Note the head bent down, the neck strained forward; the slope of the body and cock of the tail are also very characteristic.
7. The Cat-attitude (Display). The general attitude is very well represented. More white should show on the breast; and the dark portion of the wings should be grey. To represent them black lessens the effect of the real black on the crest, which in actual life is the central and most conspicuous part of the picture.
8. The Passive Pairing-attitude. Note the strange stiff appearance, the humped back, and the total closure of the crest.

PLATE II.

Fig. 9. The "Ghostly Penguin" (attitude on emergence of the diving bird in the Display Ceremony). Note the head bent down, and the forward curve of the top of the neck.
10. The same as fig. 9, side-view. Owing to the short time occupied in the action, I cannot myself be sure that all the details in figs. 9 and 10 are accurate. The general appearance, however, is well given.
11. A pair shaking. Note the erect necks, and the tails slightly cocked up.
12. Display Ceremony: the diving bird just fully emerged. (This is the only figure which is not satisfactory. It gives the positions etc. well, but does not recall reality in the vivid way done by the others.)
13. The Penguin Dance. Here again the whole ceremony takes such a short time that I cannot vouch for details; but the general appearance is very well suggested.

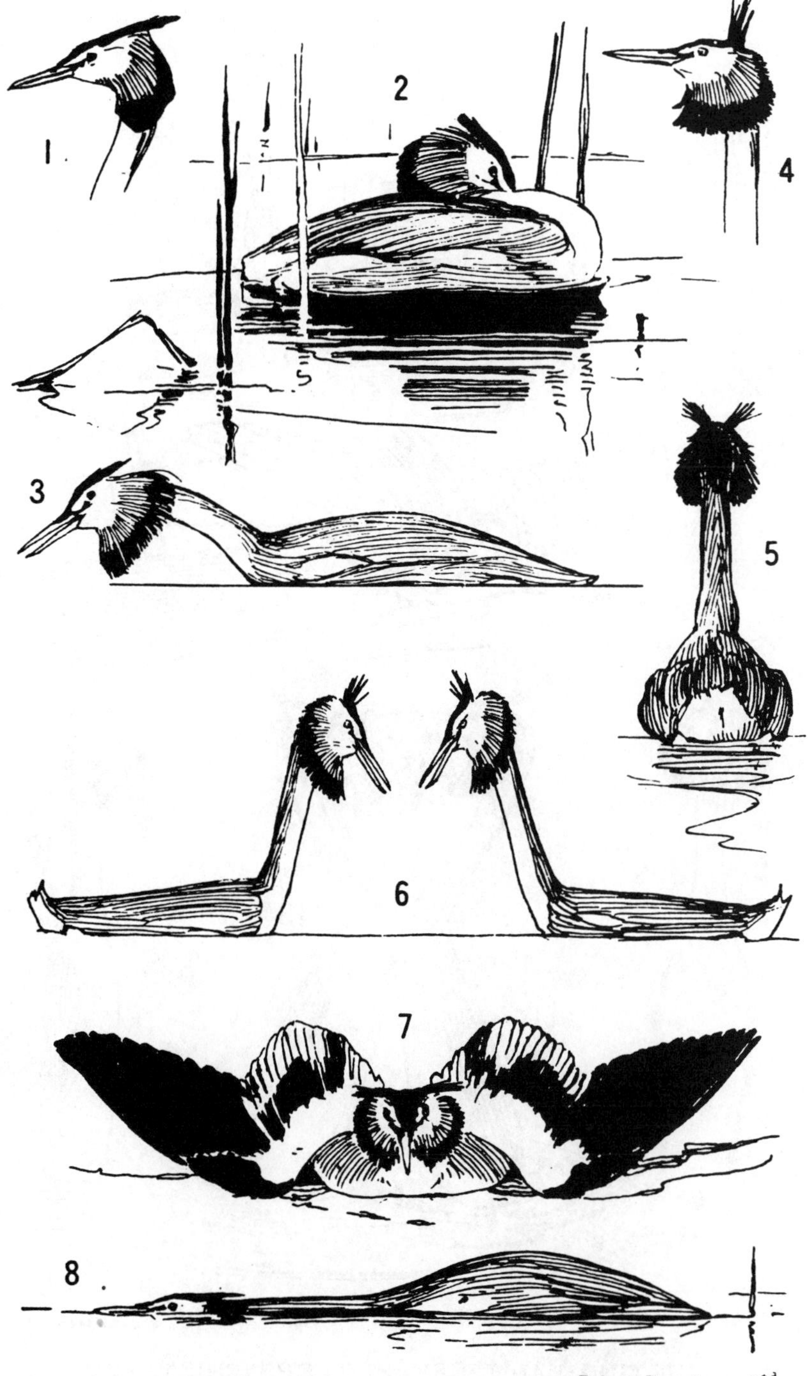

COURTING-HABITS OF PODICEPS CRISTATUS.

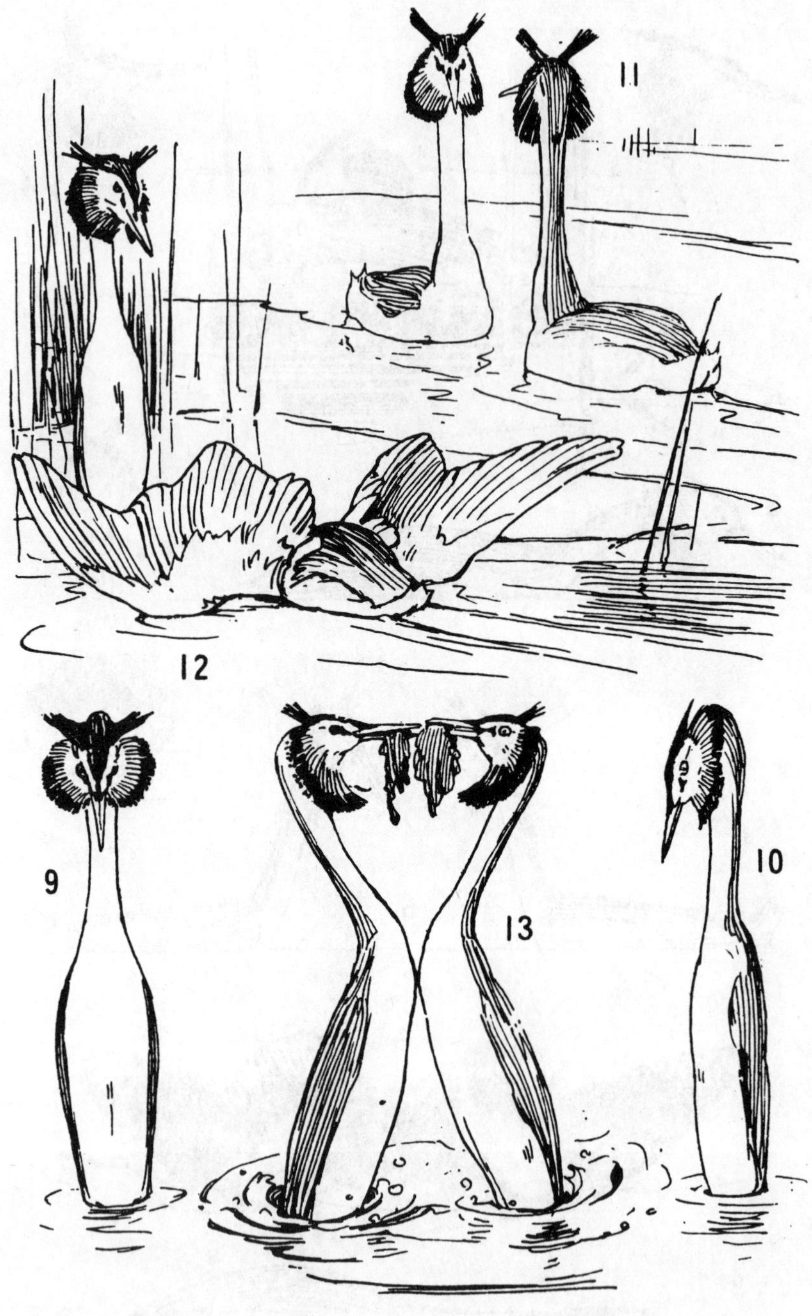

Bale & Danielsson, Ltd

COURTING-HABITS OF PODICEPS CRISTATUS.

16

ON CERTAIN PATTERNS OF MOVEMENT IN VERTEBRATES

O. Heinroth

This article was translated expressly for this Benchmark *volume by Chauncey J. Mellor and Doris Gove, University of Tennessee, Knoxville, from pp. 333–336, and 338–339 of Über bestimmte Bewegungsweisen der Wirbeltiere,* Gesellschaft naturforschender Freunde, Berlin, Sitzungsberichte *1930, pp. 333–342.*
Translation © 1985 by Chauncey J. Mellor and Doris Gove.

Whoever considers using behavior to recognize the phylogenetic relatedness of animal groups will find that peculiar more-or-less reflexive movements occur throughout certain classes, orders, or families, movements that are absent or performed differently in other forms. I am thinking here of scratching, shaking, stretching, yawning, and similar things. The purpose of this little essay is not to recount all of the known details, but to point out what one should pay attention to and what results can be achieved.

The *shaking*, familiar to anyone, of a dog coming out of the water seems to be fairly common in mammals and birds. But it seems to be completely absent in all reptiles, amphibians, and fishes. This movement is carried out with great vigor by, for example, dogs, bears, and horses, and passes through the whole body. It serves to remove foreign objects or materials like water and dust from the fur. Therefore, a kind of unpleasant stimulus extending over at least a large portion of the skin seems to precede the shaking. More-or-less naked forms such as the hippopotamus probably do not shake themselves, and one can assume that whales do not either. Sea lions and otters coming out of the water fling the water from their head and neck with a jerky, twisting motion. I know of no summary of which mammals shake as dogs do; this much at least is certain: that humans do not have this purely innate movement, and it is not characteristic of most simians, particularly the anthropoid apes. *Editor's Note:* Konrad Lorenz's copy of this paper carries on the margin his handwritten remark *"Irratum"* (erroneous statement). Heinroth seemed to favor the hypothesis that animals shake at will, and Lorenz (pers. commun.) conjectures that this explains why Heinroth overlooked the involuntary body shake in humans, which occurs, for example, at times lined to urination. (Information courtesy of Wolfgang Schleidt.)

Most birds shake extensively, which is understandable, considering their highly developed epidermal structures, the feathers. In doing this, all their feathers are uniformly and rather slowly made to stand on end; the shaking begins in the body and then generally proceeds over the neck toward the head. However, the course can also be in the opposite direction, for example, in owls. In this case, the shaking stimulus is not caused by foreign objects but by the fact that the feathers have somehow shifted or come out of position, which evidently causes unpleasant sensations on the skin. This seems to be the general case. With one flick the bird is once again neat and orderly. Some groups have a greater shaking tendency and therefore shake more than others. Most water birds tend to shake at each little provocation; for them, of course, constant plumage care is vitally important. In birds of prey, somewhat greater disturbances become necessary. Because one can observe shaking particularly often and easily in the class of birds, the issue of whether we are dealing with a reflex or a willful action is simple to solve in them. The fact that the shaking characteristic of a single species is carried out by all individuals identically, from the start, and without a model argues for reflex. However, the shaking reflex can be suppressed by other stimuli both before its inception and also during its initial stages. One can best notice this when trying to take a good picture of a more or less tame captive bird. If one takes such an animal from its enclosure and carries it to the photographic work area, then the plumage almost always becomes a bit disarrayed by the touching: the wing feathers do not lie properly, and also, in other parts of the body, the feathers have been disrupted by defensive movements. Now, if one does not want to get these disturbing irregularities on the plate, one has to wait until the bird has gotten its coat back in order. If the animal is tame and familiar with the surroundings, it will, as soon as it has quieted down, immediately shake so thoroughly that, in some species, one can hear the rustling for quite a distance, and then the picture can be taken. However, very shy birds, in their anxiety, hold their feathers stiff for a long time, and then the only thing that helps is patience. A quite gentle bristling of the small feathers occasionally runs over the body, indicating that the bird is in the mood to shake and does not dare to do so. Often, it helps then to step back a bit and look away. Only when the animals have become used to their situation do they shake more frequently and thoroughly until the feeling of well being of properly arranged feathers is achieved.

A movement corresponding, at least to a certain degree, to a shaking of body feathers, is what I call *wing flapping*. One sees it best and most often in geese and ducks. When they perform it on land or water, they generally straighten up somewhat and do a few quick wingbeats, finishing usually with an audible beat, whereupon the flight feathers disappear under the contour flank feathers. This very characteristic process occurs mainly when water has gotten on the contour flank feathers, something that occurs, for example, when non-diving ducks dive or when the animals are chasing each other on the water. It is characteristic that a domestic goose that had been born with no wings also carried out this wing flapping in its imagination, as one could see in the twitching of the breast muscles and other body movements when it had jumped from a steep bank into the water. Likewise, wing flapping is also performed by gulls, terns, cormorants, flamingoes, grebes, and coots, as well as most snipes, penguins, and alcids. So this behavior is probably performed by all bird groups that come into extensive contact with water. It occurs not only to shake off water, but also when the flight feathers have become disordered, for example, when the inner vanes have gotten over the outer ones. As in shaking, wing flapping can also be inhibited by external stimuli. Up to a certain extent, wing flapping is contagious, at least in social forms, as is yawning in humans; one can observe this especially well in a flock of geese that belong together.

In many other bird families, this wing flapping occurs only rarely, most often when the animals have bathed in water or dust. But then, in songbirds, parrots, storks, birds of prey, and others for example, it is a rather long wing fluttering or a vigorous wing beating that does not end with a distinct thump as it does especially in ducks. Gallinaceous birds assume an intermediate position, so that the flapping of a rock partridge *(Alectoris)* and a partridge *(Perdix)* is so quick that the uninitiated only notices it after it is over. In the alcids and guillemots, the very hard and conspicuous flapping is always combined with overall shaking. The extensive wing drying of cormorants that lack flank feather pockets and the standing posture of vultures with outstretched wings have nothing to do with the wing flapping mentioned above. The loudly audible wing flapping of the arrogant rooster about to crow, which of course is an expression of excitement, is likewise different, as is the repeated shaking of enraged geese.

[*Editor's Note:* Material has been omitted at this point.]

Scratching on the front part of the body also deserves special mention. Generally in mammals except man and the simians, it is performed only with the hind legs; as is known in dogs, the insides of the front paws are only used in removing objects such as muzzles from the head. Another exception is the way that cats wash themselves. Kangaroos, beavers, and various other rodents also use fingers for preening the head, chest, and abdomen. Birds scratch their heads and beaks with the claws of their front toes in two different ways: namely, birds like chickens simply reach "around the front," while almost all songbirds, for example, reach in the following manner: they raise the foot above and behind the wing to the head. We call this "around the back." These two possibilities, which, as innate actions, are each restricted to quite definite bird groups, are distributed across this class of animals in a peculiar fashion. The birds that scratch themselves around the back are: almost all songbirds (Passeriformes), swifts, goatsuckers, kingfishers, bee eaters, hoopoes, hornbills (but not toucans *(Ramphastes)*), bearded birds, (Capitonidae), and generally woodpeckers. Strangely enough, the Charadiidae, including the avocet, the black winged stilt *(Himantopus)* and the sand grouse *(Pterocles)* are also part of this group. It has been known for a long time that many songbirds have the innate habit of scratching the beak and then the head plumage after they have removed oil from the oil glands with the tip of the beak. This serves to lubricate the head plumage.

As a survey of these two forms shows, scratching around front or around behind is certainly not correlated with short or long leggedness, nor with living on branches or on the ground. Therefore, conditions of relationship must play a role, at least to a certain extent. In parrots, as Lucanus pointed out, it is remarkable that in the two groups in which the majority only rarely use the feet to hold large pieces of food (the flat-tailed parrots [Platycercidae] and the lories [*Trichoglossus*]), the around-the-back method of scratching is seen. All of the remaining parrot species scratch around the front. Consequently, in cases like this, one might look for the cause of this difference in a greater or lesser mobility of the legs and feet. However, it is strange that doves, storks, sandpipers, and ducks, which never grasp anything with their feet, scratch themselves around the front.

[*Editor's Note:* In the original, material follows this excerpt.]

Editor's Comments on Papers 17 Through 22

Whitman died in 1910, his major comparative work on pigeons and doves unpublished. His student, Wallace Craig, used some of his techniques and cited some of Whitman's observations, such as those on the use of imprinting to produce hybrid birds. Later Harvey Carr, one of J. B. Watson's most able students at the University of Chicago, edited Whitman's extensive unpublished notes and manuscripts with the help of Oscar Riddle. The excerpt presented here (Paper 17) describes the feeding of squabs by their parents, a phenomenon involving crop "milk." A half century later there was renewed and controversial interest in this behavior (c.f. Lehrman, 1955 and Klinghammer and Hess, 1964). The last section of this extract touches on imprinting. As Hess and Petrovich (1977) have documented and as has been demonstrated in prior readings in this book, imprinting has had a long history of being described, exploited, forgotten, and rediscovered. A note of interest is that Whitman's observations on imprinting were used by Kuo (1921) as a major line of evidence in his attempt to eliminate *all* notions of instinctive behavior, and even the relevance of heredity, in the study of behavior.

From 1908 until the early 1920s, Craig published several remarkable papers. In Paper 18 Craig's important distinction between appetitive behavior and consummatory acts is presented. This distinction helped resolve the dilemma involving two usages of the instinct concept: instinct as a goal-seeking urge, drive, or motivation finding expression in several outlets, and instinct as unlearned, stereotyped, usually adaptive behavior. Much needless controversy and muddled thought grew from the confusion. For example, McDougall's emphasis on the purposive aspects (e.g., Paper 8) contributed to the eventual reaction against instinct as a vague term used to cover ignorance and subvert empirical science.

Learning, Lorenz (Paper 23) later claimed, takes place during the appetitive, "searching" phases while the stereotypy is reserved for the consummatory act—the prey-killing, fighting, copulating, and suckling. Thus, for Lorenz, learning influences only the appetitve stage and not the form of the innate consummatory act. This analysis is simplistic but heuristic (Thorpe, 1979); it helped integrate both motivational and behavioristic analyses by breaking down "instincts" into a sequence. Much later controversy over the ethological rehabilitation of instinct missed the point that Lorenz used Craig's distinction to develop a middle ground beween vitalistic and mechanistic approaches: the appetitive was associated with 'motivation,' and the consummatory with a reductionistic reflexology, albeit including chains of reflexes and conditioning.

Craig corresponded with Lorenz (Thorpe, 1979) but apparently ceased active research in animal behavior by 1920. His article "Why Do Animals Fight?" (Paper 19) shows that the analysis of aggression did not need to take on the fatalistic overtones that were later attributed (somewhat unfairly) to the popular ethological writings on aggression of Lorenz, Ardrey, and Morris.

Our modern concept of social hierarchy and dominance stems from the studies on chickens of Schjelderup-Ebbe (1935, see also Schein, 1975). The modern companion concept of territory is largely due to the efforts of H. Elliot Howard (see Stokes, 1974), one of the last great amateur naturalists. A wealthy businessman, he spent many years of weekends observing birds in the English countryside. For Lorenz, Howard's observations on incomplete actions by birds in courtship, nesting, and brood care (Paper 20) were of great interest. They provided evidence that consummatory acts could wax and wane, varying in intensity for reasons other than fatigue or lack of experience. Howard also noted conflicting response tendencies in his birds, but was reluctant to view these as more than additonal mechanisms producing harmonious relations between the pair. A less Panglossian view of displacement activities and other motivational conflicts had to wait for Tinbergen and Kortlandt in 1940 (Tinbergen, 1951).

Heini Hediger is best known for his work in zoo environments through his many years as director of the Basel and then of the Zurich zoos in Switzerland. Notwithstanding his great interest in the practical knowledge of animal behavior relied on by zoo and circus personnel, he advocated careful use of ethological knowledge of animals in their display. His views on animals in captivity have been presented at length in books readily available in English (e.g., Hediger 1950).

Hediger also pioneered the careful study of spatial relationships between animals of the same and different species and applied the results to the taming of wild animals in captivity. Through his early work concepts such as individual distance, critical distance, and flight distance have entered the ethologist's lexicon (Paper 21). These were among the first ethological concepts used in the study of human behavior (e.g., Hall, 1966). These spatial concepts not only provided practical knowledge for dealing with captive animals and observing wild ones, but demanded quantitative measures of behavior. As the reader will have noticed, few numbers or quantitative measures have intruded into the descriptions and theory of preceding papers. Hediger helped mark the end of simple qualitative or crude ordinal (more or less) studies. Quantification was implicit in Lorenz's subsequent models and methodological precepts, especially the Method of Dual Quantification (Lorenz, 1950), although rarely a factor in Lorenz's own observations. Tinbergen, more than anyone else, brought quantitative description and experimentation to classical ethology (Tinbergen, 1951).

Erich von Holst was a physiologist brilliant in both theory and experiment. His studies on the endogenous rhythmicity of locomotion in a variety of animals provided Lorenz with needed physiological support in advancing his ideas on endogenous factors in behavior. These had seemed to clash with then-current views of the central nervous system. Paper 22, one of von Holst's first, was cited often by both Lorenz and Tinbergen. Some of his other important papers are also available in translation (von Holst, 1973). Shortly before his death in the early 1960s, von Holst began a seminal series of experiments on the production by brain stimulation of normal, extended sequences of behavior in domestic chickens. These contributed to the current interest in neuroethology that we are now witnessing (e.g., Ewart, Capranica, and Ingle, 1984).

REFERENCES

Ewart, J. -P., R. R. Capranica, and D. J. Ingle, eds., 1984, *Advances in Vertebrate Neuroethology,* Plenum, New York.

Hall, E. T., 1966, *The Hidden Dimension,* Doubleday, New York.

Hediger, H., 1950, *Wild Animals in Captivity,* Butterworth, London.

Hess, E. H., and S. B. Petrovich, 1977, *Imprinting,* Benchmark Papers in Animal Behavior Series, Dowden, Hutchinson & Ross, Stroudsburg, Pa., 333p.

Holst, E. von, 1973, *The Behavioral Physiology of Animals and Man,* Methuen, London, 341p.

Klinghammer, E., and E. H. Hess, 1964, Parental Feeding in Ring Doves (*Streptopelia roseogrisea*): Innate or Learned?, *Z. Tierpsychol.* **21:**338-347.

Kuo, Z. Y., 1921, Giving up Instincts in Psychology, *J. Phil.* **17:**645-664.

Lehrman, D. S., 1955, The Physiological Basis of Parental Feeding Behaviour in the Ring Dove (*Streptopelia risoria*), *Behaviour* **7:**241-286.

Lorenz, K., 1950, The Comparative Method in the Study of Innate Behavior Patterns, *Symp. Soc. Exp. Biol.* **4:**221-268.

Schein, M. W., 1975, *Social Heirarchy and Dominance,* Benchmark Papers in Animal Behavior Series, Dowden, Hutchinson & Ross, Stroudsburg, Pa.

Schjelderup-Ebbe, T., 1935, Social Behavior of Birds, in *Handbook of Social Psychology,* C. Murchinson, ed., Clark Univ., Worcester, Mass., pp. 947-972.

Stokes, A. W., 1974, *Territory,* Benchmark Papers in Animal Behavior Series, Dowden, Hutchinson & Ross, Stroudsburg, Pa. 398p.

Thorpe, W. H., 1979, *The Origin and Rise of Ethology,* Heinemann, London, 174p.

Tinbergen, N., 1951, *The Study of Instict,* Clarendon Press, Oxford, 228p.

17

Reprinted from pages 65-68 of *The Behavior of Pigeons. Posthumous Works of Charles Otis Whitman,* vol. III, H. A. Carr, ed., Carnegie Institute of Washington, Washington, D. C., 1919, 161p.

FEEDING AND CARE OF THE YOUNG

C. O. Whitman

[*Editor's Note:* In the original, material precedes and follows this excerpt.]

FEEDING OF YOUNG.

Most people know something of the way the pigeon feeds its young. There are, however, some rather extraordinary things that happen in connection with feeding that deserve mention. Suppose we place the young of a small species under the care of a larger species—a young ring-dove, for example, under a common pigeon or homer. The old birds will take care of this young ring-dove just as faithfully as they would take care of their own young. But when this young bird gets out of the nest, at a week or 10 days, he is not able to take all the food that the old birds want to get rid of; the homer can not feed this young and small bird as much as he desires to feed it, and we can imagine that this gives the homer a quite unpleasant feeling, for he behaves in a way to indicate this. When the young bird does not accept all the food that is offered to him, then the foster-parent begins to tease the young to accept it. The parent will pull the feathers of the young very gently and bite its beak, or give it a gentle pinch, just to stimulate it to feed. The little bird will respond to these things until it has so much that it can hold no more; but the old bird urges it to accept still more. Finally the parent begins to pinch the young bird's beak a little more severely, and to pull its feathers a little more strongly, and—being a larger species—behaves in a way rather rough for the smaller species, and in the end alarms the young bird, which now tries to get out of the way. The old bird follows it up—he can not give up, but is fully determined to get rid of all his food. Naturally it troubles and startles the young bird to be pursued in this rough way and the old bird sooner or later gets angry and begins to peck the squealing, retreating young very severely, sometimes pulls out its feathers, sometimes pulls off the skin of its head, neck, and back, and if the young bird is not taken away it is very often killed.

Now, I am perfectly certain that the old bird behaves in this way, not from any malicious motives, but just for its normal relief. There are a great many ways of becoming sure of this fact. This behavior, of course, is found in birds of different species. I have tried some of them in many different ways and found this view of the behavior confirmed by other behavior which is analogous. In further illustrations of this, we may consider a situation which is sometimes met with in the case of a common pigeon and her own young. When one of the two young of a brood is hatched a little too long after the first one, the first hatched is apt to get quite an extra start in growth and be much stronger than the second one; when the second one is hatched, it is fed a "little" and the other one is fed "more," and so from day to day the first one continues to outgrow the second one to such an extent that the second one has not the strength or ability to stand up and get as much of the food as the first one; so the difference increases from day to day. The second bird is more and more hungry, because the old bird is satisfied with the vigorous bearing and desire and energy on the part of the stronger bird. The second one does not satisfy the old bird and is more and more neglected and in the end it becomes a dwarf bird, or it may die.

On the question as to the part played by intelligence in these cases, I may say that it does not seem to me that the birds exercise intelligence; they simply act in accord with their feelings. They appear to be very fond of the young birds, but their fondness for the young is all determined by their feelings, or desire for relief. The moment they get over their desire to feed they also get over their fondness for the young. If the eggs do not hatch they have a desire to feed, but it is not so strong as to lead to bad results and they soon outgrow it; it does not then make them sick. In those cases, however, where parents really get started in feeding the young, the need to feed is most pressing; and if the young die in two or three days after hatching, then the old birds are not only frequently made sick as a result of not being able to feed, but frequently sick to death.

The special secretions in the pigeon's crop seems to be stimulated in part by the sight of the young and the amount of stimulation, or rather the amount of food which the parents want to relieve themselves of, and it rapidly increases in amount during the days which follow the beginning of feeding. If they do not begin actual feeding they get over it, but if they once get started it is necessary to go on; otherwise they suffer.

I have noticed in pigeons that the "weak are neglected" and the strong favored. This is seen in the feeding as noted above. To get the full benefit of the parent's ability to feed the young must *push, squeal, and flap its wings—and all this with vigor*. The moment these acts are not well performed, the parent's exertion dwindles for lack of stimulus.

The feeding process is performed—at least in many cases—purely *to get relief*. The old bird gives "milk," and at intervals this secretion must be thrown out. I have seen a parent try to induce young to feed when the latter had no desire, having been well filled by the other parent. The old bird gets into the nest and calls; if there is no response it takes hold of the beak of the young and begins to pump vigorously. If the young die, or if they are taken away, one may see after 6 to 10 hours how eager the parents are to relieve themselves of the food reserved for the young. They are ready to feed anything that will put its beak into their mouths; they will often return to weaned young (crested pigeons) and feed them for a second time. They will try to feed an adult bird if that bird's beak is only held in their mouth, and held low enough to simulate young in the nest.

On the other hand, the old birds will often try to feed when they have nothing to give, *e. g.*, when young are given to them before the regular time for hatching has come. Thus the act is performed sometimes for relief, sometimes in answer to the teasing of a hungry young.

The theory that the act was primarily one for relief from too much or from indigestible food, and secondarily turned to use in feeding, has one great difficulty, namely, to account for the "milk." I do not think that could have been at first a product of a useless nature, like ejected food. It is something that has probably been developed in conjunction with the feeding. The dove's method of feeding occurs also in some other birds.[1]

As the young get larger, their feeding becomes for the parent more and more of a dread, as they are ravenous and push with all their strength to get the food. The old bird becomes wary and makes haste to finish quickly and retire, going out of the cot often before it has full relief. It then continues to call, but hesitates to go in. The young stand at the hole and squeal. The appearance is as if the old bird wanted to entice the young outside. Later the old one grows still more cautious and the young venture out and are fed. The next time they go out quicker and the old bird retreats. It then looks as if the old bird wanted to teach the young to fly. It is evident, however, that this is no thought-out art,

---

[1] Lucas states (Mental Traits of the Pribilof Fur Seal, V, 1899), that the young of the fur seal was fed by the mother for her own relief.

but an art that grows out of the situation—young clamoring for food, old ones anxious to relieve themselves, and fear of the push of the young.

The old bird normally feeds two birds at the same time, and it seems to need the two-sided contact to get full relief.[1] If there is only one young, the old bird after feeding from one side may walk off to the other side of the nest and thus give a chance for a change in position. (SS 10, R 24, B 2b, R 7.)

[*Editor's Note:* Material has been omitted at this point.]

ANOMALOUS FEEDING BEHAVIOR.

Two young hybrids, one white (*H*) and the other blond (*J*), were with their foster-parents in a cage. The old birds were at the time out of the nest. The white young (*H*) appealed to *J* for food, whereupon the latter several times opened his beak to receive that of his nest-mate, and shook his crop and wings as if it would give up food. Soon the white young actually inserted its beak into that of *J* and the latter fed it in a vigorous manner, pumping up food liberally until I interrupted, thinking it could not be well for *J* to part with its food. *J* is only 12 days old, and it is at least 3 days behind its companion in the development of its feathers. It was a strange sight to see this unfledged bird, with only pin-feathers on its head, and much of its body bare, feeding another bird more advanced in its feathers, although of about the same age and weight. I have never before seen or heard of such a performance.

I have seen both young and old throw up food that was indigestible, especially when sick. I have seen them in good health overeat and cram themselves with more than could be readily swallowed, and then seek relief in throwing up. In so doing the movements may all be the same as in feeding the young—the head is held down, the beak open, the

[1] The parent takes care to feed both young. I am not sure that this is intentional, but it looks almost so. The parent feeds for a time, then stops, disengages, and moves around a short distance before commencing to feed again, so that it usually happens the other young gets the second feeding. (Conv. W. C.)

[2] This statement is made for the bronze-wing, which is a species that copulates on the ground. For those species which "bill" and copulate on a perch, or branch of a tree, the shifting of the billing from one side to the other would seem to be a rather difficult or awkward matter. Possibly, however, it occurs in these latter species also. It certainly sometimes occurs in the case of ring-doves when these birds copulate on the ground.—O. R.

crop is shaken, and with it the whole body, especially the wings. The power to regurgitate is, then, common to doves, old and young, and *it has come secondarily into service in the care of the young.*

What prompted *J* to feed its white companion? I think the crop was a little overloaded, and that, together with the stimulus of a teasing beak, led it to give up food. It was not, however, an act of vomiting merely; for *J* offered its open beak several times to the companion, and in a way that was unmistakable, so that I wondered at it before the feeding began. When the beak was inserted in the open beak the whole machinery of correlated movements was set in motion and the young bird behaved in all respects like an adult bird. The beak inserted in the throat may have much to do with stimulating the process.

A hybrid between a ring-dove and the European turtle was 11 days old; it had as a nest companion a ring-dove 1 week old. The ring was small and covered with pin-feathers; the hybrid had many wing-coverts and scapulars unfolding. These birds were under the care of a pair of ring-doves; both were perfectly healthful and strong. I noticed the young ring teasing the hybrid with its beak, as young birds scantily fed often do with a nest-mate. The vigorous billing about the hybrid's beak and neck led it to feel like feeding, and it opened its beak several times for the ring, and once after offering the open beak, it put its head down and shook its wings and crop as an old bird does in feeding. The young ring did not get its beak into that of hybrid, but came near doing so. (R 24.)

TRANSFERENCE OF YOUNG TO FOSTER PARENTS.

The young of any species may be transferred to foster-parents, provided they are put under the foster-parents at the time the eggs of the latter are due to hatch. It does not seem to make any difference whether the young birds are transferred when they are just hatched, or at a week or 2 weeks old, provided only that they are not old enough to stand up straight and raise the head too high. The old birds do object to taking care of any young birds that stand up and look too much like old birds. If the young simply sits down, holds its head down in the proper position for care, that is all that is required; the old birds will feed it just as well as if it were hatched out under them. If eggs are transferred one must always be sure that they are properly timed; that is, timed to hatch with the eggs which are removed; if they hatch too early, or are due to hatch too long after the time of the eggs removed, the transfer is not successful. The foster-parents are as well satisfied with the eggs of other species as with their own; they do not know the difference.

The necessity of exactly matching the time of hatching of the transferred eggs with that of the eggs removed is one of the complications that is met with in attempts, such as my own, to obtain and hatch unusually large numbers of eggs from particular pairs of birds. I have had some trouble of this sort. The common pigeons, however, are a little less particular in the matter of the exact time the eggs are due to hatch than are the wild species. Also, in crossing common pigeons with the wild species one finds that the former will usually continue the incubation a few days after their mate's usual time. The wild passenger-pigeon never waits more than 10 or 12 hours. If the egg does not hatch within that time he leaves it; even if the shell is broken and the bird is nearly ready to hatch it is deserted. If young are put into their nests before they are due to hatch they are not able to feed them; much care in this matter is necessary. "Pigeon milk," as it is called, seems to be ready at the proper moment. Everything is well-timed, and if there are young birds in the nest before this time arrives they will get no food. (SS 10.)

18

Reprinted from *Biol. Bull.* **34:**91–107 (1918)

APPETITES AND AVERSIONS AS CONSTITUENTS OF INSTINCTS.

WALLACE CRAIG,

UNIVERSITY OF MAINE.

GENERAL ACCOUNT OF APPETITE AND AVERSION.

The overt behavior of adult animals occurs largely in rather definite chains and cycles, and it has been held that these are merely chain reflexes. Many years of study of the behavior of animals—studies especially of the blond ring-dove (*Turtur risorius*) and other pigeons—have convinced me that instinctive behavior does not consist of mere chain reflexes; it involves other factors which it is the purpose of this article to describe. I do not deny that innate chain reflexes constitute a considerable part of the instinctive equipment of doves. Indeed, I think it probable that some of the dove's instincts include an element which is even a tropism as described by Loeb. But with few if any exceptions among the instincts of doves, this reflex action constitutes only a part of each instinct in which it is present. Each instinct involves an element of appetite, or aversion, or both.

An appetite (or appetence, if this term may be used with purely behavioristic meaning), so far as externally observable, is a state of agitation which continues so long as a certain stimulus, which may be called the appeted stimulus, is absent. When the appeted stimulus is at length received it stimulates a consummatory reaction, after which the appetitive behavior ceases and is succeeced by a state of relative rest.

An aversion (example 7, p. 100) is a state of agitation which continues so long as a certain stimulus, referred to as the disturbing stimulus, is present; but which ceases, being replaced by a state of relative rest, when that stimulus has ceased to act on the sense-organs.

The state of agitation, in either appetite or aversion, is exhibited externally by increased muscular tension; by static and phasic contractions of many skeletal and dermal muscles, giving rise to bodily attitudes and gestures which are easily recognized signs or "expressions" of appetite or of aversion; by restlessness; by activity, in extreme cases violent activity; and by "varied effort" (Lloyd Morgan, '96, 7, 122, 154; Stout, '07, 261, 267).

In the theoretically simplest case, which I think we may observe in doves to some extent, these states bring about the appeted situation in a simple mechanical manner. The organism is disturbed, actively moving, in one situation, but quiet and inactive in another; hence it tends to move out of the first situation and to remain in the second, obeying essentially the same law as is seen in the physical laboratory when sand or lycopodium powder on a sounding body leaves the antinodes and comes to rest in the nodes.

But pigeons seldom are guided in so simple a manner. Their behavior involves other factors which must be described in connection with appetite and aversion.

An appetite is accompanied by a certain *readiness to act*. When most fully predetermined, this has the form of a chain reflex. But in the case of most supposedly innate chain reflexes, the reactions of the beginning or middle part of the series are not innate, or not completely innate, but must be learned by trial. The end action of the series, the consummatory action, is always innate. One evidence of this is the fact that in the first[1] manifes-

[1] To see the appetitive nature of an instinct, it is necessary in some cases to observe an individual animal carefully during its first performance of the act in question. But the performance may be so quick that the observer is quite unable to analyze it. Analysis may be aided by preventing the animal from attaining the consummatory situation for a time, so that the appetitive phase is prolonged, as it were magnified. My cripple dove (example 5, p. 99) afforded just this aid to analysis. The literature is full of reports of instinctive behavior which might well be further analyzed. Consider for example the case of the young moorhen cited by Lloyd Morgan ('96, 63) which had never previously dived, but on being suddenly frightened by a puppy, dived like a flash. That act was too quick for us to analyze it. But if we could successfully impede the diving of a young moorhen so as to prolong the phases of the act, I think it probable that we should find an appetite for the consummatory situation (that of being under water) and a restless striving until it is attained; and that some details in the series of actions, details which in a normal dive are very sure to be hit upon by accident, are not innately predetermined. When one sees the first performance of an instinctive act take place very quickly and with apparent perfection, this does not prove that there is an innate chain reflex determining every detail of the act.

tation (also, in some cases, in later performances) of many instincts, the animal begins with an *incipient consummatory action*, although the appeted stimulus, which is the adequate stimulus of that consummatory action, has not yet been received. I speak of an incipient "action" rather than "reaction," because it seems clearly wrong to speak of a "reaction" to a stimulus which has not yet been received. The stimulus in question is obtained only after a course of appetitive, trial-and-error behavior. When at last this stimulus is obtained, the consummatory reaction takes place completely, no longer incipiently. Then the appetitive behavior ceases; in common speech we say the animal is "satisfied."

One may observe all gradations between a true reflex and a mere readiness to act, mere facilitation. Thus, in the dove, a stimulus from food in the crop may cause the parent to vomit the food or to feed it to young: there are all gradations from an immediate crop-reflex, in which the food is vomited upon the ground, through intermediate cases in which the parent is much disturbed by the food in his crop, but appetitively seeks the young and induces them to take the food; to other cases is which the parent is only ready to feed the young if importuned by them; and finally to cases in which the stimulus from the crop does not even cause facilitation, and the parent does not disgorge the food at all, even if importuned by the young.

While an appetite is accompanied by readiness for certain actions, it may be accompanied by a distinct *unreadiness* for certain other actions, and this is an important factor in some forms of behavior. It is altogether probable that this unreadiness is due in some cases to the fact that the activity of certain neurones *inhibits* the activity of certain other neurones. It is now well-known, too, that unreadiness may be due to the condition of the internal secretions. And the mutual exclusion of certain forms of instinctive behavior is inevitable, due to the incompatibility (Washburn, '16) of their motor components.

Unreadiness may be accompanied by aversion, and vice versa; but either of these may occur without the evident presence of the other. An aversion is sometimes accompanied by an innately determined reaction adapted to getting rid of the disturbing stimulus, or—this point is of special interest—by two alternative reactions which are tried and interchanged repeatedly until the disturbing stimulus is got rid of (see example 7, page 100).

The escape from a disturbing situation or the attainment of an appeted one is accomplished, in case of some instincts, far more surely and more rapidly after one or more experiences. In the first performance of an appetitive action, the bird makes a first trial; if this fails to bring the appeted stimulus he remains agitated and active, and makes a second trial, which differs more or less from the first; if this fails to bring the appeted stimulus he remains still active and makes a third trial; and so on until at last the appeted stimulus is received, the consummatory reaction follows, and then the bird comes to rest. In later experience with the same situation, the modes of behavior which were followed immediately by the appeted stimulus and consummatory reaction are repeated; those which were not so followed tend to drop out.

If a young bird be kept experimentally where it cannot obtain the normal stimulus of a certain consummatory reaction, it may vent that reaction upon an abnormal or inadequate stimulus, and show some satisfaction in doing so; but if the bird be allowed at first, or even later, to obtain the normal stimulus, it will be thereafter very unwilling to accept the abnormal stimulus. That this is true of the sex instinct has been shown in a former article (Craig, '14). It is true also of the appetite for a nest. Thus a female dove which has never had a nest, nor material to build one, lays eggs readily on the floor; but a dove that has had long experience with nests will withhold her egg if no nest is obtainable. The male dove similarly, if he has never had a nest, goes through the brooding behavior on the floor; but an experienced male is unwilling to do so, and shows extreme anxiety to find a nest. These examples illustrate the fact that the bird must *learn* to obtain the adequate stimulus for a complete consummatory reaction, and thus to satisfy its own appetites.

There is often a struggle between two appetites, as when a bird hesitates, and it may hesitate for a long time, between going on the nest to incubate and going away to join the flock, eat, etc. By watching the bird one can predict which line of behavior it will follow, for each appetite is distinguished by its own expressive signs (consisting partly of the incipient consummatory action), and one can see which appetite is gaining control of the organism.

These outward expressions of appetite are signs of physiological states which are but little known. Since my own observations have been on external behavior only, I say little about the internal states. They are probably exceedingly complex and numerous and similar to the physiological states which in the human organism are concomitants of appetites,[1] emotions, desires. They doubtless include stimulations from interoceptors and proprioceptors; perhaps automatic action of nerve centers; perhaps readiness or unreadiness of neurones to conduct. It is known that some of the periodic appetites are coincident with profound physiological changes. Thus Gerhartz ('14) found that during the incubation period in the domestic fowl the metabolism of the body as a whole is at a low ebb. In some cases a stimulus from the environment is the immediate excitant of an appetite; especially, stimulation of a distance-receptor may arouse appetite for a contact stimulus, as when the sight of food arouses appetite for the taste of it. But probably in every case appetite is dependent upon physiological factors. And in many cases the rise of appetite is due to internal causes which are highly independent of environmental conditions, and even extremely resistant to environmental interference.

Appetitive behavior in vertebrates is evidently a higher development of what Jennings ('06, p. 309) calls the positive reaction in lower organisms; aversive behavior in vertebrates corresponds to what Jennings (p. 301) calls negative reactions.

The attempt to distinguish between instinct and appetite, as in Baldwin's Dictionary ('01), is not justified by the facts of behavior. Baldwin says : "Appetite is distinguished from instinct in that it shows itself at first in connection with the life of the organism itself, and does not wait for an external stimulus, but appears and craves satisfaction." These characteristics,

[1] Hunger furnishes a typical case of appetitive behavior (Carlson, '16; Ellis '10, 198–199). Carlson makes a distinction between hunger and appetite. The distinction he finds is certainly real, but the use of words is unfortunate, for hunger is clearly one kind of appetite.

here ascribed to appetite, are the very ones which I have observed in the instinctive behavior of pigeons. The instincts of pigeons satisfy Baldwin's further description of appetite in that each appears first as a "state of vague unrest" involving especially "the organs by which the gratification is to be secured"; and "a complex state of tension of all the motor . . . elements whenever the appetite is aroused either (*a*) by the direct organic condition of need, or (*b*) indirectly through the presence or memory of the object." This last point is illustrated, *e. g.*, by doves learning to drink (example 1, page 97), in whom the sight of the water-dish at a distance aroused the drinking actions by associative memory. I have observed appetitive behavior as Baldwin describes it in nearly all the instinctive activities of doves, and I think that sufficient observation will reveal it in all their instincts.

The most thorough attempt to distinguish instincts from appetites and to show the logical consequences of such distinction, in all the literature to which I have access, is in an old article by Professor Bowen ('46). This article is still worth study, to suggest the conclusions to which one is logically led if he denies that instincts contain any element of appetite. These conclusions, taken almost literally from Bowen, may be summarized as follows: (1) (P. 95) "If the name of instinct be denied to these original and simple preferences [appetites] and aversions, there will appear good reason to doubt whether man is ever governed by instinct, whether all his actions are not reducible to passion, appetite, and reason." (P. 115) The "passions" of man can not be concomitants of instinct. (2) (P. 117) "Instinct is not a free and conscious power of the animal itself. It is, if we may so speak, a foreign agency, which enters not into the individuality of the brute." (P. 118) Instinct "has no effect on the rest of their conduct, which is governed by their own individuality." (3) Bowen contends with logical consistency that if instinct contains no appetitive factor, the ends toward which instincts work, as seen by an observer, are not ends for the agent; that therefore the agent has no power to make the instinctive behavior more effective. In short, instinctive behavior is not susceptible of improvement by intelligence. (4)

Bowen concludes that the intellect and the "passions" of man are not products of evolution. (5) It may be added that even Bowen, strive as he did to separate appetite from instinct, was compelled to admit that the attempt at such separation leads one into difficulties and disputed cases. In contravention of Bowen's conclusions I contend: (1) That much of human behavior is instinctive. (2) That Bowen's description of instinct as "a foreign agency, which enters not into the individuality" is true of reflex action, such as coughing or sneezing, but is not true of instinctive behavior, which is extremely different from such mere reflexes. (For a fuller statement on this point, see below, page 106. See also Hobhouse, '15, 98–99.) (3) That, of the useful results toward which instincts tend, some, not all, are ends for the agent. For they are the objects of appetites, and the animal strives and learns to attain them. (4) That human conative behavior evolved from the instinctive appetitive behavior of lower animals.

In another article I hope to publish soon a further discussion of the literature.

Examples.

1. The case of doves learning to drink, as described in detail in a former article (Craig, '12), illustrates appetite. The observed appetitive behavior was aroused by stimulation of distance-receptors, such as the sight of the water-dish being brought to the cage, and of the man bringing it; these acted as appetizers. Each dove, as soon as it had learned to associate such stimuli with the drinking situation, responded to these stimuli by making drinking movements (incipient consummatory action) at once without going to the water dish. The first drinking movements failing to bring water, the dove repeated these movements again and again, sometimes walking a few steps, sometimes turning round, until after many trials and many errors it did get its bill into the water, received the stimulus from water in the mouth (appeted stimulus), whereupon the drinking movements (consummatory reaction) were made not incipiently but completely, the water being swallowed, after which the bird rested and appeared satisfied.

2. A good example of appetitive behavior is seen in the way in

which a young male dove locates a nesting site for the first time. The first thing the observer sees is that the dove, while standing on his perch, spontaneously assumes the nest-calling attitude, his body tilted forward, head down, as if his neck and breast were already touching the hollow of a nest (incipient consummatory action), and in this attitude he sounds the nest-call. But he shows dissatisfaction, as if the bare perch were not a comfortable situation for this nest-dedicating attitude. He shifts about until he finds a corner which more or less fits his body while in the tilted posture; he is seldom satisfied with his first corner, but tries another and another. If now an appropriate nest-box or a ready-made nest is put into his cage, this inexperienced dove does not recognize it as a nest, but sooner or later he tries it, as he has tried all other places, for nest-calling, and in such trial the nest evidently gives him a strong and satisfying stimulation (the appeted stimulus) which no other situation has given him. In the nest his attitude becomes extreme; he abandons himself to an orgy of nest-calling (complete consummatory action), turning now this way and now that in the hollow, palpating the straws with his feet, wings, breast, neck, and beak, and rioting in the wealth of new, luxurious stimuli. He no longer wanders restlessly in search of new nesting situations, but remains satisfied with his present highly stimulating nest.

3. Fetching straws to the nest is apparently due to an appetite for building them into the nest. The dove has an innate tendency to pick up straws, and an innate tendency to build them into the nest (consummatory reaction); but it has apparently no innate tendency to carry a straw to the nest, no innate "chain" of reflexes. When an experienced bird finds a straw he seizes it repeatedly and toys with it, sometimes making movements resembling those by which he would build the straw into the nest. He seems thus to get up an appetite for building the straw in, and when this appetite is sufficiently aroused he flies to the nest, guided by associative memory, and performs the consummatory reaction completely. A young female, no. 70, which I observed picking up a straw for the first (?) time, on her 54th day, showed the lack of a "chain reflex." For she continued toying with the straw an excessively long time, not carrying it at all,

though she happened to be very near the nest. This was the more remarkable as she had a well-formed habit of going to the nest on all occasions. At length she did go to the nest with her straw, and made well-ordered movements to build it in.

4. The male and the female dove take regular turns in sitting on the eggs. The male is seized by the appetite for brooding about 8 or 9 A. M., and the female about 5 P. M., the state evidently being brought on in each case by physiological causes which are part of the daily physiological rhythm. When either one, *e. g.*, the female, comes to the side of the nest prepared to enter and sit, she already has somewhat the attitude of the sitting bird, the body sunk down on the legs and the feathers fluffed out (incipient consummatory action). If her sitting appetite be thwarted, as by her mate refusing to budge from his position, she shows restlessness and makes intelligent efforts to obtain possession of the nest. When at last her mate yields his place, she steps exultingly into the center of the nest and settles herself on the eggs with many movements indicative of satisfying emotion (complete consummatory reaction).

A broody hen of course illustrates the same principle.

5. It is an interesting fact, exhibited in a variety of instincts, that a young bird may make feints of performing actions which it has never yet performed. Thus the young dove makes feints of flying before it has ever flown. This was illustrated in a peculiarly instructive manner by one of my young doves, no. 46, which developed cripple wings and was unable to fly. When placed in a box with sides 3½ inches high it was just able to jump on the edge. Nevertheless, when its roosting instinct developed, it endeavored strenuously every evening to fly to the perch which was some inches above its head. It looked at the perch and aimed at it with perfect definiteness, opening its wings and making feints of flying. In the evolution of birds, there can be no doubt, flying developed gradually from jumping. The new movements of flying were gradually intercalated into the interval between the initial action, leaping from the ground, and the final action, landing again upon the feet. The young dove to this day shows *first* the incipient end action, aiming at the perch to be alighted on, and only after it has launched itself

toward this end situation does the "chain" of flight reactions take place.

6. In the pigeons the order of activities culminating in the sexual act is, first display, second billing, third copulation, with numerous details each finding a place in the succession. Yet the sexual tendency is mainfestly present from the beginning of the "chain," and the preliminary steps are directed, with much guidance by experience, toward securing the stimulation required for discharging the sexual reflex. In absence of the normal stimulus to the consummatory reaction, the instinct manifests itself in marked appetitive behavior, and, especially in inexperienced birds (Craig, '14), in those imperfect consummatory reactions known as perversions and auto-erotic phenomena. The behavior of the sexual appetite is now so well known that it may be cited as the type of appetitive behavior; and to readers who are familiar with modern analyses of the sex instinct I may make my whole article clearest by saying that all the appetitive mechanisms I have mentioned, and I believe all the instincts of the dove, behave in the same manner as that of sex, in regard to appetitive manifestations and anticipation of the consummatory reaction.

7. I shall take space to describe only one example of aversion—the so-called jealousy of the male dove, which is manifested especially in the early days of the brood cycle before the eggs are laid. At this time the male has an aversion to seeing his mate in proximity to any other dove. The sight of another dove near his mate is an "original annoyer" (Thorndike, '13, Chap. IX.). If the male sees another dove near his mate, he follows *either of two* courses of action; namely, (*a*) attacking the intruder, with real pugnacity; (*b*) driving his mate, gently, not pugnaciously, away from the intruder. When he has succeeded either in conquering the stranger and getting rid of him, or in driving his mate away from the stranger, so that he has got rid of the disturbing sight of another dove in presence of his mate, his agitation ceases. If we prevent him from being successful with either of these methods, as, by confining the pair of doves in one cage and the third dove in plain sight in a contiguous cage, then he will continue indefinitely to try both methods. If we leave all three

doves free in one pen, the mated male will try the mettle of the intruder and conquer him if he can; if he fails, he will turn all his energies into an effort to drive his mate away from the intruder. Or if in former experiences he has learned to gage this individual intruder, if he conquered him before he will promptly attack him now, but if defeated by him before he will now choose the alternative of driving his mate away. In sum, the instinctive aversion impels the dove to thoroughly intelligent efforts to get rid of the disturbing situation.

8. In some cases the seeking of a certain situation involves both appetences and aversions in considerable number. Thus, when the day draws to a close, each dove seeks as its roosting-place a perch that is high up, with free space both below it and above it, with no enemies near, with friendly companions by its side, but these companions not too close, not touching (except in certain cases of mate, nest-mate, or parent). The endeavor to achieve this complex situation, to secure the appeted stimuli and to avoid the disturbing ones, keeps the birds busy every evening, often for an hour or more.

Cycles.

Instinctive activity runs in cycles. The type cycle, as it were a composite photograph representing all such cycles, would show four phases as follows.

Phase I.—Absence of a certain stimulus. Physiological state of appetite for that stimulus. Restlessness, varied movements, effort, search. Incipient consummatory action.

Phase II.—Reception of the appeted stimulus. Consummatory reaction in response to that stimulus. State of satisfaction. No restlessness nor search.

Phase III.—Surfeit of the said stimulus, which has now become a disturbing stimulus. State of aversion. Restlessness, trial, effort, directed toward getting rid of the stimulus.

Phase IV.—Freedom from the said stimulus. Physiological state of rest. Inactivity of the tendencies which were active in Phases I., II., III.

Some forms of behavior show all four phases clearly. The following are examples.

Sex.—(Phase I.) The dove, either the male or the female, shows sexual appetite and invites the mate to sexual activity. Gradually they lead up to (Phase II.) the consummatory sexual act. (Phase III.) After the sexual act, in some cases one bird shows marked aversion, *e. g.*, by striking at the mate. Either the male or the female may show aversion. In some species, signs of aversion after the sexual act seem to be a normal and regular occurrence. In other species they are shown only by a bird whose mate, having failed of satisfaction, invites to further sexual activity. (Phase IV.) The pair usually become sexually indifferent for a considerable time after each copulation.

Brooding.—(Phase I.) The dove shows the brooding appetence, goes to the nest, and, if need be, struggles to obtain possession of it. (Phase II.) It sits throughout its customary perood, during which it often resists efforts of the mate to relieve it. (Phase III.) At the end of this period, in contrast, it comes off at a slight sign from the mate, runs about, flaps its wings, and thus shows its joy in being off. This may be interpreted as a sort of mild aversion for the nest. (Phase IV.) It goes away and becomes temporarily indifferent to the nest.

In other cases, one or other of the phases is not clearly present, so that there are various sorts of incomplete cycles, such as the following.

(*a*) When the bird shows appetitive behavior but fails to obtain the appeted stimulus, the appetite sometimes disappears, due to fatigue or to drainage of energy into other channels; in which case, Phase II. is not attained.

But many instinctive appetites are so persistent that if they do not attain the normal appeted stimulus they make connection with some abnormal stimulus (see page 94); to this the consummatory reaction takes place, the tension of the appetite is relieved, its energy discharged, and the organism shows satisfaction. This is of course *compensation*, in the sense in which that word is used in psychiatry. But the abnormal stimulus is usually inadequate or incomplete, the relief or discharge is imperfect, the satisfaction is marred by the fact that some of the constituent elements of the appetite, failing to receive their appeted stimuli, are still in Phase I. and abnormally active, while at the same time other elements have already reached Phase III., aversion.

(*b*) Some forms of behavior consist of appetite and satisfaction which are not, in ordinary cases, followed by any distinct aversion. For example, the drinking cycle shows clearly: (Phase I.) appetite for water; (Phase II.) the drinking reaction, with expression of satisfaction; (Phase IV.) indifference. The dove when it finishes drinking shows no distinct sign of aversion (Phase III.) except withdrawing the bill from the water. But if the observer takes this dove then gently in the hand and re-submerges its bill in the water, it shows marked aversion, struggling to withdraw the bill and to shake the water out of it.

(*c*) On the other hand it may seem that there are some forms of behavior, *e. g.*, fear, in which Phases I. and II. are lacking; that there is no appetite for the fear stimuli and no satisfaction in them; that when the slightest of these stimuli is received it at once arouses (Phase III.) aversive behavior. Yet it is an interesting fact that even in these cases a slight degree of appetite and satisfaction may be present. Children seek and enjoy a little fear. A dove, when it hears the alarm cry from other doves, at once endeavors to see the alarming object. Even pain is (in man) to some degree, sought and enjoyed.

In actual life the cycles and phases of cycles are multiplied and overlapped in very complex ways.

For example, when a certain satisfaction has been attained, this, instead of leading at once to a state of surfeit and aversion, may lead to further appetite, which leads to a second satisfaction, and so on. Thus Phase I. and Phase II. continue to alternate, constituting a "circular reaction" (Baldwin). I have seen a pair of house sparrows copulate thirteen times in immediate succession, and know by the sound of their voices that I did not see the beginning of the series. In many cases such circular reaction serves to rouse the organism to a high state of appetite and readiness for action.

Smaller cycles are superposed upon larger ones. For example, when a female bird is building a nest, so long as she is in the nest she is in a certain nest-building attitude, a high state of satisfaction, which constitutes the consummatory reaction (Phase II.) of a large cycle. But each time she reaches for a straw, seizes it, and tucks it into the nest, she exhibits thus a little cycle containing a little appetence followed by its own satisfaction.

The time occupied by a cycle varies extremely, from cycles measured in seconds to those that occupy a year or even longer. The relative duration of the phases also is extremely variable. In some cases the appeted situation is attained without delay, and Phase I. thus passes so rapidly as to be overlooked by the observer. In other cases the bird strives hard to overcome great obstacles which stand in the way of the attainment of the appeted stimulus, consequently Phase I. is of long duration. Phase II. may last, in the case of drinking, about one second; in the case of incubation, about three weeks.

It should be stated, too, that the phases are not sharply separated; each passes more or less gradually into the next. Thus, from Phase IV. of one cycle in a series to Phase I. of the succeeding cycle, there is often a gradual rise of appetite; active search for satisfaction does not commence until a certain intensity of appetite is attained. This is what is known in pedagogical literature as "warming up." This gradual rise of the energy of appetite is followed (Phase II.–III., or II.–IV.) by its sudden or gradual discharge. This rise and discharge are named by Ellis ('03), in the case of the sex instinct, "tumescence" and "detumescence." They are important phases in the psychology of art, in which sphere they are named by Hirn ('00) "enhancement" and "relief." The discharge (Phase II.) is also exemplified in "catharsis" in art and in psychiatry.

The cycles in the behavior of birds are fundamentally the same phenomenon as the cycles in human behavior. Human cycles are enriched by an intelligence far surpassing that of doves, but this is a difference of degree only. If the dove's cycles are determined largely by instinct, habit, physiological conditions, and not intelligence, so are some human cycles, as those of sleeping, eating, drinking, sex. F. H. Herrick ('10, 83) emphasizes the fact that a bird may scamp one cycle in order to begin another. Thus, birds may abandon young which are not yet weaned, because their appetence for a new brood has set in. But the same principle works, though not quite so crudely, in human life; as in the case of a mother who grows indifferent or even somewhat hostile toward her older children each time a new child is born. Herrick emphasizes also the fact that when anything disturbs the bird in the progress of a cycle, she very often gives up that cycle and begins a new one. Thus, a cedarbird who has just completed her nest one day finds a man examining it; she forthwith abandons that nest and begins to build another. But, again, the same phenomenon appears in human behavior. A man begins to build a house; when he has progressed far with the building he meets some horrible experience in it which "turns him against" it, and nothing will induce him to proceed with that house; he abandons it and begins to build elsewhere. The cedarbird has had a, to her, horrible experience which has turned her against her nest; that nest has lost its *value* for her; the sight of it now, instead of arousing her appetence, arouses aversion.

C. J. Herrick ('15, p. 61) says that many of these cyclical activities of birds are "simply complex chain reflexes." The reason he gives for this statement is that "each step in the cycle is a necessary antecedent to the next, and if the series is interrupted it is often necessary for the birds to go back to the beginning of the cycle. They cannot make an intelligent adjustment midway of the series." But all this, in some degree, is true of the behavior of human beings toward their mates, their nests, and their young. This has been illustrated in the preceding paragraph, and a few illustrations are here added. As to mates: When the cordial relation between a husband and wife is, by some mischance, broken, the pair may make an "intelligent adjustment" if the difficulty is not too great. But birds also make such adjustments constantly, when the difficulty is not too great. And with human beings, as with birds, the difficulty may be insurmountable; in which case, the husband and wife separate for a week, a month, or a year, after which period of rest (Phase IV.), they can commence a new cycle with Phase I., courtship. As to their nests: The fact of homesickness proves that the behavior of a human being toward his or her home runs in a series which conforms to Herrick's statements. As to behavior toward the young: The inability of human parents to make "an intelligent adjustment midway of the series" is shown by the fact that they cannot arouse the fullest degree of parental behavior toward an adopted child unless they adopt the child in its infancy. These facts do not prove that the human behavior in question consists of mere chain reflexes. Neither do the similar facts as to avian cycles prove that the avian behavior consists of mere chain reflexes.

The birds in their cycles exhibit attention (using this and all the following terms in a strictly behavioristic sense), intelligence, memory, intensely emotional behavior, conflict of tendencies, hesitation, deliberation (of course an elementary sort of deliberation), rise, maintenance, and decline of appetences, behavior conformable to certain laws of valuation. All these forms of behavior function in bringing about the consummatory situations of the cycles. Thus the instinctive behavior of birds, so far from consisting of mere chain-reflexes, and having no relation to "individuality" (Bowen, vide ut supra, p. 97), is in reality very highly integrated, and is the very core of the bird's individuality.

All human behavior runs in cycles which are of the same fundamental character as the cycles of avian behavior. These appear in consciousness as cycles of attention, of feeling, and of valuation.

This description is true not only of our behavior toward objects specifically sought by instinct, such as food, mate, and young, but also of our behavior toward the objects of our highest and most sophisticated impulses. Consider, for example, the course of a music-lover's feelings and attention in the case of a symphony concert. Before the concert, if his internal state is favorable (Phase I.), he is all eagerness, desire, interest. He goes to the concert-hall, chooses a good seat for hearing, and in every way shows appetitive behavior. (Phase II.) The music begins, he pays close attention, and feels satisfaction. (Phase III.) If the concert continues too long, he is surfeited, his pleasure diminishes, he even feels some unpleasantness, and his attention turns away, which is of course a form of aversion. (Phase IV.) When the music at length ceases he feels restfulness, relief, and his attention goes elsewhere. This cycle of the whole concert is overlaid by a complex system of epicycles, each extending through one symphony, one movement, or a smaller division, down to the measure and the beat. This is only one illustration of the fact that the entire behavior of the human being is, like that of the bird, a vast system of cycles and epicycles, the longest cycle extending through life, the shortest ones being measured in seconds. This view helps us to understand the laws of attention; for example, the law that attention cannot be held continuously upon a faint, simple stimulus. For as soon as such a stimulus is brought to maximum clearness, which constitutes the consummatory situation, the appetite for it is quickly discharged and its cycle comes to an end. This familiar fact shows that we, like the birds, are but little able to alter the course of our behavior cycles.

BIBLIOGRAPHY.

Baldwin, J. M.

'01 Dictionary of Philosophy and Psychology. New York. Art. "Appetite."

Bowen, Francis.

'46 Instinct and Intellect. North Amer. Review, 63, 91–118.

Carlson, A. J.

'16 The Control of Hunger in Health and Disease. Chicago.

Craig, W.

'12 Observations on Doves Learning to Drink. Jour. Animal Behav., 2, 273–279.

'14 Male Doves Reared in Isolation. Jour. Animal Behav., 4, 121–133.

Ellis, Havelock.

'03 Studies in the Psychology of Sex. Vol. 3, Analysis of the Sexual Impulse, etc. Philadelphia.

'10 Studies in the Psychology of Sex. Vol. 6, Sex in Relation to Society. Philadelphia.

Gerhartz, H.

'14 Ueber die zum Aufbau der Eizelle notwendige Energie (Transformationsenergie). Pfluger's Arch., Bd. 156, 1–224.

Herrick, C. J.

'15 An Introduction to Neurology. Philadelphia.

Herrick, F. H.

'10 Instinct and Intelligence in Birds. Popular Science Monthly, 76, 532–556, 77, 82–97, 122–141.

Hirn, Y.

'00 The Origins of Art. London.

Hobhouse, L. T.

'15 Mind in Evolution. Second edition. London.

Jennings, H. S.

'06 Behavior of the Lower Organisms. New York.

Morgan, C. Lloyd.

'96 Habit and Instinct. London.

Stout, G. F.

'07 A Manual of Psychology. Second edition. London.

Thorndike, E. L.

'13 The Original Nature of Man. (Educational Psychology, Vol. 1.) New York.

Washburn, M. F.

'16 Movement and Mental Imagery. Boston.

19

Copyright © 1921 by The University of Chicago Press
Reprinted from *Int. J. Ethics* **31:**264–278 (1921)

WHY DO ANIMALS FIGHT?

WALLACE CRAIG.

THE fact that all animals fight has attained immense importance in our day, because it is used as an argument in favor of the doctrine that men also should fight one another, that warfare ought not to be abolished. This doctrine I shall speak of for convenience as militarism; which is far preferable to calling it Nietzscheism or Treitschkeism or Prussianism, for all such names are invidious and more or less unjust. Militarists are at work in every nation, and in every nation they emphasize what they call the "biological" argument for war. They paint lurid pictures of "nature red in tooth and claw," dwelling on the many instances of rapacity, cruelty and destruction which undoubtedly occur in nature. They claim that theirs is a true picture of the life of animals, and also of the natural life of man. Their argument looks plausible, and it furnishes entertaining reading for the populace. But I believe it to be fallacious, partly because it exaggerates the cruel facts in nature, but far more because it misinterprets their meaning.

The attempts to refute the biological argument for war, so far as I have seen them, have been inadequate, some of them even absurd. Some pacifists have claimed that "No animal fights its own kind." Now, if the word pacifist means a person who longs to see war abolished, and who is willing to labor to the very best of his ability toward that end, I am myself a pacifist. But I believe that the cause of truth is more fundamental than the cause of any one man's theory as to how to make peace. Let us tell the truth, regardless of consequences. And the essential truth in this matter is that every animal fights its own kind. If we wish to discover any biological support for a policy of pacification, we must not seek to do it by asking "Do animals fight?" That question is not worth investigation, because the answer is already known. We must change the question and ask "Why do animals fight?"

The militarists are ready with an answer. They claim that the means by which a race progresses, or even maintains its present standard, is the killing of the less fit in the struggle for existence. And if the struggle to the death should cease, the race would degenerate. This presupposes the truth of the theory of the all-sufficiency of natural selection. Now, I shall not allow myself to be drawn into the debate between the rival theorists in genetics. I favor no one proposed solution of the problem of heredity and the method of evolution. On the contrary, I emphasize that this problem is not solved and that, therefore, the militarist has no right to base an argument for war upon one particular theory as to the origin of species. Our knowledge of genetics is not sufficient, and we doubt whether it ever will be sufficient, taken by itself, to settle the debate either for or against the biological argument for war. The pacifist is right in saying that the argument for war is "not proven," but he cannot say that it is positively disproved, by what we know as to the method of evolution. To find the evidence on which to base a positive verdict, we must turn away from the theories of evolution—which are all highly speculative—and examine at first hand the facts as to fighting among animals.

To understand why an animal fights, we must watch its fighting behavior, and also study the relation of its fighting to its other behavior, to its life history—in short, to its whole economy. This cannot be done by the old-fashioned method of surveying the entire animal kingdom, collecting therefrom a mixture of fragmentary and miscellaneous information. On the contrary, our problem can best be attacked by an intensive study of one species, or one group of related species. And if it can be shown that even one flourishing group of animals has evolved into its present prosperous state without its members engaging in internecine strife, that is enough to prove that warfare is not necessary for evolution.

My own specialty, as a student of behavior, is the Columbidæ, and I shall use pigeons as my chief examples. But I have studied other birds and mammals sufficiently to be sure that the statements made in this paper have a wide and general application among them. The birds and mammals are the most important animals for our problem. And the pigeons are a properly representative group: because, first, their behavior is typical; they quarrel and fight just about as much, or as little, as do the majority of birds. A healthy pigeon never allows another to trespass on his territory, or in any way interfere with his interests, with impunity. And, secondly, the pigeons are a "dominant" group; that is to say, the pigeon family is found all over the world, it has evolved into a large number of species, and the number of its individual members is enormous. All signs indicate that the pigeon family is in the most flourishing condition and in a state of rapid, progressive evolution. If the members of such a group live and act in a manner contradictory to the militarist theory, this is sufficient to prove that the militarist policy is not necessary for the welfare or the evolution of a race.

A friendly critic has asked me whether the generalizations presented in this article are true of carnivores such as the lion. In reply I would say that I believe they are. The published accounts of leonine behavior, so far as I have read them, indicate that our knowledge of the lion is fragmentary and incomplete. But the lion is only a large cat. Feline behavior is best known in the common cat. All the generalizations set forth in the present article are true of cats. Some of them are less true of cats than of pigeons, but, on the other hand, some of them are more true of cats than of pigeons. For example, the felines "space out" much farther than pigeons do, and consequently come less into conflict with each other. The terrible powers of the carnivores are exercised chiefly upon their prey, not upon their own kind. I never knew of a cat being killed by another cat in a fight. And our problem concerns only fighting between animals of the same species.

With reference especially to the higher vertebrates, we shall maintain and defend the two following theses: I. Fundamentally, among animals, fighting is not sought nor valued for its own sake; it is resorted to rather as an unwelcome necessity, a means of defending the agent's[1] interests. II. Even when an animal does fight, he aims, not to destroy the enemy,[1] but only to get rid of his presence and his interference.

I. The animal fights in order to gain or to retain possession of that which is of value to him, such as food, mate or nest. With animals, as with men, the cause of a quarrel is very commonly a coveted territory. My dog drives away other dogs from my house and yard. In general, each agent drives away other animals from his own nesting place, his chosen place for sleeping at night, his place for basking in the sun, or other territory which he can appropriate and use. Two animals fight only when their interests conflict. This is the fundamental fact in regard to infra-human fighting.

Animals do not enjoy fighting for its own sake. Unless his anger is aroused, the agent's behavior indicates that he has no appetence[2] for the fighting situation; he does not seek it; when in it he does not endeavor to prolong it; and he reveals by his expressions that he does not enjoy it. On the contrary, fighting belongs under the class of negative reactions or aversions;[2] it is a means of getting rid of an annoying stimulus. As McDougall says, the stimulus of the instinct of pugnacity is the thwarting of some other instinct. If the animal's instincts are not thwarted, if annoying stimuli do not thrust themselves upon him, he will never fight. He fights only when he is attacked, or threatened, or his interests are interfered with. He does not go in active search of a fight, except in play, in which

[1] We shall use the term "agent" for the animal whose behavior we are studying. His opponent we shall speak of as the "reagent" or the "enemy." The word enemy is thus used, as it is used by writers on the science of war, in a non-invidious sense.

[2] Craig. Appetites and Aversions as Constituents of Instincts. *Biol. Bull.*, 1918, Volume 34.

no injury is done to either side. Of course it is true that when necessity compels him to fight, he shows eagerness to attack, and joy when he achieves the victory; these are necessary in order that he may be a good defender of his rights. But when we try the experiment of keeping a bird in such a peaceful environment that he never tastes the joys of battle (except in harmless play), he shows no sense of loss, but is manifestly happier than the birds that fight. The pigeon, unless his temper is aroused, has no appetence for a battle. He has appetence for a great many other objects; as, water, food, mate, nest: if kept without such appeted objects he shows distress, tries to get out of his cage, and in every way makes clear to us that he is seeking the appeted object. But when he is kept without enemies, he never manifests the least appetence for them. This is true of all birds and mammals, so far as I know them. I am sure that it is true of the common fowl, although the cock is one of the most celebrated of fighters. It is a popular error to suppose that a bird such as the game cock, which shows unyielding bravery when in a fight, must necessarily be an aggressive bird, seeking the joy of battle for its own sake. The fact is that in the poultry yard the game cock is not a quarrelsome bird; if neighboring cocks do not unwarrantably thrust their presence upon him he lets them alone and attends to his own affairs.

At this point the question naturally arises, Why do animals fight so much as they do? For it is undeniable that under certain conditions there is a very great amount of fighting among them. In seeking an answer to this question it is important to notice the circumstances under which the excessive amount of fighting takes place. One of these circumstances is that of caged animals which are crowded so closely together that they constantly fall afoul of one another. Pigeons, if thus crowded in quarters that are too small for them, fight to a degree that is cruel and distressing. Since each pair of birds insistently drive trespassers away from their nest and from a certain territory around their nest, if the pigeon-keeper crowds the nest-boxes too closely together constant fighting must inevitably result. Such a state of affairs is not natural. It does not exist in any species in a state of nature. And it has given to some theorists an exaggerated notion as to the frequency of fighting among animals. In a great many wild species the amount of fighting is, I am sure, extremely slight.

Indeed, anyone who has watched wild animals with a philosophical eye must have been struck by the beauty and delicacy of the adjustments through which they avoid collisions with their fellows. Thus, in a flock of flying geese each goose maintains with astonishing precision the standard distance between himself and the goose ahead of him. When a great number of robins are searching for worms on a large lawn, each keeps with a pretty faithful constancy a distance of two or more yards between himself and any other robin. This tendency to "keep one's distance" is so widespread among animals and so various in its manifestations that it constitutes a study in itself. It is known to naturalists as "spacing out." Birds "space out" their nests with similar regularity, the distance varying with the species. The bank swallow burrows within a foot or two from the burrow of his neighbor; the kingfisher prefers a distance which I should estimate at about a half-mile. The interesting fact about these myriad cases of spacing-out is that nearly all of them (of course, not quite all) are adjusted without fighting. Evidence indicates that in a vast number of cases the animal seeking a nesting-place, and finding one that is already occupied, peacefully passes on and looks elsewhere. These quiet cases of adjustment are likely to be overlooked by the observer; the cases of fighting are conspicuous and thus seem to be a greater proportion than they really are.

Notwithstanding all these facts, there remain cases in which two animals, even among those that are in a free and natural state, may be observed to quarrel and fight throughout a whole day, and to renew the quarrel when they wake the next morning. The reason for such excessive fighting may be stated briefly as follows:

The adjustment of conflicting interests requires intelligent co-operation and some degree of social organization. Most animals, because of the low state of their intelligence and of their social organization, have a narrowly limited power of co-operating. That is why their differences must be settled so often by fighting. The following illustration, though simple, shows the point clearly. If we set up a new pigeon cote containing several compartments, each with its own door, and allow the pigeons to choose compartments for themselves, it may happen that two males will choose different doors from the very first, in which case they may live side by side in peace. But it may happen that on trying the new dove cote both males become enamored of the same door and each tries to enter it and make it his own. If these birds were endowed with reason, one of them would address the other in this wise: "Friend, in this dove cote there is plenty of room for you and me and for our families. Let us agree that you shall use the right door and I the left." But since pigeons are not endowed with reason they cannot make such a conceptual agreement; if both birds have chosen the same door they can adjust the difficulty only by fighting for it. In short, the reason why animals fight is that they are too stupid to make peace.

That this is the true explanation is indicated by the fact that if we lend the birds our reasoning power, if we act as arbitrator and settle their disputes for them, they gladly accept our adjustments and live in peace. Thus, in the case of the two pigeons fighting for possession of the same compartment in the dove cote, if I take one of the birds and keep him in a different compartment until it becomes "home" to him, then I can let him out, he will return to his new home and leave his neighbor at peace in the one which he had first chosen. This is an illustration of what we said before (p. 268), that if a man who keeps a flock of birds acts as a "benevolent despot" among them, administering justice and successfully resolving all conflicts of interest, the birds under his rule show no desire to fight, and are happier than those that do fight.

The amount of quarreling among animals varies (other conditions being equal) with the degree of their stupidity. Some individual pigeons are much more quarrelsome than others. The truculence of some individuals is due to the fact that they were reared under unnatural conditions which kept them in constant brawls. If as adults they are allowed to live a free, normal life, they outgrow their excessive quarrelsomeness. As their experience increases they tend more and more to adjust their differences with their neighbors without fighting. This fact is of great importance for our problem; because it proves that as viewed by the birds themselves fighting is not a good, it is a necessary evil.

The law that aggressiveness does not pay is conspicuously true among animals. The contentious pigeon, as every fancier can testify, brings disaster not only to his neighbors but also to his own cherished home and family. Hence it is not surprising that in the course of evolution the birds have become as gentle and peaceful as we know them to be, and that even the individual bird strives, with his limited intelligence, to adjust the difficulties that arise between himself and his neighbors and to avoid actual fighting.

II. These last observations naturally lead up to our second thesis, which is, that even when an animal does fight he aims, not to destroy the enemy, but only to get rid of his presence and his interference. This point is important, for if the militarist theory were correct, that the function of fighting in the economy of nature is the elimination of the "weaker" individuals or groups, then we ought to find that the behavior of the fighting animal is directed toward the extinction of the enemy's line of descent, as, by the destruction of the reagent himself and also his eggs and young. But the behavior of animals does not conform to this theory. Especially among the higher vertebrates, observation shows that they do not follow the militarist policy, and without it they live and thrive and progress.

The only animals which in any degree follow the militarist policy are, I believe, those of parasitic habit. Some of the ants, bees and wasps have become "robbers," systematically destroying their congeners, even those of their own species, in order to appropriate their stores of larval food. It is a sad and discouraging fact that wherever in the animal kingdom some members have developed sociality and co-operation and thrift, their very prosperity has furnished opportunity for the breeding of a race of despicable parasites. But there is a grain of consolation in the fact which has long been known and has recently been emphasized by Professor Wheeler,[3] that the parasitic habit leads, in the course of time, to the destruction of the parasite itself. A specialist on the hymenoptera could probably write a positive refutation of the militarist theory, by showing that those animals which do follow a militarist policy thereby lead to their own extinction.

But I am not a specialist on insects. Among the higher vertebrates, parasitic behavior is found only in rare cases, as in that of the European Cuckoo. We are told that the young cuckoo works systematically to push its nest-mates out of the nest and thus destroy them. Such a systematic attempt to destroy its rivals is, I am sure, not to be found in any non-parasitic species. The young cuckoo exhibits the behavior that would be found in other animals if they followed a militarist policy, and makes clear to us by contrast that they do not follow a militarist policy. The non-parasitic bird or mammal aims, as we said before, not to destroy his rival, but only to free himself from his presence and his interference.

To prove this thesis we shall present an analysis of the animal's fighting behavior. We shall inquire especially as to the behavior of that individual animal who is the better fighter of the two in the contest, the one destined to be the winner if the battle be fought out. With such an individual as agent, the reagent may do one of three things: namely, (a) flee; (b) submit; (c) persist in fighting. We shall describe the behavior of the agent in these three cases.

[3] Wheeler, W. M. The Parasitic Aculeata, a study in evolution. *Proc. Amer. Philos. Soc.*, 1919, Volume 58.

(a) If the reagent flees, the agent does not pursue him indefinitely and seek to destroy him. On the contrary, he pursues him only far enough to eliminate him from the field of interest about which the battle is being fought. Thus, when a dog has driven an intruding dog out of the territory over which he claims sovereignty, he does not pursue him farther (unless there be some other cause for the pursuit), but by barking and other expressions of triumph and satisfaction he shows that his end has been fully accomplished. So it is with all animals that I have ever observed driving others away from their territory or their mates or their food. That is all that the agent cares to accomplish. He shows no tendency to pursue the enemy to the death.

That this generalization is true and fundamentally important is evidenced by the fact that in many groups of animals, indeed in probably all animals in some degree, there has been an evolutionary change from destructive forms of fighting to forms of fighting which are merely expressive or ceremonial, which drive away the reagent by threatening or warning him without doing him any injury. This is a part of what Hocking[4] has named "the dialectic of pugnacity." The pigeon warns his enemy by a display which is highly ceremonious, consisting of elaborate cooing and gesturing. The celebrated mock-battles of the capercaillie and of a great many other birds furnish other instances of the same trend in evolution. All birds and all mammals are endowed with instincts to threaten the enemy, to make feints, to hiss or growl or roar, or to vent their anger in other expressions which serve to warn the reagent and often cause him to flee without a blow being struck. It is important to notice that these attempts to drive away the enemy are used first, before the physical combat. If the aim were to destroy the enemy he would be attacked silently and by stealth: but in most species he is not so attacked; he is first warned, and given every opportunity to withdraw from the field. In a great majority of the

[4] Hocking, W. E. *Human Nature and Its Remaking.* New Haven, 1918. Chapter XXIV.

conflicts among animals, the ceremonial combat is all that is needed and all that is used: the reagent may withdraw as soon as he is threatened; or he may at first make a counter-display, but withdraw on discovering that the agent is more determined than he. The physical combat is resorted to only after the ceremonial has been tried and has failed to settle the dispute.

(b) If the enemy submits, the agent ceases fighting. In pigeons this is witnessed again and again. In the heat of battle the agent may rush upon his enemy, jump on his back, peck him with all his might, and pull out his feathers. But if the reagent lies down unresisting, the agent's blows quickly diminish into gentle taps, he jumps off his prostrate foe, walks away, and does not again attack the enemy so long as he is quiet. This behavior is typical, and it proves that the pigeon is devoid of any tendency to destroy his rival.

Further study of this behavior indicates that it is not merely negative, not merely the absence of an impulse to destroy. The bird has a positive impulse to quit fighting a non-resisting bird of his kind. One explanation of this impulse is to be found in the mode of instinctive sex recognition. When a male meets a stranger belonging to his own species, provided this male has not learned by experience to discriminate the sexes, the only discrimination he shows is this: if the stranger fights, the agent treats it as a male; if the stranger refuses to fight, the agent treats it as a female; if the stranger first fights, then submits, the agent treats it first as a male, then as a female. This mode of sex recognition is so widespread in the animal kingdom that it seems to be fundamentally ingrained in the nature of the male. Audubon tells us that when he watched a battle between two wild turkeys, when one of them had been defeated, he was surprised to see that the victor, instead of injuring him, showed toward him the amorous behavior which is generally accorded to a female. Audubon need not have been surprised. Behavior of this sort is now known to be characteristic of a great many animals ranging all the way from the lower invertebrates to the Primates. Thus we see that in the male animal there is a fundamental trait which tends to prevent him, and in most cases does prevent him, from doing any injury to a non-resisting member of his species.

We showed under (a) that the enemy who flees is not pursued to the death. We have now shown under (b) that the enemy who submits commonly finds his life spared. The non-destruction of the enemy is more certain under (a) than under (b). An enemy who has fled from the field of operations is perfectly safe. It is but natural that an enemy who has submitted is not perfectly safe; the pugnacity may smolder, and may break out again into violent fighting. The facts are too complex to be treated in this short article. Suffice it to say that in those rare cases in which an animal, or a group of animals, kills a non-resisting member of the same species, such killing is in various ways exceptional or accidental;[5] it is not a policy, not a common and regular form of behavior, and very far from being a systematic pursuit.

(c) When the reagent refuses to flee and refuses to submit, the agent is obliged to resort to physical force. It is extremely interesting to notice that even in this case the physical force used is often of a form which serves merely to rid the agent of his enemy without doing him any hurt. Thus, when the common pigeon quarrels with his neighbors on one of the high ledges on which they like to perch, his principal method of dealing with his opponent is to seize him by the nape of the neck, drag him to the edge of the ledge, and hurl him off into space. The bird that is thus hurled off spreads his wings and flies without injury.

From what has been said thus far, it is clear that when a pigeon deals with a rival pigeon, his behavior is directed first toward inducing the reagent to flee voluntarily, then toward forcing him off the field. Only when these means have failed, when the reagent refuses to flee, refuses to sub-

[5] See, *e. g.*, W. H. Hudson. *The Naturalist in La Plata.* London, 1892. Chapter XXII.

mit, and is too powerful to be hurled off the ledge, only then does the agent endeavor to the utmost to injure his opponent. Then the two pigeons meet in a grim, silent, unrelenting, physical struggle. This brutally physical struggle appears in extreme contrast to the more common pigeon fights, which are highly ceremonious. Yet even these fiercest struggles, unless they are protracted for a very long time, do not result in the death of either combatant. And at any time when either combatant feels that he has had "enough," he needs only to leave the field in possession of the victor; he thereby saves himself from further injury. For even after the most prolonged and painful battle it still remains true that if the enemy flees or submits the agent ceases fighting.

In conclusion, we can give only a few brief statements as to the bearing of the facts of animal behavior upon human problems and upon philosophy.

1. No bird or mammal follows a policy of non-resistance. And we find no trend of evolution toward a policy of absolute non-resistance. Even in the peaceful settlement of disputes each animal asserts his rights by expressing his determination, and often by making feints or threats. Further, each individual and each social group is prepared to resort to force, and to exert force to the utmost, if necessary, in order to defend its interests. Defensive fighting pays.

2. On the other hand, aggressive fighting does not pay. Among animals, as among men, fighting is a wasteful and harmful means toward the attainment of the ends sought by the contestants. In adaptation to this fact, we find that both in the history of the race and in that of the individual there is a trend away from destructive forms of fighting, toward the adjustment of disputes by harmless means. Progress in this direction involves the development of intelligence, of self-control, and of a technique for the adjustment of difficulties. Such progress has been made by all the higher animals, in varying degree. Birds and mammals strive to control their angry impulses. They

cope with the problem of pacification, for their interests depend upon it. We human beings, when we strive toward world peace, are only travelling farther in the same line of progress in which our infra-human ancestors took the first steps. Whether we shall ever achieve world peace, I do not predict; because I do not know whether we can ever develop our understanding, our self-mastery and our political technique to a degree of perfection sufficient to cope with the immense difficulties of world organization.

3. The third conclusion, which we shall present here chiefly as a criticism of militarism, is, in its broader aspects, a criticism of the whole "biological"[6] philosophy of certain schools of current thought. This conclusion is that a distinctively "biological" need for fighting does not exist. The reasons why animals fight are substantially identical with the reasons why men fight. These reasons all pertain to the subject matter of our first two conclusions, and they may be summed up in one sentence. Animals and men fight because they must conserve their interests, and their technique for the adjustment of conflicting interests is too imperfect to adjust all cases of conflict. This, we believe, is the only true argument for war. The militarist denies that this is the whole argument. He says that if we ever should achieve a world organization which would adjust all conflicts peaceably and abolish war, the race would degenerate, because war is a "biological" necessity. His "biological" argument takes two forms; viz., (a) the definite; (b) the indefinite.

(a) The definite "biological" argument is based upon the theory of the all-sufficiency of natural selection. In reply to this argument, I do not need to offer any opinion as to the method of evolution. All I need to say is that, whatever the method of evolution may be, it cannot be in contradiction with the facts of animal behavior. And the facts of animal behavior prove that fighting to the death

[6] This word is put in quotation marks to make clear that our criticism is aimed, not at biology proper, but at the misuse of biological theory by certain philosophers and writers on human affairs.

is not necessary for the welfare or for the evolution of a race.

(b) The indefinite form of the argument pictures a mysterious, irresistible "biological" necessity or "destiny," which threateningly overhangs our human efforts and will cause the extinction, or at least the degeneration, of any people that lives up to its humanitarian ideals. In answer to this, I have given evidence to prove that the animals also are humanitarians. The higher animals strive to avoid destructive fighting, and some large and important groups of species have so far achieved this result that they have reduced fatalities to a negligible quantity. These great groups of animals have been evolving their pacific régime, and thriving under it, for millions of years, and are today in a state of progressive evolution, the very opposite of degeneration. Therefore, we may brand as false and contrary to the evidence of facts, the militarist statement that degeneration threatens to overtake us if we should put into practise our humanitarian ideals.

UNIVERSITY OF MAINE.

20

Copyright © 1929 by Cambridge University Press

Reprinted from pages 58-60, 85-86, 88, and 92 of *An Introduction to the Study of Bird Behavior,* Cambridge University Press, Cambridge, 1929, 136p.

THE WHOLE HAS VALUE, THE PARTS BY THEMSELVES HAVE NONE

H. Elliot Howard

As a result of analysis we are left with a number of different modes of behaviour strewn about without much coherence. True they follow one another in ordered sequence; but of the relation of one mode of behaviour to another, of the behaviour of the male to that of the female, of the behaviour as a whole to an environment seldom stable–of unity, nothing so far has been said.

All these diverse forms of behaviour are commonly spoken of as 'instincts'–aggressive, acquisitive, sexual, nest-building, parental. But to regard them as the product of abstract forces capable of being studied in compartments has two objections. The first, that we have no notion what the forces are, nor where in organisation they lie; and hence are no whit nearer interpretation when we attribute a certain mode of behaviour to a certain instinct. The second, a graver objection still, that by thinking of behaviour in terms of instinct we lose sight of the unity of the whole. So I avoid the word and speak throughout of reactions.

What do I mean by a "pattern of reactions"? I mean not only that there is inherited *structural provision* for these particular reactions, but such unity of structure that the same intrinsic excitation brings all the different parts functionally into being, or the same stimulation that elicits one reaction (its appropriate one) thereby raises temporarily the susceptibility of others.

It may seem strange to seek unity where stimulation is constantly pouring in upon the bird from the outside world exciting alternative reactions–alternative in the sense that now one now another is prepotent where one or other at any moment may be of major importance. Clearly there must be some provision for singleness of action. I spoke of rhythmical excitation, and gave some examples to show the sort of way that reactions wax and wane, promising to return to the matter. I must do so now, for the rhythms are an expression of what may prove to be a system of control.

The waning of a reaction under prolonged excitation can be observed in all the reactions that go to form the pattern, though more marked in some than in others. My attention was first called to it by the way a male fights in defence of his territory or to acquire it. Knowing how furiously birds fight, one expects a battle to rage until the issue is decided by injury or by death. Injury often follows, death sometimes, but expectations are rarely fulfilled; and a conflict pursued to the bitter end without pause is seldom seen. Well, that is not perhaps very remarkable; a bird must suffer muscular fatigue, must pause to carry on. But it does not merely grow tired, nor merely await an opportunity to attack while gaining breath; a change comes upon it at intervals, a change of deep-seated origin that temporarily alters its attitude. The very birds that a minute ago were fighting in deadly earnest now seek food side by side; and presently they will renew the contest with former energy, passing from hostility to friendship and from friendship to hostility without any outward change to account for it.

So I turned to other reactions—to those concerned with nest and young, to the sexual—and found that each likewise waxed and waned under prolonged excitation, periods of intense activity alternating with periods of calm. A bird builds; she gathers material, flies to the nest, weaves—gathers, carries and weaves again and again; her energy is centred on the nest as if it were urgently necessary to finish, nothing in the outside world disturbs her, nothing interests her. But gradually small irregularities appear; she becomes less active, the material in her very beak is a burden and she lets it fall.

So, too, when feeding her young; she collects food, delivers it, cleans the nest, and repeats the routine again and again. The reaction then wanes and the brooding reaction takes its place; she brings food, sits on the rim of the nest, watches the young stretch their necks but cannot respond and so swallows the food and broods.

All this, it may be said, is natural enough. When a bird fights or builds or is sexually excited it must suffer loss of energy which alternating periods of calm alone can redress. But the change that comes upon it under prolonged excitation is not the sort of change that weariness produces; not merely a period of rest but a change of conduct, as the attitude of rival males shows; not merely loss of activity but diversion of it. And this diversion of activity into other channels shows that the waning of a reaction has nought to do with muscular fatigue; for a reaction that takes the place of a waning reaction may employ the same muscles and require even greater muscular effort.

There is no doubt that these reactions in their functioning resemble the reflex. They suffer, if not the so-called fatigue of the reflex, a condition approximating to it.

But how does this provide a system of control? "One obvious use attaching to it is the prevention of the too prolonged continuous use of a 'common path' by any one receptor. It precludes one receptor from occupying for long periods an effector organ to the exclusion of all other receptors. It prevents long continuous possession of a common path by any one reflex of considerable intensity. It favours the receptors taking turn about. It helps to insure serial variety of reaction. The organism, to be successful in a million-sided environment, must in its reactions be many-sided. Were it not for such so-called 'fatigue', an organism might, in regard to its receptivity, develop an eye, or an ear, or a mouth, or a hand or leg, but it would hardly develop the marvellous congeries of all those various sense-organs which it is actually found to possess." That, in the opinion of Sir Charles Sherrington, is the biological value of fatigue in the reflex. Apply his words to the reactions that form the pattern concerned in reproduction and at once we reach the heart of the matter. Were it not for the waning of a reaction a bird might, in regard to its receptivity, develop a mechanism to fight, to build, to feed its young and brood them, but it would hardly develop the harmony without which the common end would not be attained. For it is plain to anyone who studies the behaviour that the whole has value, but that the parts by themselves have none; that one reaction is neither more nor less important than another; that failure of one means failure of all.

Now the agencies which constitute adequate stimulation to various reactions constantly pour in upon the bird. What does it do: respond in this direction or in that at haphazard, oscillating between different modes of behaviour so that nothing is accomplished? Of that sort of antagonism there is none; none in the sense that one reaction interferes with another—to its harm.

[*Editor's Note:* Material has been omitted at this point.]

Strangeness, suddenness, summation of successive stimuli—these are the principal features of posturing: they are different from the normal forms of stimulation, but lead to fertilisation where these others fail.

I mentioned that prooestrum might be a reason why, in polygamous species, a female seemed to choose one male in preference to another. But in judging of her behaviour we must also take into account summation of successive stimuli and the waning of a reaction. The story of the Hedge-Sparrows shows that a male and female may pair, live together, enjoy sexual flight, build and yet fail to procreate their kind because the reaction of one or other, or of both, stands at such a low level of intensity that effective synchronisation cannot be reached. So, too, in polygamous species those individuals in which intensity of reaction or of stimulation stands habitually low must produce fewer young.

Posturing is an accompaniment of postponed reaction. What brings about the postponement? Something in the state of the bird to which the posturing is directed—that is what we feel tempted to say. The evidence is conflicting. A female assumes the attitude appropriate for conjugation—raised tail, quivering wings, trembling body—yet the male hesitates and postures; and his posturing may increase with hesitation or fade away. On the other hand, a female flies to her mate, is excited, but does nothing at first; he postures and worries her until she assumes the appropriate attitude when conjugation takes place. It is difficult to attribute his postponed reaction in the first instance to the state of the female. To what then is it due? We can only attempt explanation. She assumes the appropriate attitude, postures, waits—yet a period intervenes between the application of stimulation and conjugation. The postponement is evidently due to the low intensity of his reaction; just as, in the second instance, her reaction stands at too low a level of intensity to enable her to respond. In both instances, however, summation of successive stimuli elicit reaction. Yet it seems strange at first sight that he should posture if his reaction is of low intensity. But only strange if outward manifestation were always in ratio to intensity of reaction. Of this we found little evidence; we found, rather, that perfect synchronisation of rhythms was preceded by little nervous discharge into muscular activity.

Whitman suggested that one purpose of posturing is self-stimulation. We seemed on the point of reaching an opposite view regarding its purpose in the second phase—that it was self-exhausting. Can it serve the purpose of self-exhaustion in one phase, self-stimulation in another? That probably depends upon the extent of it, its duration and strength at the start. Synchronisation of rhythms which is required for successful fertilisation would scarcely be facilitated if posturing drained the reaction in one member of a pair, even though, by increasing intensity of stimulation, it provoked response in the other. Consider a Pied Wagtail—how, with trailing wings and raised feathers, he runs round and about the female, and how at length conjugation takes place. The postponement must be due to the state of the female or to the state of the male. Supposing it was due to hers, his preliminary behaviour by providing intense stimulation might elicit her reaction; but, at the same time, by draining his own reaction it might leave him less susceptible to stimulation at the appropriate moment. Hence what was gained in one direction might be lost in the other. Supposing, on the other hand, it were due to his—which means that the intensity of his reaction at the start stood at a low level—and the escape of nervous discharge into muscular activity lowered the intensity still further; even though her posturing heightened stimulation it is unlikely that the balance would be redressed. But if Whitman is right and posturing, while heightening stimulation, raises the level of intensity of the reaction, more effective synchronisation will without doubt be obtained.

[*Editor's Note:* Material has been omitted at this point.]

At the beginning of a sexual cycle conjugation seems to fail more frequently, and delay is then more prolonged. Help may come later from learning and from revival in the form of imagery reinforcing a weak reaction. The female's hurried return to a place where conjugation has occurred before, even though the place may not be within her visual field, seems to point to revival. But whatever the explanation of the later facility may be, we must remember that even then the completion of the act is only easy in contrast with the perceptible difficulty of synchronisation at the beginning–a difficulty which sometimes, as happened with the pair of Hedge-Sparrows, never is overcome; and which only can be overcome by grading the intensity of stimulation. Normal extrinsic stimulation comes from voice or appearance, either of which is sufficient to establish a reaction at the peak of intensity. But a weak or waning reaction needs intense stimulation. The arched neck, depressed tail, and stately walk combine to provide strange and sudden stimulation: stimulation so imperative that though reaction is feeble in the male he nevertheless mounts her, stands upon her back, slides off and pecks her violently, executes all the movements except the consummatory one; and though reaction in the female is too feeble for her to assume the appropriate attitude, she nevertheless mounts him and stands helpless upon his back.

The escape of nervous discharge from a postponed reaction is thus turned to good account, pressed into the service of the organism, and used on the one hand for grading the intensity of stimulation, on the other for fortifying a failing reaction. And the actions of male and female combine to form a harmonious whole, beautifully adapted to bring about appropriate synchronisation of rhythms.

From start to finish those pairs, or those males and females, in which the rhythms of the reactions combine harmoniously will score a larger number of successful conjugations: from start to finish those individuals whose organisation by providing various forms of stimulation can establish a weak reaction, and in which sexual reaction, while not holding the field for too long and so preventing other reactions from functioning at the appropriate moment, stands habitually at a high level of intensity, will naturally be selected.

[*Editor's Note:* Material has been omitted at this point.]

Now sexual reaction, in common with other reactions, wanes under prolonged excitation; moreover, its intensity depends upon nutriment, upon temperature, humidity, atmospheric pressure or perhaps upon combination of the three, and in so far as dependent must suffer from the changes to which these external agents are subject. All this means that fertilisation is difficult, for a reaction of weak intensity forms a barrier to mutual synchronisation of rhythms. Bearing the reflex in mind we sought for some means of intensifying stimulation and so of redressing the balance between stimulation and response. As far as we could discover, the particular stimuli to which sexual reaction is susceptible are voice, appearance, mode of approach, motion, imagery—some we could not detect. These are normal and adequate when a reaction is in a state of full excitability. But there are times when normal stimulation fails to bring about conjugation, and a period accompanied by much posturing seems necessary for completion of the act. The suddenness of posturing and the summation of successive stimuli which it provides combine to form intense stimulation, thereby provoking response from a reaction of sub-maximal intensity.

An escape of nervous discharge into muscular activity cannot be without effect on the reaction. It must either heighten the threshold of excitability or lower it; the former would make synchronisation of rhythms still more difficult, the latter would make the birds more sensitively reactive to one another. Hence we agree with Whitman when he speaks of posturing as self-stimulating.

21

ON THE BIOLOGY AND PSYCHOLOGY OF FLIGHT IN ANIMALS

H. Hediger

This article was translated expressly for this Benchmark *volume by Chauncey J. Mellor and Doris Gove, The University of Tennessee, Knoxville, from pp. 21–22, 26–27, 29, 34–35, 40, of Zur Biologie und Psychologie der Flucht bei Tieren,* Biologisches Zentralblatt, *54:21–40 (1934). Translation © 1985 by Chauncey J. Mellor and Doris Gove.*

Introduction. In contrast to many complex phenomena in the way of life of animals (reproduction, brood care, food intake, etc.), astonishingly little attention has been paid up to now specifically to the most banal life phenomena such as flight. It seems therefore worth the effort for once to devote oneself somewhat more closely to this phenomenon. To be sure, it is not possible here to discuss all the observable facts in connection with flight; rather, we shall first limit ourselves merely to the flight of a (higher) animal from man or a biologically equivalent enemy. Since with few exceptions (such as vampire bats, piranhas, certain sharks), all higher animals tend to draw away from man, that is, to flee from him at his approach, then flight is that phenomenon of life that man is most likely to find the opportunity to observe. The following remarks are far from claiming completeness. It can only be a matter of drawing attention to flight as a life phenomenon with certain lawful regularities. Since this paper had to be completed shortly before the beginning of a trip, it bears the character of a preliminary communication and represents the beginning of an attempt to investigate the most important behaviors of animals in their biological and psychological connections with respect to man. Later the phenomenon of tameness and training will be treated similarly.

1. **The Flight Reaction.** At all times, but now perhaps more strongly than ever, the most diverse animals have been influenced in their way of life by man—according to Kopstein (p. 73), *Birgus latro*, the coconut robber crab on Christmas Island, is a quite diurnal animal that often can be found there in great numbers because it is not hunted. But on islands where this large crustacean is eaten, it is very shy and only leaves its hiding places at night. In the Coconut Islands, Darwin became acquainted with this animal as a diurnal one; however, Forbes, by 1879, found it to be decidedly nocturnal there (Kopstein). According to Lutz (p. 541), the *Hydrochoerus capibara* in Brazil in heavily hunted regions has also become a typical nocturnal animal, whereas it is otherwise also active by day. According to older reports, the hippopotamus was occasionally found far inland crossing high and steep mountains (Erbach, p. 284). Today, under the influence of human

persecution, it is scarcely to be termed more than "amphibious," but in many cases lives almost totally in the water. "Therefore, the life style of hippopotamuses, just as in other wild animals, is closely connected with the degree of their persecution," Erbach says.

However, at this point, we cannot pursue these interesting connections any further. From the abundance of life phenomena we can now pick just one for further investigation: flight reaction. Upon closer examination of these flight reactions, some peculiarities and regularities immediately stand out, especially the relative constancy for the individual species. The specific aspects in these various flight reactions are in part so striking that, in my paper on the herpetofauna of New Britain (1933), for a quick characterization, I immediately arrived at expressing the flight reactions of a certain species schematically in formulae.

On the outer wall of a hut in Gasmata (New Britain), for example, every evening a group of geckonids appeared. Now as I approached this geckonid colony, the following occurred. Some animals fled upwards rather vertically and disappeared in gaps between boards and under roof beams, etc. The rest fled directly down to the ground and hid there under stones and tufts of grass. After some captures, it was evident that I was dealing here not with a uniform population but with one consisting of two species, and that individuals fleeing up always belonged to the one species, *Gehyra oceanica*, and those fleeing down belonged to the other species, *Gymnodactylus pelagicus*. It is interesting in this connection that *Gehyra* possesses a fully developed clinging apparatus on the undersides of its toes, whereas such a thing is totally lacking in *Gymnodactylus*. In order to prove the consistency of this remarkable flight reaction, several tests were tried; *Gehyra* always fled up, and *Gymnodactylus* always fled down. By the way, Mell (253) also established similar things with three species of Chinese geckonids that lived together, each with its specific flight reaction.

[*Editor's Note:* Material has been omitted at this point.]

2. **Flight Distance.** After having clearly pointed out in the introduction that, with very few exceptions, all higher animals tend to withdraw before man, that is, to flee from him, a second rule can be added here: that the wild animal flees only to the extent and only to the distance from man as appears necessary in the interest of the preservation of the individual and the species. By the expression flight distance (abbreviated F.D.) we understand not that distance that an animal in flight puts behind it in fleeing from man, but rather that distance to which a person must approach a free wild animal to cause it to flee, merely by the person's approach.

It would be completely useless to want to introduce this concept for long if it were not more precise, at least for certain cases, than the already current expression, shyness, or some such, and if the F.D. were not subject in certain groups to quite definite lawful regularity, and above all if it would not show striking constancy and a value subject to measurement. Obviously, this value is not absolute, but is nevertheless specific despite all deviations for the individual species and its subcategories, and for this reason it deserves quite special interest.

Pearse (p. 128) reports on a Philippine winter crab *(Uca)* that always fled into its holes at the approach of a human to 15 m. In other words, the F.D. of this crab amounts to 15 m. For myself, the nature of the F.D. never became clearer than at the moment, when, at the mouth of the Regreg river in Morocco, I approached a swarm of winter crabs, *Uca tangeri*: at a certain distance from me—likewise around 15 m—all the crabs disappeared into their holes. At the spot which just a moment before was alive with crabs, there was nothing to be seen. After I had gone on for a stretch, the animals behind me came out of their burrows; I stood at the center of a roundish or oval zone of approximately 30 m (2 × F.D.) diameter that was completely free of crabs, whereas, on the other side of that strikingly sharp border, it was crawling with winter crabs. Wherever I moved, this zone always shifted, always keeping me in the center. In this special case, to be sure, the F.D. may coincide with the threshold value for certain sensory stimuli. However, we shall see below that the F.D. also represents a quite independent value that is not identical with thresholds (as used in sensory physiology) for optical and other stimuli.

[*Editor's Note:* Material has been omitted at this point.]

Deer have frequently become accustomed to the lack of danger of railway trains thundering by (Alverdes, *Tierpsychologie,* p. 73), so that they are not disturbed in the least by the trains while grazing. Similar things are also reported about African wild animals in the vicinity of the great railway lines. "Animals' fear of man is traditional and not purely instinctive," Alverdes writes (*Tiersoziologie,* p. 123), "for all travelers who have sought out areas hitherto not visited by humans agree in their description that the animals there were completely trusting. Only slowly did they learn to flee their most grievous enemy, and this knowledge is passed on from generation to generation by the example of the older animal in each case." It will be shown later that "completely trusting" parents can have the "shyest possible" young and vice versa. Also that it is often quite difficult to distinguish between instinctive and traditional, or rather inherited or learned, "trustfulness" or "shyness." It seems as if the number of generations plays a very important role in this, yet this leads to the problem of the transmission by inheritance of acquired characteristics, and, as a matter of principle, we must forego here the opportunity to deal with that issue.

[*Editor's Note:* Material has been omitted at this point.]

3. **Defensive Distance.** In the example of the hermit crab and the hedgehog it was already shown that flight does not simply continue indefinitely when the fleeing animal is pursued, but that in these cases, a special reaction appears when the distance between the pursued and the pursuer diminishes to a certain size. This reaction is a conversion of flight into defense. Animals that have certain defensive and protective devices at their disposal now put them to use. The hermit crab pulls back into its house, the hedgehog rolls up into a ball; antler and horn bearers set their weapons at the ready. All the well-known, many-faceted defense measures are taken at this instant. This is also the moment where the many typical fear-producing effects (such as the baring of teeth, the exposing of garishly colored body parts, the unfolding of the hood in cobras, etc.) are brought forth.

This defensive reaction is, so to speak, a last warning to the pursuer. If he now withdraws, then the confrontation has no further consequences, but if he still does not leave the pursued in peace, then a fight (the unloading of poisonous or smelly secretions, etc.) will ensue. In an earlier essay (1928, p. 407), I described the behavior of the Berber skink *(Eumeces algeriensis),* which suddenly "makes its stand if one comes too close to it in chasing it over the steppes." The hooded adder *(Macroprotodon cucullatus)* on cultivated areas of Moroccan farms even makes a stand against tractors on their rounds that come too close to such a snake in flight. Every observer must know numerous examples of this from experience, so it is unnecessary to cite further ones.

Let us be brief. The nature of the defensive distance (D.D.) can then be observed when a wild animal living freely is not only caused to flee but continues to be pursued while in flight. By D.D. we understand that distance to which the pursuer must approach the pursued in order to set off the defensive reaction, that is, the change of flight into defense. The analysis of flight phenomena will lead straightaway to the recognition of this D.D. as a specific behavioral characteristic similar to F.D. Likewise, it also represents a measurable value, which, to be sure, varies within certain limits (tiring or exhaustion, previous stimulations, length of chase, formation of the environment, peculiarities of the pursuer, etc.), but nevertheless in no way lacks specific qualities.

[*Editor's Note:* Material has been omitted at this point.]

4. **Critical Distance.** The normal reaction of the free-living wild animal to the approach of humans, as we have seen, is flight. Various circumstances can inhibit the course of this reaction: captivity, wounding, gravidity, peculiarities of habitat formation, etc. All these factors can prevent the animal from flight and consequently cause the transformation from flight

to attack or defense. The "attacks" taking place under such conditions always have the character of self-defense. It is completely wrong to speak of "vindictiveness," as often occurs.

The phenomenon of critical distance (C.D.) has a certain similarity to that of D.D., but the difference consists in the fact that in D.D. there is nothing in the way of further flight, whereas the C.D. is only observed when flight is somehow inhibited. By C.D. we understand therefore that distance to which a person must approach a wild animal (wild as opposed to tame) hindered (in reality or imagination) in its flight so as to set the animal off to attack merely through the human's approach. This attack here completely bears the character of defense.

Like F.D. and D.D., C.D. is also dependent on the individual's experiences, momentary moods, and other such factors. But beyond that, it stands in a quite definite connection not only to the size of the animal, but also to its type of weaponry, to its defensive devices, etc. But here it can be clearly established that despite multifarious variational possibilities, relatively constant species specific characteristics do obtain in the C.D.

[*Editor's Note:* In the original, material follows this excerpt.]

LITERATURE CITED [in Hediger's Bibliography]

Alverdes, F. 1925. *Tiersoziologie.* Leipzig.
———. 1932. *Die Tierpsychologie in ihren Beziehungen zur Psychologie des Menschen.* Leipzig.
Erbach-Fürstenau, R., Graf zu. 1912. Beobachtungen über das Tierleben in Ost-und Zentralafrika. *Sitzber. Ges. natf. Freunde Berlin* 5, 271–299.
Hediger, H. 1928. Die Tierwelt auf einer marokkanischen Farm. *Bl. Aquar. Terrarkde. Stuttgart* 39, 406–408.
———. 1933. Beitrag zur Herpetologie und Zoogeographie Neu-Britanniens. *Zool. Jahrb.* (in press)
Mell, R. 1929. Grundzüge einer Ökologie der chinesischen Reptilien und einer herpetol. *Tiergeographie Chinas.* DeGruyter Berlin u. Leipzig.
Pearse, A. S. 1912. The Habits of Fiddler-Crabs. *Philipp. Journ. Sci.* 7, 113–132.

22

ON THE PROCESS OF CENTRAL NERVOUS COORDINATION

Erich von Holst

This article was translated expressly for this Benchmark *volume by Chauncey J. Mellor and Doris Gove, The University of Tennessee, Knoxville, from pp. 149–153, 157, 158 of Über den Prozess der zentral-nervösen Koordination,* Pflügers Archiv f. d. Gesante Physiol. *des Menschen und der Tiere 236:149–158 (1935).*

Translation © 1985 by Chauncey J. Mellor and Doris Gove.

A) *PROBLEM.*

One of the most characteristic properties of the nervous system is its ability to order movements and positions of various body members and parts with respect to one another in such a way that they make up a harmonious whole. This property is also one of the most difficult to interpret. As far as the explanation of this coordinating function has been attempted at all, ideas generally fall into the framework of the usual reflex schemata. People assume, say, that one extremity carrying out a certain motion sets another extremity into motion by reflex. This activates a third or reactivates the first, etc. Up to this point, where their validity has been tested experimentally, these ideas have not proven to be useful.[1] The reflex theory of coordination brings us no closer to the understanding of the coordination process. This knowledge and the fact that we are dealing with a very complex function that can be influenced from various directions, not all of which the experimenter surveys, let alone controls, have on occasion recently led to the view that these and other higher central functions are subject to certain effects that we cannot comprehend with mechanical methods.[2] This opinion seems to me questionable because it relegates a purely physiological question to the realm of philosophical speculation. That it could come to this can, in my view, be attributed to the lack of a useful methodology permitting us to say quantitatively precise things about the process of central coordination. And I therefore made it my task to find a new method.

[1]See for worms, v. Holst. 1932–33 [6.], for fishes, 1935 [*1.*]; in addition, the papers of Bethe on the coordination in arthropods and others, 1931 [2.], 1936 [3.]
[2]Thus, for example, Auersperg 1934 [*4.*] speaks of the "physical energetical indeterminate accomplishments" of the central nervous system, of the "metamechanical character of coordinative accomplishments," and of the fact that this "totality of accomplishment" includes an "immanent functional order inaccessible to purely analytically oriented research" and that it "appears impossible to grasp the events occurring intra-centrally with methods of measurement."

B) *METHOD.*

The views and information obtained recently on the theory of plasticity provided the point of departure. They provide a certain insight into the problem of locomotory coordination in arthropods, and simultaneously contain the first beginnings of a new approach to the solution of this problem. We know today that in these animals, the system of movement is not fixed by rigid reflexes, but that the legs can move in any mutual relationship of position. But we also know that a "system gradient" exists that keeps the legs constantly in a certain coordination equilibrium, and we know in addition that this equilibrium can be changed by various central and peripheral interventions. Only when a rather large number of legs in sequence is removed in diplopods and chilopods does this equilibrium no longer assert itself. The legs located before and after the gap move in a different rhythm. However, the question remained open here whether some sort of mutual connection might exist despite this allorhythmia.

This question caused me to look for a more suitable subject in which organs of locomotion operate at different frequencies, but also in which a certain mutual influence remains despite this independence. This influence must be expressed in certain deviations of frequency and amplitude. The suitable subject is *Labrus festivus*, a bony fish of the Gulf of Naples. It usually swims not by undulation of the body but by wave form oscillation of various fins that are moved in the same way as the tail fin of the goldfish in the "propeller reflex."[*5.*] In the main, *Labrus* uses the pectoral fin and the posterior broadened part of the dorsal fin. Occasionally, the anal and caudal fin can also participate. In mere observation one sees that the dorsal fin oscillates more quickly than the pectorals, which, for their part, are generally moved alternately, but on occasion in synchrony. One can further note that the dorsal fin oscillates regularly only when it alone is active. If the pectorals join in the action, then certain deviations appear in the dorsal fin rhythm that apparently have a connection to the pectoral fin rhythm.

Now the next task was to record these motions in such a way that as many things that could have a changing effect on them were excluded. This way, the mutual connection of the rhythms is unambiguously expressed. Above all, therefore, the brain, with its constantly changing effects on the activity of the spinal cord, must be excluded. On the other hand, the active movement of the fins must be fully maintained. I finally found that these conditions are fulfilled when a very careful section is made through the upper medulla of the fish, specifically behind the exit of the oculomotorius and optimally in front of the exit of the vagus nerve. With this section, the influence of the brain on the spinal cord is lost. Respiration ceases, and the rhythmic fin motions are continued for hours and days with the greatest consistency.

Animals of 15–25 cm of length were used; the operation was done with urethane narcosis. A transverse slit is cut into the skull with a scalpel in the appropriate place. Through this slit, a small honed tab made of razor blade steel is gradually pushed down. The tab is just wide enough to fill the skull cross section at this place. In this way the medulla is carefully severed. Loss of blood is minimal. The fish is put in a small tank of seawater where its mouth is attached to a tube that brings water in (see Illustration 1a). The body rests on a lead fork shaped to fit (1b). The rhythm of the *dorsal fin (D-rhythm)* and of the *pectoral fin (P-rhythm)* are recorded. Since the pectoral fins always oscillate alternately in this condition, the recording of one fin, the left, suffices. The balanced recording arms (h) are made of straw, the writing nibs of paper. The part that moves in the water is made of a thin celluloid tab. So as not to interfere with the movement of the fins, the arms are not directly connected to them, but are instead connected with fine wires (d) about 2 cm long that are flexibly connected to the arm on one end and on the other are connected with a little hook to two or three fin rays. Attachment was always made on the two uppermost rays of the pectoral fin approximately at the midpoint. On the dorsal fin, it was made on the fourth to sixth ray, counting from the back, two-thirds of the way along its length. The axis of the writing arm ran parallel to the axis of the fish.

Usually the fish survives several days after the operation (up to six). Soon after waking from anesthesia, the brain again begins to function. The fish begins to move its eyes rapidly back and forth. After some time they begin to protrude severely so that without entirely losing their mobility, they stick out far from the head after a few days. Also, many internal organs often collect substantial amounts of fluid. Since these things occur only after the incision, in my observation, I believe that here in the medulla there lies an especially important area for the regulation of water balance—a question worthy of further investigation.

The spinal cord regains irritability in the course of several hours (2–10). At first, there is no response at all to stimuli. After a while, one can elicit reflexive fin movements, especially by pinching the front part of the anal fin. Later, a mere touch at this point suffices to produce sustained pectoral and dorsal fin rhythms. Finally, the movements appear by themselves. At first, they are weak with little amplitude. In the course of the following hours, the rhythms grow stronger, and finally take their course in full natural amplitude and frequency with the greatest uniformity. In the process, dorsal and pectoral fins produce a stream of water directed down and back. Many animals show hardly any interruption of motion. In others, small pauses occasionally occur. Either all the fins are at rest or the dorsal fin rests alone. Also, one or both pectoral fins stops for a while.

C) *FINDINGS.*

Frequency and amplitude of the P-rhythm are always uniform regardless of whether the D-rhythm is active or not. The D-rhythm, by contrast, is only uniform as long as both pectoral fins are at rest. If they start being active, then the dorsal rhythm always shows strong deviations, as shown in the example in Figure 2.

Now we must first counter the objection that these deviations are of mechanical origin, occurring, for example, through the movements of water caused by the pectoral fins. If one divides the tank into two compartments with a cross wall that touches the body of the fish on both sides such that the pectoral fins move in the front compartment and the dorsal fin moves in the back compartment, then the irregularities of the D-rhythm persist to the full extent. Or: the dorsal rhythm can be inhibited for a time by brief touching of the dorsal fin or by pinching the upper part of the caudal fin once without disrupting the pectoral rhythm. The dorsal fin then remains completely motionless, as shown in Figure 3. Finally, to cite a third proof, it can be easily calculated that the deviations of the D-rhythm, if they were being mechanically caused by the pectoral fin motion, would have to be the opposite of the actual deviations. The result emerges therefore with complete certainty that some sort of central nervous process lies at the base of these irregularities of the D-rhythm.

[*Editor's Note:* Material on the measuring of curves has been omitted at this point.]

Basically, the comparison of the three curves of the same animal yields the result that *the influence proceeding from the P-rhythm does not fundamentally change with a changing frequency ratio of both rhythms.*

Before we can ask more detailed questions about the nature of this influence itself, it first seems important to decide from whence it comes. Two possibilities are open. Either they are reflexive excitations originating through the motion of the pectoral fins that gain an influence on the rhythm formation of the D-rhythm. Or we are dealing with an effect of the central process eliciting the P-rhythm itself on that process that generates the D-rhythm. The latter is by far the more probable. The following experiment decides the question. Two small levers are attached to the ends of both pectoral fins in a suitable way and are connected to one another in such a way that their motion sets the pectoral into a passive alternating oscillation that corresponds to a great degree to the normal active oscillation. The point of attachment itself is made insensitive with a small transverse cut made through the fin rays in question quite close to the attachment place. As mentioned above, the P-rhythm occasionally stops for a short time. One can also bring about this cessation artificially, for example, by momentary pressure at the back of the pectoral fins. While the pectoral fins are still, the D-rhythm is always completely uniform. Now, if the pectoral fins are passively set into rhythmic alternate oscillations, then the uniformity of the D-rhythm is always fully maintained. No deviations of any sort comparable to those described above appear. Only when the passive motion becomes very violent can certain stimuli arise. These stimuli, however, lead to a partial inhibition of the D-rhythm that lasts for a certain period of time (see Fig. 8). If any receptors stimulated in the act of pectoral fin movement were the cause of those regular deviations in the

D-rhythm, the same influences would have to proceed even from merely passively moved pectoral fins, since the stimulation of the exteroceptors in passive and active oscillation of the fins is the same, and we know the same about the proprioceptors, for example, through Sherrington's well-known experiments on "plastic tone." *Consequently, receptors and reflexes play no role here. The effects proceeding from the P-rhythm must be of a central nature*.

This result stands in highest agreement with earlier experiments on the coordination of other fishes. In a quite different way I have been able to show [*1.*] that the harmonious cooperation of the individual segments in swimming motions does not, as hitherto assumed, rest on the chain reflex mechanism, but that we are concerned with a purely central process, up until now completely unknown in its nature. Now for this process, a quantitative measure is obtained for the degree of the effect of the P-rhythm on the D-rhythm. And thereby the possibility is obtained to begin its closer analysis.

In this, the first paper, we have become acquainted with a hitherto unknown form of mutual coordination of movement. I suggest calling this type of coordination "relative coordination" in contrast to "absolute coordination" in which a definite mutual relationship of the movement phases is always maintained—regardless of whether it can change (dogs' or horses' gait) or not (bird flight). Absolute coordination can either be facultative (arthropods) [*3.*] or obligate (higher chordates, with certain exceptions to be discussed later).

In a paper soon to follow, I will attempt to show, among other things, that this relative coordination is none other than the precursor of absolute coordination—the transition from no harmony of movement at all to complete harmony of movement.

SUMMARY

Dorsal and pectoral fins of *Labrus festivus* oscillate independently of one another in a rhythm of changing frequency ratio; in the process, a continual influence proceeds from one, the "dominating" rhythm (pectoral fins) onto the other rhythm, which is expressed as a quantitative increase or decrease of the beat duration and beat velocity of the dorsal fin rhythm.

A method is communicated by which, after excluding the influence of the brain, which constantly influences the activity of the spinal cord, a precise recording and thereby quantitatively precise determination of this coordinating influence becomes possible.

The duration and velocity tables in Figures 5 and 7 [omitted in this excerpt] show examples of it for the same animal in the changing frequency relationship of both rhythms.

This effect, extending from one rhythm to the other, is not of a reflexive, but of a purely central origin.

The present new type of movement coordination is contrasted to the hitherto known "absolute" coordination and is called "relative" coordination.

LITERATURE CITED

1. Holst, E. v.: Pflügers Archiv **235,** 345 (1935).
2. Bethe, A. and E. Fischer. *Handbuch der normalen und pathologischen Physiologie* (Handbook of normal and pathological physiology), Vol. 15, p. 1045 (1931).
3. Holst, E. v.: *Biol. Reviews* **10,** p. 234 (1935).
4. Auersperg, A.: *Pflügers Archiv* **233,** 549 (1934).
5. Holst, E. v.: *Z. vergl. Physiol.* **20,** 582 (1934).
6. Holst, E. v.: *Zool. Jb. Physiol.* **51,** 547 (1932); **51,** 67 (1933).

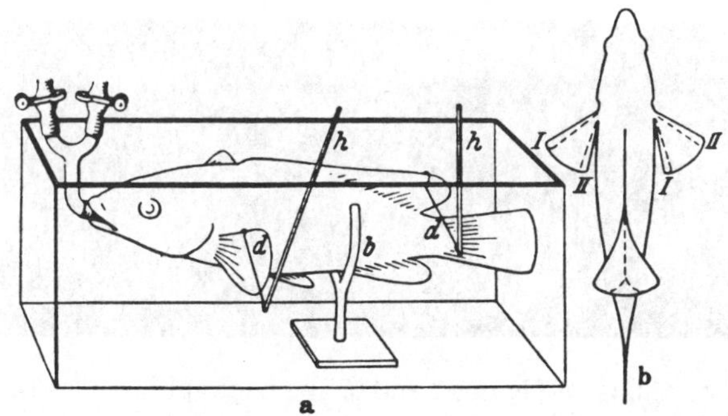

Figure 1a and b. a. Experimental set up—detailed description in text. b. *Labrus festivus* seen from above; the oscillating range of the dorsal and pectoral fins shown schematically.

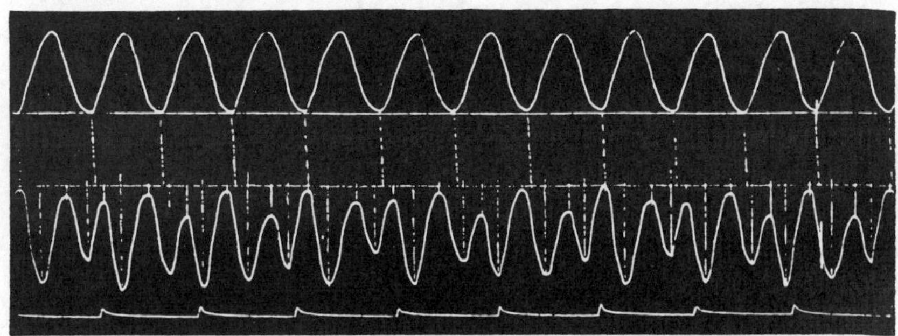

Figure 2. *Labrus festivus*, sample of curves. Above: Left pectoral. Below: dorsal. Time: (as in the other curves) in seconds. (Animal XXIII, 18.)

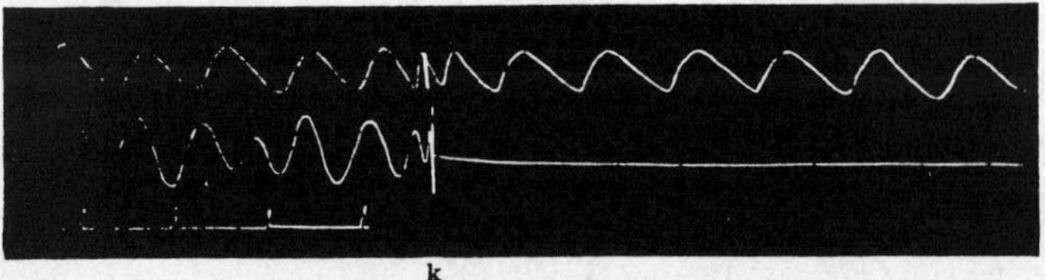

Figure 3. Inhibition of the D-rhythm by single pinching (k) on the tail. (Animal XXI, 14.)

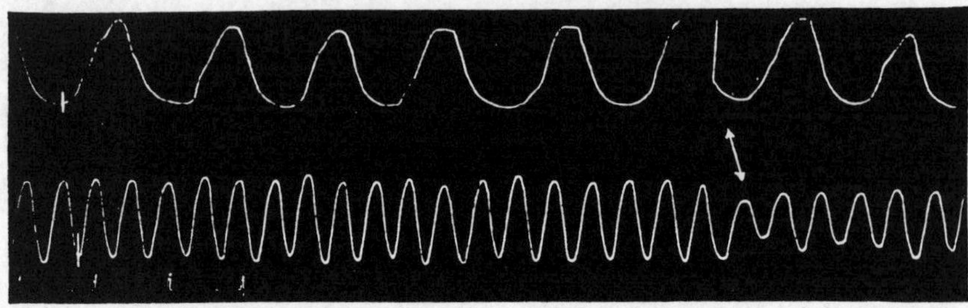

Figure 8. Same animal as Figures 2 and 6. The pectoral fins have been passively set into oscillation here. The arrow designates an inhibition of the D-rhythm caused by a too violent movement of the pectoral fins. (XXIII, 20.)

Part V

THE SYNTHESIS ATTAINED

Editor's Comments on Paper 23

23 LORENZ
On the Concept of Instinctive Action

The preceding papers provide much insight into the views of animal behavior held by a variety of thoughtful persons. Konrad Lorenz, after years of raising and observing many species (delightfully recounted in Lorenz, 1952 and 1954), developed his insights and those of others into a comprehensive approach. It relied largely on evidence from vertebrate animals, and this undoubtedly facilitated its eventual serious consideration by psychologists.

Some aspects of this paper have been discussed in the Introduction. Only a few final points will be added here. First, the term 'instinctive action' proved confusing to many people since instinct had a history of being used for more than just consummatory acts or inherited motor coordinations, the two senses in which Lorenz used the term in this paper. Thus fixed action pattern became the accepted term (Tinbergen, 1951) for the latter. In fact, the proper translation of the German word central to Lorenz's analysis here, *Instinkthandlung,* is not trivial. In translations of other papers by Lorenz from this period it has been rendered 'instinctive act,' 'instinctive behavior pattern,' as well as 'instinctive action,' used here. 'Instinct action,' as a parallel with reflex action, is an alternative, but, unlike reflex, use of instinct as an adjective is not standard English. We have striven for an accurate, literal translation as Leyhausen has pointed out (in Lorenz and Leyhousen, 1973) that Lorenz's ideas at this time were still hesitant and even contradictory.

Second, it should be emphasized that Lorenz himself emended his system in several important respects since the late 1930s. Some of his examples were also corrected. For instance, in Paper 23 Lorenz claims that shrikes need to learn how to orient themselves to the thorn on which they impale insects and other prey. Later work showed that when more properly reared, shrikes do know how to impale prey on thorns when confronted with them for the first time. This type of result made Lorenz wary about any evidence for experiential effects based on deprivation experiments (see Lorenz, 1965). This constrasts with the reverse strategy later employed by Hebb and others in trying to establish the effects of early experience (e.g. Hebb, 1949). We now know that Lorenz's skepticism was well-founded. Lorenz's point that one could establish what was innate from deprivation experiments has been attacked in many quarters; nevertheless, carefully executed and interpreted deprivation experiments are of continued importance in ethology and psychology. Lorenz (1965) later redefined innate to refer to genotypic sources of 'information' rather than the phenotypic characteristics themselves, and broadened his treatment of learning and experience, incorporating genetic constraints on these as well (Lorenz, 1969). However, this latter point had also been anticipated (c.f. Morgan, Papers 7A, 7C). The 'good for the species' terminology was also played down in later writings. The break with reflex theory announced here was final.

Third, this paper shows the diverse contemporary and international sources used by Lorenz in the 1930s. Many workers influencing Lorenz could not be represented in the source documents. These include important writings, not yet available in translation, by O. Koehler, Kramer, and Verwey, a large number of Americans, both psychologists and biologists. This international eclecticism contributed to the enduring vitality of the classical synthesis. American laboratory animal psychology was in large measure reducing its adaptability by a narrowing of the factors considered important. On the other hand, ethology was not uncritically eclectic; it went beyond mere platitudes on the need to consider natural behavior, diverse species, different sensory mechanisms, and evolutionary processes. The ethologists outlined a specific set of mechanisms that appeared to have wide scope. They could be tested and modified, but as with Darwin's theory of natural selection, the theory proved resilient enough to withstand many a "fatal" objection.

Those familiar with Lorenz's writings only from his popular works, might find this reading unexpectedly complex; but careful study in conjunction with the source readings should be rewarding.

REFERENCES

Hebb, D. O., 1949, *The Organization of Behavior,* Wiley, New York, 335p.

Lorenz, K., 1965, *Evolution and Modification of Behavior,* University of Chicago, Chicago, 121p.

Lorenz, K., 1969, Innate Bases of Learning, in *On the Biology of Learning,* K. Pribram, ed., Harcourt, Brace, Jovanovich, New York, pp. 13-93.
Lorenz, K., and P. Leyhausen, 1973, *Motivation of Human and Animal Behavior: An Ethological View,* Van Nostrand Reinhold, New York, 423p.
Lorenz, K. Z., 1952, *King Solomon's Ring,* Methuen, London, 202p.
Lorenz, K. Z., 1954, *Man Meets Dog,* Methuen, London, 199p.
Tinbergen, N., 1951, *The Study of Instinct,* Clarendon Press, Oxford, 228p.

23

ON THE CONCEPT OF INSTINCTIVE ACTION*

Konrad Lorenz

This article was translated expressly for this Benchmark *volume by Chauncey J. Mellor and Doris Gove, The University of Tennessee, Knoxville, from pp. 17–49 of "Über den Begriff der Instinkthandlung,"* Folia Biotheoretica, *Series B, No. II:* Instinctus, *E. J. Brill, Leiden (1937). Page numbers are indicated in brackets in the text.*
Translation © 1985 by Chauncey J. Mellor and Doris Gove.

But there must be a
concept to the word.
Goethe

TABLE OF CONTENTS

* Revision of the lecture given at the Second Symposium of the Prof. Jan van der Hoeven-Foundation for Theoretical Biology on November 28, 1936 in Leiden. The subject of this symposium was the concept of instinct.

[18]

INTRODUCTION

As pleased as I was at the time to accept Professor C. J. van der Klaauw's request to speak at the symposium on animal psychology in Leiden on the problem of instinctive actions, I am now equally pleased to accept his invitation to extract the purely conceptual material from my recent paper on this topic in the "Naturwissenschaften" and to compile it in a manner appropriate for the "Folia Biotheoretica." Of course, I will not pass up the opportunity to make corrections in those places where I think I have learned something new from the lively discussion that followed the appearance of that treatise.

Instinct is a word, nothing else. And the attempt to define "instinct" too often just goes around in a circle: pigeons have a homing instinct because they fly home. They fly home because they have a homing instinct. Connecting the word "instinct" in this way with the specification of an activity can become detrimental when it is considered to be an explanation of the process. Such illusory explanations then often obstruct the way to further analytical research. Thus, in most cases, one resorts to the assumption of an "instinct" when the behavior of an animal cannot be explained by the psychic accomplishments we know from our own experience; for example, an organism's capacity to learn and its "insight" clearly are not sufficient to control purposeful actions it executes for the survival of its species. Or an animal may carry out such actions in a completely perfect way with no prior experience at all. Thus, first and foremost, every *innate* purposeful action is judged to be an outgrowth of an "instinct." But a more precise analysis shows that even among types of innate behavior there are two categories of movement patterns that can be distinguished with exceptional clarity: the *orientation reactions* or *taxes,* and the movement patterns based on *inherited coordinations.* Taxes are innate insofar as they represent *norms of reaction to direction-giving stimuli.* The norms are independent of the individual experience of the animal. They orient the animal in space, and the movements guided by them directly adjust the locomotor [19] movements of the animal's body to the spatial circumstances of its surroundings at the time. The *form* of the movement depends directly upon the given external stimuli in each case.

The coordination and sequence of the muscle contractions are not rigid, but are adaptively influenced by the individual stimulus situation. The hereditary coordination, by contrast, is *not a reaction norm,* but a *movement norm*—a sequence of movements of certain muscles and muscle groups established once and for all in the species involved. The coordination and rhythm of contraction of these muscles and muscle groups are *not* influenced by the type and direction of the external stimuli in effect at that moment. The *form* of the movement is independent of the stimuli. These stimuli are only capable of determining whether and with what *intensity* the genetically coordinated movement formula runs its course. The regular gradations of the phenomena associated with various intensities of the same reaction will be discussed elsewhere.

The further we proceed in the analysis of animal behavior, the deeper the gulf appears to be between the movement patterns governed by hereditary coordinations and those controlled by the higher psychic achievements—capacity to learn and insight. By contrast, one cannot draw a clearcut externally detectable boundary in the behavior of organisms between the orientation reactions on the one hand and the learned and insightful actions on the other. Indeed, this becomes more difficult, the more sharply and objectively one tries to formulate these concepts. It is precisely this circumstance that seems to justify drawing a conceptually precise distinction between types of behavior based on hereditary coordinations and *all* others and contrasting such types with the orientation mechanisms as well as with learned and insightful actions as something of a different nature. The possibility of really establishing a concept justifies, in turn, selecting a term, and since the movement patterns that are fixed in the hereditary coordinations quite certainly constitute the essence of what people used to regard as the effect of "instinct," I have termed these patterns instinctive actions.

For the time being, and perhaps even in principle, it is not possible to give an intrinsic definition for any sort of biological phenomenon. Its extrinsic definition can be based only on an enumeration of a very large number of its characteristics, and an attempt to enumerate the facts most clearly characterizing the instinctive action will be made in the present treatise. To the extent the separation appears at all justified, the separation of this observationally founded version of the concept of instinctive action [20] from all other types of behavior must be included in the concept itself. Similarly, an attempt will be made to separate the concept of instinctive action from that of reflex. A special section will be devoted to discussing and distinguishing those elements of animal behavior that, according to the version of the concept of instinctive action advocated here, should be considered separate from instinctive action, but which are considered to be instinctive under older conceptualizations. The justification of the new conceptualization we present here for discussion will show how these distinctions provide further opportunities for an analysis of animal behavior.

I. THE CHARACTERISTICS OF THE INSTINCTIVE ACTION

A) THE BEHAVIOR OF THE INSTINCTIVE ACTION IN ZOOLOGICAL SYSTEMATICS

The most striking and perhaps most important characteristic of the instinctive action for anyone thinking in terms of biology is its behavior in zoological systematics. It is one of the most remarkable facts that there are movement patterns which are not only present in all individuals of a species in the same manner, but which also can be characteristic of much larger categories: of genera, orders, indeed, even of entire classes. Under certain circumstances these movement patterns exceed most morphological characteristics in taxonomic value. For example, when one looks up the diagnostic characteristics of the order of doves in a modern zoology text, one finds the following compilation of morphological characteristics: "carinate, altricial bird with weak, soft-skinned beak that has a blister-like distension in the area of the nostrils; with medium-long, tapered wings and short feet either cleft or for squatting." Not even such a combination of various characteristics suffices to define the group without exception. The crowned pigeon *(Goura)* is not truly altricial; *Didunculus* has an entirely different kind of beak; in *Goura* again, short, round wings similar in every respect to those of gallinaceous birds are found; and a whole series of terrestrial pigeons does not share the characteristic of short legs at all. But, on the other hand, if one says that the coordination of drinking movements in doves requires rhythmic sucking in of water with the beak held still and constantly immersed during drinking, then this fact perfectly defines the group. There is as yet no known kind of dove equipped with a [21] different drinking movement, nor any other kind of bird that might be confused with doves through having the same movement pattern. To the extent that they drink at all, all other birds drink either in the well-known manner of the domestic chicken, by dipping and scooping and raising the beak, or by movements of the tongue, as in hummingbirds and sunbirds. Consequently, the drinking movement pattern in the order of doves proves to be phylogenetically more conservative and thereby taxonomically more significant than any single characteristic listed in the morphological diagnosis; indeed, it is even more so than the sum of these characteristics.

The demand for phylogenetic-comparative investigation of instinctive actions arises from their taxonomic utility, a demand that Whitman summed up as early as 1898 in the sentence: "Instincts and organs are to be studied from the common viewpoint of phyletic descent." Heinroth, in his paper "Beiträge zur Biologie, namentlich Ethologie und Psychologie der Anatiden" [Contributions to the Biology, especially the Ethology and Psychology of the Anatidae], which did not appear until 1910, demonstrated that taxonomic consideration of instinctive actions in many cases justifies statements about phylogenetic connections that a comparative morphologist is hardly ever free to make with equal certainty. His results were strikingly confirmed by the investigations of Poll, who investigated the various disturbances in the maturation of sperm in male hybrids of various forms of anatids and employed the degree of completion of their spermatogenesis as a useful and extremely precise measure of the relatedness of the two parent species. In all cases where both researchers departed from the traditional systematics of anatids, they were in agreement with each other. Later, Heinroth extended the phylogenetic approach to the instinctive actions of far larger systematic groups and showed in his paper "Über bestimmte Bewegungsweisen bei Wirbeltieren" [On Certain Patterns of Movement in Vertebrates] that a whole series of hereditary coordinations can be shared characteristics of whole classes, indeed, virtually of entire phyla. From the rich content of the above-mentioned treatise, let us just take the behavior of a single coordinated movement in the zoological system. The *scratching movement* is fundamentally the same in salamanders, lizards, and mammals. The scratching hind leg is brought laterally to the head past the foreleg that is standing on the ground, as everyone has seen in the domestic dog. This coordinated movement behaves [22] in a particularly interesting way in the class of birds. The anterior extremity no longer "stands on the ground" in birds, but is situated far dorsally and not in the way of the scratching posterior extremity at all. Despite this, in a large number of species, the original spatial relationship of the extremities is reproduced before the movement formula of scratching takes place. The wing is lowered, and the leg is brought laterally past it to the head. In other birds, we find this evidently more primitive movement pattern changed in such a way that the wing remains in its rest position and the foot simply moves around forward in the direction of the head. The distribution of scratching "around front" and "around back" in the class of birds corresponds entirely to systematic relationships and proves to be completely independent of the proportions or other anatomical properties of the individual species and genera. Systematically different birds that have, by convergence, become exceedingly similar in anatomical form and in other movement patterns can show opposite scratch reactions. And, on the other hand, forms that are extremely different anatomically often scratch themselves in the same way. Both these facts indicate without question that phylogenetic relationships exclusively determine the issue of around front or around back and that functional factors play no role at all.

A certain group of instinctive actions deserves a particularly high taxonomic value. The species-preserving value of these instinctive actions consists in their signalling effect, that is, in the release of certain other instinctive actions on the part of a conspecific. These movement patterns acting as signals are particularly useful in systematics since their special form is not directly influenced by factors in the animals' surroundings and is determined almost purely by historical factors, as are, for example, the signs of the Morse code alphabet. It is pure "convention" that, for example, the tail wagging of many canid predators signifies a peace offering, while a similar movement in felid predators means hostile tension. As far as the functions of both signals are concerned, they could just as well be reversed. From these circumstances it follows that the similarity of releasing instinctive actions in various species nearly always means homology, since, of course, their precisely identical elaboration by convergence in two different animal forms is virtually impossible.

The incontestable systematic utility of the instinctive action is of great importance in two different respects. First, it proves beyond doubt that instinctive actions, with their phylogenetic mutability, behave entirely like organs; indeed, in many cases, [23] they conserve the details of their particular form even more tenaciously than organs do. Taken by itself, this organ-like behavior during phylogeny suggests that the form of a movement is closely *linked* to *structures* in the central nervous system. Second, the fact that it is possible to use instinctive actions as taxonomic characteristics argues very strongly against assuming that the individual experiences of the animal change them. Where would Whitman and Heinroth (both of whom raised the smallest and finest details of movement forms to taxonomic characteristics of the highest dignity) have gone with their conclusions if the form of an instinctive action were influenced by the personal experience of the individual creature? Both made their observations on captive animals, and, in most cases, on ones actually reared from an early age. Nevertheless, their conclusions have been so convincingly confirmed.

B) BEHAVIOR OF THE INSTINCTIVE ACTION IN ONTOGENESIS

If, from personal experience and on the basis of a sufficiently large number of individual observations, one is familiar with the above-mentioned behavior of instinctive actions in the zoological system, then one finds the following characteristic of the instinctive action discussed by most authors as the primary, most important, and most peculiar, to be quite obvious: the instinctive action can develop in a young animal without prior experience and without the example of an older conspecific. Indeed, the action can appear quite suddenly in its species-preserving perfection. Nevertheless, the definition of instinctive action by Driesch requires a qualification. When this author says, "Instinct is a reaction that is perfect from the beginning," then one must counter with the fact that the instinctive action, quite like any organ, goes through a maturation process during the ontogeny of the animal, a process that can be complete at the birth of the animal, but, in reality, is certainly not so in all cases. The instinctive action can mature at the same time as the skeletal and muscle system whose movements it coordinates. In such a case, one cannot decide immediately whether the incompleteness of the movement or the incompleteness of the organ is responsible for the still deficient functioning. However, the maturation of the movement coordination can far outstrip that of the organ, and this sometimes leads to quite strange phenomena. For example, although the very small chicks of many anatids [24] do have tiny, completely useless wings in relation to their other body parts, they nevertheless exhibit from their first days of life all the movement patterns of these organs that we observe in the adult bird. This is particularly conspicuous in the fighting reaction when the opponent is held with the bill and struck with the bend of the wing. When little chicks perform this movement, they do not reach their enemy with their wing at all because the coordinations, which are attuned to the body proportions of the adult bird, cause the bill of the young bird to hold the opponent much too far away from its body. In such cases, the independent maturation of the instinctive action can be demonstrated very clearly. But the opposite case, in which the maturation of the organ outstrips that of the movement coordination, can also occur. So if we see a young animal with organs of movement that are already entirely functional performing uncoordinated movements, whose coordination is gradually becoming more perfect in the course of ontogeny, then we cannot decide at first glance whether we are witnessing a process of *maturation* or of *learning*. Authors who believe in a mutability of the instinctive action by individual experience often take such observations quite uncritically as arguments for this assumption. This question—maturation process or individual learning—can only be decided experimentally, and the few experimental papers that have investigated the ontogenetic completion of innate and easily definable movement patterns argue unambiguously for the assumption of a process of maturation. Carmichael used prolonged narcosis to prevent amphibian larvae from performing their natural swimming movements for quite some time. Grohmann raised young pigeons in very confining boxes in which they could not open their wings or practice any flying movements. In both cases, the ontogenetic completion of the movement coordinations suffered no impairment compared with that of control animals growing up normally.

Considering the other similarities to organs that the instinctive action shows in ontogeny, it seems not insignificant that, under certain circumstances, instinctive actions show a tendency to recapitulate phylogenetically old forms in the course of development, just as organs sometimes do. Thus, in ravens, pipits and larks that, as adults, stride instead of hopping with both feet like other passerines, one finds in juveniles a narrowly limited time during which they show the latter movement pattern as a "palingenetic" characteristic. All the same, such [25] ontogenetic recapitulation of movement patterns of ancestral forms is not very common.

C) THE REGULATIVE PHENOMENA OF THE INSTINCTIVE ACTION

One peculiarity of a number of instinctive actions is their capacity to be regulated in certain ways. This regulative capacity has often been used erroneously as an argument for obliterating all boundaries between instinctive actions and learned types of behavior. Now, these regulative phenomena appearing in instinctive actions are similar in many ways to those also observable in the external form of body organs. For example, the simplest and least highly differentiated actions possess a far higher capacity for regulation than the more highly developed ones. Bethe, for example, investigated the regulative capacity of the relatively simple coordinations of walking movements in arthropods. In decapod crustacea, various leg amputations were able to cause a nearly inexhaustible number of various regulative coordinations—always integrated—of the walking movements of the remaining legs.

Let me point out explicitly that the regulative phenomena established by Bethe are essentially independent of learning and experience. In most cases in which a regulation appeared at all, it was present in a completely finished state immediately after the operation, especially in arthropods. In a dog, whose two ischiatic nerves were sewn together across its lower back, a complete regulation was established in the motor domain, but remained absent in the sensory domain. Though normal in its movement patterns, the dog reacted to pain stimuli applied to one hind leg only with the other hind leg, confusing the sensory excitations coming from the two legs in accordance with the crossed nerves. The experiment shows that in this case, experience, which, of course, should have taught the animal the very opposite thing, played no role in the formation of the motor regulation.

D) INTENSITY DIFFERENCES OF THE INSTINCTIVE ACTION

An instinctive action very often becomes noticeable in the behavior of the individual animal as a subtle suggestion of the movements initiating the chain of movement based on an inherited coordination. However, at "low intensity" of the entire reaction, the animal breaks off the initiated chain after the first introductory links. [26] In the opposite case, at high intensity, it is precisely these introductory movements of an action chain that can be skipped and left out. Actions begun and left unfinished often tell us that the state of excitation associated with a particular instinctive action is beginning to awaken in an animal, thus revealing the animal's "intentions" to us. For this reason, these actions are often called intention movements. Between such barely indicated beginnings, perceptible, if at all, only to one knowledgeable about the instinctive action involved, and the full performance that fulfills the species-preserving function of the movement formula, there exists every single imaginable intermediate stage. One can observe entire intensity scales particularly clearly in the annual reawakening of certain movement patterns of the reproductive cycle, for example, in the nest building actions of very many birds. Since the manifestations of a reaction that are attributable to the individual intensity levels blend with each other in an unbroken transition, one can assert that these manifestations are present in an infinite number and can cite this number as an argument against the assumption that a reflex runs a fixed course. On the other hand, in these very intensity scales such a rigid, or one is tempted to say, machinelike regularity is expressed that one quite involuntarily reaches for a metaphor of physics to describe them and speaks of inner "excitation pressure" and the like. The difference in intensity with which the same reaction responds to the same stimuli is by no means subject to a haphazard willfulness of the animal, but instead, obeys its own laws. But these laws cannot be derived from the stimulus-reaction model of the reflex and therefore require a special explanation for a person inclined to view an instinctive action as a reflex process.

E) THRESHOLD LOWERING AND VACUUM ACTIVITY

One of the most peculiar phenomena we can note regarding the course of instinctive actions is that there is a regular variation in the threshold value of the stimuli releasing them. The rule governing these variations could be briefly summarized by saying that the instinctive action can be released more easily the longer the period since it last ran its course. If the appropriate stimuli normally releasing an instinctive action remain absent for a longer interval than generally corresponds to the circumstances in the natural habitat, then we note the animal's gradually increasing readiness to respond with the reaction concerned to other similar stimulus situations that are not [27] normally responded to with the performance of this instinctive action. The longer the time that release does not occur, the lower the selectivity of response of each instinctive action. And furthermore, in cases where the strength of the individual stimuli can be roughly quantified, the *threshold value* of the required stimulus strength also declines. In the extreme case, the two phenomena can lead to the "spontaneous" performance of the instinctive action, that is, *without* a demonstrable stimulus. This performance of the entire movement sequence, independent of releasing stimuli and independent of the total situation in which the instinctive action fulfills its species-preserving meaning, is of the greatest theoretical significance in several respects.

This "vacuum activity" [Translators' Note: "Leerlauf" means engine running in neutral, getting nowhere. English "vacuum activity" has become current and is used for Lorenz' "Leerlauf*reaktion*" as well.], as I have termed the emergence of a sequence of hereditarily coordinated movements independent of a stimulus, cannot be derived from the stimulus-response model of the reflex. Indeed, the essential quality of a "reflex," and the quality that determines its name, that is, the reflecting out of something streaming in, is totally absent here. In quite specific instinctive actions, one might be led to think of inner stimuli whose intensity increases with the length of time an instinctive action is "dammed up." The movement sequences of ejecting feces, sexual products, etc., as well as the movement sequences of food and water intake, can be dependent on such inner stimuli, and certainly are in many cases. But it is of the greatest importance to establish that threshold lowering and vacuum activity are, in all probability, *typical of all instinctive actions*, that is, also typical of those for which such an easily identifiable chemical or physical inner stimulus can be excluded with certainty, as, for example, in the flight reactions of very many animals, reactions especially prone to emerge as vacuum activity. Two things go hand in hand: (1) The regular rise in the readiness of the organism for a quite definite sequence of movements, and (2) an identical rise of the intensity of an action when it is finally released or erupts as vacuum activity. Both of these facts suggest the idea of an inner accumulation of a reaction-specific excitation substance that is destroyed or made ineffective when the movement runs its course. We will have to return to these matters in the section on instinctive action and reflex.

Another reason why vacuum activity is important is that it shows in a particularly clear way that the sequence of movements established by a hereditary coordination is wholly independent of any sort of additional stimuli by which the animal might be able to guide itself. In his book "Purposive Behaviour in Animals and Men," Tolman presents all animal behavior as purpose-directed behavior, and actually even uses the word [28] "behaviour" in this sense. Tolman says, in the discussion of such stimuli that support behavior ("behaviour supports"), "behaviour cannot go off in vacuo." But one could imagine no more conspicuous characteristic of instinctive action than the capacity "to dissipate in a void" in the absence of the stimuli that accompany its normal, species-preserving performance, nor could one imagine a better designation of vacuum activity than "to go off in vacuo." In his attempt to show it is absurd to assume the existence of a behavior that is not subjectively purpose-directed, Tolman demands as proof for such a behavior the very thing that we can provide in the form of an instinctive action spontaneously running its course. To be sure, the instinctive action is purposeful in an species-preserving way, but it is not a behavior that is subjectively purpose-directed.

One also finds examples in the literature of the phenomena of threshold lowering and vacuum activity. Eliot Howard, for example, pointed out at an early date that the intensity of the reaction and the readiness of the animal for the reaction in question were subject to regular variations, and he was also the first to surmise that the part of the reaction immediately affected by the lowering of the threshold was perception. In fact, human perception changes when a reaction has been dammed up for a relatively long time: For example, a very hungry person perceives a roast about to spoil as smelling good; the same roast would repel a person who was sated. The experiments of Lissmann, who used models to investigate the releasability of fighting reactions of the fighting fish *(Betta splendens),* can be evaluated quantitatively regarding the subsequent rise of a lowered stimulus threshold.

F) APPETITIVE BEHAVIOR

The increasing of intensity and the lowering of thresholds of releasing stimuli are, however, not the only factors that cause the organism to discharge an instinctive action. One of the most important and peculiar aspects of the instinctive action is that the organism does not passively wait for stimuli leading to the release of the action, *but actively seeks these stimuli*. Wallace Craig was the first to show that the animal tries to bring about the discharge of its own instinctive actions by a purpose-directed behavior that he calls "appetitive behaviour." Striving after a certain instinctive action is known to all of us in the form of an appetite. [29] Because of the rather narrow meaning of this term [*Appetit* = hunger] in German, I have translated Craig's expression as "*Appetenzverhalten*." Appetitive behavior is a behavior that pursues a purpose invested in the organism as subject, a typical case of *purpose-directed behavior*. By this term we understand along with Tolman all those types of behavior that, while the goal striven after by the subject remains the same, show adaptive variability in order to attain that goal. When, for example, an animal first strives for a given spatial goal in one way and then, finding the first avenue blocked, attempts another, then we have here the objective manifestation of the subjectively desired purpose. This possibility of defining the subjective purpose in the terminology of objective behavioral theory is very important for us.

The fact that the discharge of an instinctive action represents a goal for purpose-directed behavior throws open a question that is not easy to answer. The release, or perhaps better stated, the unblocking, of the discharge of an instinctive action has, in many cases, so much in common with reflex processes, as we shall show in the next section, that we are accustomed to regard at least this part of the reaction as an unconditioned reflex in the proper sense of the term. For the time being, however, it is altogether inexplicable how it happens that the organism as subject strives for the release of these very reflexes, something that it by no means ever does with respect to all its reflexes. It would occur to no human to make a special effort to seek out the stimulus situation in which the patellar reflex is released, whereas almost all organisms strive for the release of the reflexes leading to the discharge of sexual products, even in cases where the safety of the individual is put at risk. It is actually integral to the notion of the reflex that it lies constantly ready like an unused machine and only becomes active when certain key stimuli impinge on the organism. It is not in the nature of the reflex that the reaction, so to speak, announces its presence, thus disquieting the animal and causing it actively to seek these stimuli.

I would like to use the term *drive* for this disquieting of the animal, which is the direct cause of its seeking a stimulus situation that unblocks a certain instinctive action. In so doing, I am aware that this word has already been used with a greatly different meaning. Here, only a search directed exclusively toward *one specific reaction* is viewed as the effect of a drive. Without doubt, there are *two* factors that cause the animal [30] to strive for the release of a certain instinctive action. First, the factor just termed drive, and second, after prior experience, those subjective feelings of pleasure that accompany the impingement of the releasing stimuli and the discharge of the instinctive action itself. This "sensual pleasure" linked with exteroceptive and proprioceptive perceptions is quite certainly the actual purpose, from the subject's standpoint, of all behavior striving for the release of an instinctive action. Though we cannot go very far in explaining these subjective phenomena causally today, we are nevertheless very tempted to interpret them as final [causes] with regard to species preservation. The organism is "pushed and pulled" by the restlessly goading drive and by the "enticing pleasure" to discharge species-preserving inherited movement patterns.

Several authors espouse the view that there are two types of "instincts," those of the first and those of the second order. The difference between them is seen in the fact that the first order instinct, which is also called a superordinate instinct or "first order drive," avails itself of a "subordinate instinct" as a means to an end. Thus, a "parental instinct" (McDougall) allegedly makes use of the instincts of brooding, feeding, etc., and one finds the criterion of purpose-directedness of the superordinate instinct in this means-ends relationship. But superordinate instincts in this sense cannot be demonstrated. We would only be justified in assuming their existence if we were to see a factor in action (1) that has a goal-setting, unifying effect, (2) that goes beyond the capacity of individual reactions to be regulated, and (3) that coordinates these individual reactions for a common effect. Now, such a factor cannot ever be found, and the disruption of a single, only seemingly subordinated, instinctive action usually shatters the mosaic combination of instinctive actions that contribute to a specific species-preserving function, as can be shown in very many experiments. Each component action is carried out for its own sake, is sought after by its own appetence. The only planful and unitary aspect is the harmonization of a large number of autonomous appetences and instinctive actions working toward the objective goal of species preservation, a harmonization uniquely developed in the evolution of each species.

Nevertheless, we can speak of superordinate and subordinate instinctive actions, though in a fundamentally different sense: that is, there are some movement patterns based on central hereditary coordinations, but distinguished from other instinctive actions in that they are generally not performed [31] for their own sake, that is, for the sake of their own pleasurability, especially not in the normal habitat of an animal. For example, in a great majority of cases, the coordinations of locomotion are used *in the service of appetences of other kinds*, used, in fact, like an implement, as a means to an end. A wolf runs *in order* to catch and eat a deer. A bird flies *in order* to weave a grass stalk into its nest, etc. But we must never forget that these locomotor coordinations are not in such cases the implement of another *instinctive action*, but rather of a *purpose-directed behavior*. This purpose-directed behavior can be a pure taxis or a pure insightful action, or, in a special case, can even be an appetitive behavior directed towards another instinctive action, as in the above-mentioned examples. When that is the case, one can in fact say with some justification that the implement reaction is subordinated to the instinctive action that represents the desired conclusion of the subjectively purpose-directed chain of action. But we must not slip into the anthropomorphism of assuming that only "hunger and love" hold the workings of the world together, and that all other instinctive actions are subordinate implement reactions of these two appetences. To be sure, McDougall says expressly that a sated predator does not hunt, but every dog owner knows that the state of nourishment of a dog has no influence at all on its appetite for the instinctive actions of hunting. Quite similar things seem to hold for nearly all implement reactions. In some animals, it is exactly these hereditary coordinations of locomotion that need to be relieved of just a bit of their load serving other appetences to reveal most clearly a lowering of the threshold of the stimuli releasing them and to cause appetitive behavior of the animal. This is also true of the performance of other implement reactions such as the pecking and scratching of chickens, the delousing behavior of monkeys, etc. The captive animal strives for these reactions with true elemental force, and the animal is satisfied with the most improbable substitute stimuli. After all this, we are quite certainly not justified in generally terming any instinctive action as a subordinate instinctive action; much less can we separate it off conceptually as "innate aptitude" ("innate skills or dexterities" of American authors) from other instinctive actions. *Any* hereditary coordination can in principle become the goal of its own autonomous appetitive behavior.

[32] II. THE RELEASE OF THE INSTINCTIVE ACTION

The conditions that can lead to the release of a single instinctive action are variable to the same extent that the movement formula of that action is rigid and void of all goal-directed variability. The release of one and the same movement pattern can be dependent one time on an unconditioned reflex and another time on something learned or insightful. Learning and insight can decide the question whether and at what intensity a certain instinctive action is released in an individual case, but they never, even in the slightest, change the *form* of the movement pattern ultimately executed. Even in this, the instinctive action behaves once again like an organ, which, like the beak of a bird or the hand of a monkey, can, to be sure, be used in various ways under the influence of experience and insight, but cannot be changed in form. Such a course of action is in and of itself rigid, varying only along a scale of intensity that is fixed from the very beginning. Its release, caused by a whole series of different, and above all, differently ranked centers [Instanzen], will now be discussed in greater detail.

In very many cases, probably more often in lower animals than in higher ones, the process releasing the course of the instinctive action is an unconditioned reflex. Now, such unconditioned reflexes are especially interesting because the stimuli, or, better said, the stimulus combinations to which they respond, usually are of a quite particular nature. Among the many potential stimuli of the objective overall situation, in which the chain of movements concerned must be discharged in order to fulfill its species-preserving meaning, we find relatively few stimuli that act as characteristic features. However, we always find stimuli that characterize the situation sharply enough to make an "erroneous" occurrence of the action very unlikely in a situation other than the biologically correct one. Thus, the biting reaction of the common tick (*Ixodes ricinus* L.) responds to the combination of a heat stimulus of 37 degrees and the chemical stimulus of butyric acid. As simple as this stimulus combination is, it nevertheless suffices to adequately characterize a mammal, the appropriate host of the parasite. Adapting von Uexküll's terminology, I have elsewhere called such receptive correlates of very specific key stimuli "innate releasing schemata." The unblocking of instinctive movement chains by the unconditioned response of the innate releasing schemata is by no means found only in lower animals. In particular, social instinctive actions of birds very [33] often respond exclusively to highly specialized innate schemata. In such cases, special differentiations for the transmission of key stimuli of acoustic and optic nature often correspond to these highly differentiated receptive correlates. The sole function of a very large number of bright colors, striking structures, emitted sounds, and peculiar "ceremonial" movement patterns is to transmit such releasing stimuli.

In the presence of an innate releasing schema, the releasing process of an instinctive action corresponds in this respect to an unconditioned reflex. However, in many higher animals there are very many instinctive actions whose release, as in the case of a "conditioned reflex," is dependent on a stimulus situation, the "knowledge" of which must first be acquired by the individual. Since this acquisition is equivalent to true learning, that is, a complex and not yet satisfactorily analyzed process, not to be confused with the ultimate occurrence of the instinctive action, I have spoken in such cases of an "instinct-training interweaving." And I have done so even where the training is limited exclusively to the receptive aspects. Another term might be the expression "conditioned instinctive action." Such a releasing mechanism can be found in very many instinctive actions that involve a specific object, without its characteristic features being given to the animal in the form of an innate schema. For example, a young shrike (*Lanius collurio* L.) acquires the knowledge of a thorn necessary for the successful performance of its impaling reaction by self-training after a relatively long search by the principle of trial and error. Frequently, the acquisition of such knowledge of an object is guided by an innate schema consisting of few and simple signs. Thus, for example, for the reactions of prey catching in young birds of prey, an innate schema of the prey animal is present that is made up of only such signs as mobility, approximate size, and similar things, but which is then supplemented and perfected by learned characteristics. Such an addition of acquired characteristics, which release actions in a "conditioned" manner, to the characteristics of the innate schema, which act in an unconditioned manner, is extraordinarily common. Indeed, one could say that it is demonstrable in higher animals in almost every instinctive action. Thus, for example, the defensive reaction of a mallard duck mother, immediately after the hatching of her young, responds fully to the distress call of *any* mallard chick, but, a few weeks later, responds only to the distress calls of her own young, which she now knows individually.

The addition of acquired material to the releasing factors [34] of an instinctive action always means a very substantial increase in the *selectivity* of its response. One of the most striking and important differences between innate schemata and acquired, "conditioned" action-releasing stimulus situations is that the schemata are virtually always *very poor in distinguishing characteristics*, while the conditioned stimulus situations are generally "complexes" very rich in distinguishing characteristics. In the highest mammals, it does of course occur, as Verlaine has shown, that only isolated features, as arbitrarily chosen characteristics so to speak, are "abstracted" from the stimulus situation that has been entrained. In intellectually lower animals, it generally occurs that very many details are equally effective in an entrained releasing stimulus situation; indeed, it is perhaps even the case that, in principle, any perceived details whatsoever are equally effective; all of them being submerged in the overall quality of what has been perceived, and all simultaneously determining this quality, as Volkelt in particular has emphasized.

The instinctive actions called implement reactions above, generally carried out in the service of other appetences, rather than for their own sake, show a particularly wide variety of forms of release. The release of typical implement reactions, such as locomotion, is subject to, among other things, the most central and least analyzable parts of the central nervous system: the "voluntary control" of the animal, as we usually say. Such a movement pattern is then dependent on quite different centers [Instanzen] in the nervous system. Thus, the chain of movements of a bird taking off, which is in and of itself quite rigid, can take its course as vacuum activity after long damming up. But it can also be released by an unconditioned reflex, for example, by the response of an innate releasing schema, as when a greylag goose spots an eagle for the first time. But the same instinctive action can also be set in motion by a conditioned reflex, as when the same goose spots a man with a rifle, a sight which it must have learned to fear on its own. Finally, the reaction in an insightful action of the animal can also be unblocked "voluntarily," as when the goose flies over a fence in a detour experiment. The action is always identical; the movement formula is as rigid as any can be, but the centers [Instanzen] capable of unblocking it are extremely various. It is precisely the absolute sameness of the movement pattern that, despite the variety of the processes of the nervous system capable of governing it, is extraordinarily characteristic of the simple instinctive action of locomotion.

[35] III. INSTINCTIVE ACTION, REFLEX, AND AUTOMATIC-RHYTHMIC STIMULUS GENERATION

Since the instinctive actions prove to be so extraordinarily similar to anatomical structures in their ontogenetic and phylogenetic behavior, then the idea appears to follow that their special form might be rigidly tied to such anatomical structures. A connection of certain movement patterns to certain structures in the central nervous system is nothing new in physiology, which, after all, assumes an anatomical substrate for all *reflex movements* in the form of a reflex arc. Now it is an incontestable fact that every conceivable transitional stage exists between instinctive actions and those processes that are generally viewed as reflexes. This caused Ziegler to define the instinctive action quite simply as a chain reflex and even to put the histological-anatomical aspect of his definition in the foreground. Now, a series of objections against this view arises merely out of the described properties of the instinctive actions. These objections seem to me worthy of special discussion, since von Holst has shown that quite similar and very compelling arguments suggest that very many movement sequences that are generally considered to be reflexes are based on processes that have nothing to do with reflexes. Von Holst' statement holds quite generally both in physiology and in animal psychology:

> Virtually everything that we know about the accomplishments of the central nervous system is concentrated on this question: stimulus and response to stimulus. One effects an external or internal *change* of condition and determines the ensuing reaction. The onesidedness of this view *necessarily* led to the view still prevailing today that the accomplishments of the central nervous system were strictly limited to the response to and the evaluation of stimuli.

The totally unexperimental observation of captive animals that I had at first kept only as a fancier made me aware of the curious emergence of instinctive actions that was biologically senseless and independent of stimuli. This certainly represents the most weighty argument against viewing the instinctive action as a reflex. Now, as early as 1916, Graham Brown demonstrated that *automatic-rhythmic processes* in the spinal cord of mammals underlie the movement coordinations of locomotion; in other words, they underlie processes that in our view are true instinctive actions. For worms and fishes, E. von Holst has demonstrated that, after the complete exclusion of all receptors that [36] can be considered to excite locomotor reflexes, well-ordered and antagonistically rhythmic impulses continue to issue from the central nervous system. These impulses lead to the execution of locomotor movements that are typical and that can also be observed in the intact animal in a completely similar manner. Now, in this case, when both the rhythm and the coordination (that is, the functional alternation of antagonistic impulses) prove to be a manifestation that the central nervous system is independently producing stimuli, this fundamentally sets the locomotor rhythm of the animals concerned on the same level as the rhythm of breathing and cardiac activity, whose automatic nature has long been known. In the spinal cord of fish, von Holst found a whole series of phenomena that had previously been known solely from the automatic rhythm of cardiac activity. These phenomena would be completely incomprehensible under the assumption of the reflex nature of locomotor coordination. This and a series of other findings on the manner of coordination of antagonistic movements, the discussion of which would lead us too far astray here, have led him

> to the opinion that the process of nervous coordinations is of a purely central nature, that is, that it can fundamentally dispense with peripheral receptors just as the rhythm of individual movements can. This assumption of an automatic (= self-activating) coordination mechanism *does not* of course exclude the release of coordinated action by peripheral stimuli.

It is a fact that the assumption of centrally coordinated basic processes automatically producing stimuli can easily explain a whole series of properties of the instinctive action. And it is also a fact that the correctness of this assumption is *proven* for a number of locomotor coordinations that ethologically *show all the essential characteristics of the instinctive action*. These characteristics were discussed in the first part of the present treatise. It will have to be one of the most pressing tasks of future research on instinct to pursue the question of the extent to which *one time* movement sequences, that is, those instinctive movement sequences that are not rhythmically repeated, as locomotor movements are, behave differently. Let us mention here briefly a series of parallels that to a considerable extent might be taken as arguments for the presence of automatic stimulus production in all instinctive actions.

First, let us consider the independence of the form of movement from any receptive processes whatsoever. This independence becomes evident in the vacuum activity of captive isolated animals, a situation that is quite analogous to the "deafferentiated" spinal cord preparations of von Holst. Second, the phenomena of threshold lowering and [37] reaction intensification after a relatively long damming up should be mentioned. We find a hypothetical explanation for these phenomena, which arranges all the details in a most attractive way, in the assumption of an inner accumulation of reaction-specific excitation material. Quite analogous phenomena are found in the fish spinal cord in the form of what Sherrington called "spinal contrast." Thus, for example, it can be shown quantitatively in the spinal preparation of the seahorse, *Hippocampus*, that the intensity of the spontaneously appearing movement sequence of erection and beating of the dorsal fin is directly proportional to the length of time that this reaction was suppressed by specific inhibiting stimuli, and was thus "dammed up." A third point of comparison lies in the altogether similar way the release of a completely identical movement formula, both in the spinal cord automatism and in the instinctive action, can be dependent on quite different centers [Instanzen]. Thus, in the wrasse, *Labrus,* there are two movement patterns of the pectoral fins, an alternate beat and a simultaneous beat. The first keeps the fish hovering in place, and the second moves it forward. Both of these locomotor coordinations are subject to the "voluntary control" of the intact fish. However, they can regularly be released in the spinal preparation by changes in the oxygen concentration. Finally, I would like to mention the fact that the instinctive action, like the automatic process in the spinal cord, is subject to inhibition from a central nervous source. It seems to me very likely for the instinctive action as well that the release of the movement formula is not directly caused by superordinate centers [Instanzen], but merely that free rein is given to the automatism that is in fact always active. In particular, the reaction-specific susceptibility to fatigue of the instinctive action seems to me to argue along these lines. When a bird pair once, twice, or three times performs the movement pattern of feigning injury when a person threatens its nest, and then, despite continued presence of the stimulus situation that has just released that reaction, the pair begins to look for food or preen, then one immediately gets the impression that the reaction is exhausted and, for lack of internal excitation production, does not run its course, despite full "unblocking" by the receptors and all higher centers [Instanzen] of the central nervous system.

Today, it is certainly quite premature to make definite statements about the possibility of equating the instinctive action with the automatic rhythms found by von Holst. However, for a correct evaluation of such a hypothesis, it is nevertheless worth emphasizing that, on the basis of such an assumption, it is precisely the most peculiar [38] and, given any other point of view, the most contradictory characteristics of the instinctive action that not only become comprehensible, but are virtually *required*. Neither the view that the instinctive action is a reflex process nor the concept that it is purpose-directed behavior can explain the following facts without the greatest difficulties: (1) the inability of all receptive processes, of experience in the broadest sense of the word, to influence the movement form of the instinctive action, and (2) the tendency of instinctive action toward spontaneous eruption. On the other hand, with the assumption of automatic rhythmic and centrally coordinated, stimulus-producing fundamental processes, it is precisely these most remarkable salient characteristics of instinctive actions that become self-evident. In any case, it can be regarded as established that all the objections von Holst raised against the all too confident assumption that the reflex is the basis, and the elemental unit, of all animal movement processes still stand, especially in opposition to the definition of instinctive action as a chain reflex. In my paper in the "Naturwissenschaften," on which the present treatise is based, I defined instinctive action as a chain reflex for lack of differences that could be defined between instinctive actions and those processes generally considered to be reflex processes, and I defined it as a "sought-after chain reflex" in accordance with the prominent feature of the instinctive action that it forms the goal of appetitive behavior. Now, von Holst has subjected the concept of the reflex to very sharp scrutiny, so that today, knowing his results, I no longer dare give the above definition. To be sure, the sought-after *release* of the instinctive action, especially where this comes about by the response of an innate schema, still appears to be a genuine reflex process in the proper sense of the term. It is precisely these innately releasing stimulus situations that are often the goal of an appetitive behavior in an especially unmistakable manner, so that the definition of the "sought-after reflex" characterizes very well this form of the release process. But to what extent the movements following this release process at a later stage are of an automatic nature and only are freed by this release of the central inhibition otherwise preventing their result, or to what extent they in fact may be considered reflex processes, will have to remain subjects for future research.

Now, whether the movement processes of the instinctive action are based on reflexes or on processes that automatically produce stimuli, the *fact that their release is sought after* must remain, in any case, the main characteristic of the instinctive action. This being sought is in both cases equally peculiar and in both cases presents the same difficulties. One can, [39] to be sure, (see page 29) define the concept of goal-seeking purpose-directed behavior in the terminology of objective behavioral theory, as Tolman does. As valuable as this fact is, one must nevertheless not forget that the *true* goal that the animal subject seeks with its appetitive behavior is those subjective phenomena that accompany both the perception of the releasing stimulus situation and the occurrence of the movement formula itself. There are logical and philosophical difficulties in directly connecting such heterogeneous concepts as that of a striving after subjective things and that of a physiological process such as the automatic process or the reflex. And our definition has something of the naivete of the view of Descartes, who considered the pineal gland to be the focal point of all connections between mental and physical processes. But this is precisely what characterizes the difficult position (and perhaps consequently the important and enlightening position) that the instinctive action deserves not only in biology and physiology, but also in the theory of human and animal feelings and desires.

IV. NON-INSTINCTIVE BEHAVIOR

A) THE UNANALYZED REMAINDER

It is one of the greatest mistakes that natural scientists can ever make to commit themselves too early to definitions. Such premature definitions almost always signify an arrogant overestimation of our knowledge of the defined phenomenon. The untimely defining of "instincts" by theoreticians who were only very superficially familiar with the fundamental properties of the innate types of behavior, especially of the hereditary coordinations, that is, of the instinctive actions in our sense, led to the formation of clumsy, indeed, virtually useless concepts of instinct. These concepts have been and remain very great impediments to the advance of analytic research. We must take care not to lapse into a similar error with regard to the non-instinctive types of behavior. The fact that we have, in my opinion, analyzed and rather cleanly isolated from the complex of animal behavior those sequences of movement that I have here called instinctive actions does not give us any right to make any statements about what is left. "The highest philosophy of the researcher of nature consists precisely in tolerating an incomplete world view and preferring it to a seemingly complete but inadequate one," as Mach says. Therefore, I would prefer to be very careful in my statements about whether all [40] animal behavior that is not instinctive is in the proper sense of the term purpose-directed behavior. By Tolman's definition mentioned on Page 29, every animal's action is purpose-directed that, in reference to a constant goal, shows adaptive variability in attaining that goal. This definition, which captures only the ultimate function of the reaction, and which says nothing about its causation, certainly seems well suited to make "purpose-directed behavior" conceptually synonomous with "non-instinctive behavior," as of course, we did above in the interest of a preliminary classification. But we must remain clear about the fact that, of the movement patterns falling within this concept, scarcely any have been researched even slightly as to the analysis of their causes. Indeed, very few lend themselves at all to experimental causal research at present. From the quite large number of purpose-directed types of behavior in Tolman's sense, there are essentially *two* fundamentally different types of processes about which something more is known. These are the orientation reactions and the conditioned reflex. The apparently fluid transition from the simpler and more easily analyzable orientation mechanisms to insightful behavior will be discussed later. The question arises whether something similar exists between the simple conditioned reflex and the highest accomplishments of learning.

B) THE "CONDITIONED REFLEX" (THE CONDITIONED REACTION)

Pavlov and his school, as well as the American "behaviorists," are of the view that a pathway type "association," which is to be conceived of mechanically, between two and subsequently many centers [Zentren], alone suffices to explain absolutely all animal behavior, however complex and, specifically in the sense of the preservation of the species, purposeful it is. Now it is quite certainly correct that through a temporal coincidence of one stimulus, which is at first meaningless to the organism, with another that innately releases an "unconditioned" reaction, be it reflex or instinctive action, a connection of some sort is made between the originally ignored stimulus and the performance of the reaction such that from that time on, the stimulus gains a significance and is capable of replacing the innately appropriate stimulus more or less completely by producing the same releasing effect. But it is just as certain that the simple principle of stimulus substitution cannot explain a whole series of phenomena. No matter how understandable it might be to us from human psychology, that the first researchers on association, [41] in their joy at a phenomenon that could be analyzed well and indeed perhaps even causally explained, attempted to explain *everything* on the basis of this phenomenon, nevertheless a persistence in this attempt today seems to us to be blind to the facts. Quite apart from the fact that the hereditary coordinations a priori cannot be built up from conditioned reflexes, there is also very much in the purpose-directed behavior of animals and still more of humans that puts insurmountable difficulties in the way of a purely associative explanation. Such difficulties rest in the necessity, which in fact exists, for the conditioned and unconditioned stimuli to impinge in a definite chronological sequence, and above all, in the fact of "experimental extinction," that is, the disappearance of the reaction to the substitute stimulus if this reaction several times in a row is not followed by the unconditioned stimulus. But, if one conceives of the reaction of an animal as subjectively purpose-directed behavior, as Tolman does, these very phenomena find a quite unforced explanation that takes all of the details into account astonishingly well. Considering that the experiments on the conditioned reflex were set up in a time of mechanistic thinking, and for precisely the purpose of substantiating the mechanistic theories of the behaviorists in the stricter sense, then it is really quite amazing how uncommonly well subjectifying interpretations can be harmonized with the facts found at that time. A large number of phenomena that present particularly great problems for the simple principle of stimulus substitution are easy to understand when one assumes that the entrained stimuli are signals leading the organism to the goal of satisfaction of its subjective needs. The anticipation of the unconditioned reaction in response to the entrained stimulus, a process characterizing every conditioned reflex, becomes comprehensible in this way. But when it is maintained by Tolman and his school that the signal stimulus always releases such a reaction that must be understood either as a preparation for a stimulus that the animal seeks, or as an avoidance of an unpleasant stimulus, then this is, in my opinion, a generalization that is not altogether admissible. The only proof against the assumption of a conditioned reflex as purpose-directed purely in a subjective manner would be attained, according to Hilgard, if an association were to come about under such totally senseless and bizarre conditions that the reaction would be neither preparation nor avoidance, and consequently not accessible to explanation as a subjectively purpose-directed behavior. Now I believe that the conditionability of reflexes never appearing in the environment [Umwelt] of the subject, as, for example, the conditionability of a person's pupil reflex, cannot [42] be explained in this way. I also believe I have some examples of observations fulfilling Hilgard's requirement and appearing to demonstrate a truly purely mechanical association of events that occur successively. For example, I know of a dog that was repeatedly shot with buckshot who now begins to limp at the sound of a shot. This behavior can neither be called a preparation for being shot, nor an avoidance of it, nor a purpose-directed behavior at all, nor even objectively purposeful. Birds, whose learning seems in general to be on a lower plane, exhibit similar "senseless" associations more frequently than mammals do.

While noting the qualifications, certainly necessary, that the theory of conditioned reflex has experienced rather recently, one must not forget that cases of the simplest association *really do exist*, that is, of the simplest stimulus substitution, as the earliest researchers of these phenomena understood it, and that these very cases are the most essential ones for future research because they are the easiest to analyze. I believe we should conceive the notion of learning narrowly and in the sense of purely associative learning; and where this principle is not enough, we must simply take the effects of heterogeneous factors into consideration. I find it very possible that neglecting the simple process of stimulus substitution and its conceptual blending with higher processes, which are only understandable in terms of ultimate causes ("insight learning" of Russell), signifies a mistake quite similar to many of the extensions of the concept of instinct that are so much of an impediment to analytical research.

C) ORIENTATION REACTIONS

As mentioned at the beginning, the orientation reactions are reaction norms to direction-giving stimuli, which, as such, may also be innate, and therefore are frequently lumped together with instinctive actions, which are innate norms of movement independent of such stimuli. We usually call simple direction-giving turnings taxes. When a frog, before discharging its snap reaction, first focuses its eyes on the fly, and then aligns its whole body with this fly in little steps, we can certainly best describe this movement in terms of the theory of taxes and investigate it with the methodology of taxis research. However, we must not forget that fluid transitions exist from such simple spatially orienting turns to [43] the highest forms of insightful behavior. At least from the standpoint of objective behavioral theory, no sharp borderline can be drawn between these simplest of orientation reactions and the highest "insightful" solutions to detour tasks that we find in the "smartest" mammals. There is also nothing preventing us from crediting the frog with "insight" into the spatial relationships between it and the fly. I definitely do not want to assert that all "insight" is fundamentally "the same thing" as the forms of taxis that are simpler and already quite analyzable today. This word "insight" seems to me rather to designate an unanalyzed remainder of accomplishments leading to orientation, about whose mechanism nothing is known. In terms of the psychology of experience, in my opinion, no sharp distinction can be made between true taxes and insightful behavior either. The "aha experience," as Bühler so aptly calls the subjective correlate of insight, must surely always occur whenever the condition of disorientation yields to that of orientation. When one is thrown unexpectedly into cold water while half asleep and, in the first seconds, is completely disoriented spatially, so that one literally does not know which way is up and which down, the regaining of the simple tropotactic gravity orientation is connected with the most intense, utterly liberating aha experience. In higher insights of humans also, the aha experience accompanies only the initiation, not the presence of orientation. For example, one says "aha" when one has just grasped the nature of a logarithm, but not every time one uses it in calculation. Consciousness or unconsciousness, therefore, cannot be cited for differentiating taxis and insight, and, in this connection, one can perhaps say that all taxis is a form of insight, but that is not to make any sort of statement about whether and how the higher forms of "genuine" insight are to be explained as processes similar to taxes, or whether and how they lend themselves to the methods of taxis research. The problematic issue involving taxis and insight is one of those areas where one must be especially cautious not to confuse psychological and physiological concepts. The fact that a process such as attainment of orientation in our experience can be determined in one way or another says nothing at all about whether it lends itself to physiological causal explanations and, if so, to which ones. On the other hand, just these kinds of orientation mechanisms seem to lend themselves best to a precise physiological examination. And a parallel investigation in physiology and the psychology of experience is perhaps one of the most promising avenues of research for the establishment of the bridge between physiology and psychology [44] representing the only truly exact path toward the creation of an Animal Psychology in the proper sense of the term.

V. THE COMBINED EFFECTS OF INSTINCTIVE ACTION AND NON-INSTINCTIVE BEHAVIOR

A) THE OBLIGATION TO ANALYZE

The concept that we have given above of the instinctive action is very narrow compared to the conceptualizations used by most other authors. The justification for formulating a new concept deviating from the traditional one exists only when it brings us closer to an understanding of the processes under investigation, and also, above all, appears useful for future research into causes. In this sense, it seems to me absolutely necessary to single out the sequences of movement based on hereditary coordinations as sharply delineatable units of animal behavior, for, without further analysis of their own construction, these internally consistent units not only *can*, but *must* be used conceptually in the analysis of the total behavior of the animal. One can, of course, capriciously formulate all kinds of concepts of instinct. For example, one can designate everything innate in the individual as instinct, thereby including all innate orientation reactions. One can also call anything more or less unconscious "instinctive," as Wundt has done, who calls habitual movements that occur unconsciously "acquired instincts." One can even attribute all purposefulness in animal behavior that is not explainable by capacity to learn or achievement of insight to an "instinct," as many vitalistically oriented animal psychologists do. But there is a question of whether such a collective term for unanalyzed remainders can be at all useful for the progress of research. It seems to me that a neglect of the causation of an action because one connects its ultimate biological success with the word "instinct" would have to have precisely the opposite effect. It is just all too easily forgotten that the demonstration of an ultimate purpose does not mean a solution to the problem, but rather indicates the existence of a problem for the field of causal analysis. But, however one wants to define "instinct" or instinctive action, one will always have to come to grips with the existence of the movement sequences that were called instinctive actions here—the movement sequences that are sought after by appetitive behavior and fixed on the basis of central hereditary coordinations. [45] Even if one considers it better to introduce a new term for them, this still holds completely true.

We owe our knowledge of these instinctive actions primarily to an approach borrowed from comparative morphology and first used in a logically consistent way by Whitman. Nevertheless, it has by no means become common currency in psychological research that these inherited movement patterns exist at all. Historical factors play a strong role in this, especially the clash of dogmas between mechanism and vitalism. Mechanistically inclined researchers of animal behavior have completely overlooked instinctive actions in their attempts to explain all actions of animals by the conditioned reflex. On the other hand, a number of vitalistically oriented animal psychologists show an attitude toward the presence of instinctive action that seems to me to justify M. Hartmann's reproach saying that a misunderstood vitalism could become an impediment to research "by its prematurely putting what is not resolvable and not rationalizable where such a barrier does not yet exist, thereby blocking the path to causal research." Thus, McDougall argues against any theory that sees automatism in the instinctive action, just as if this assumption were equivalent to a confession of faith in the mechanistic dogma and signified the conception of the animal as a machine. This attitude is completely mistaken because, by the very fact that these theories might succeed in laying some of the movements of an animal open to a causal explanation (which means here the same as a physiological explanation), the animal is no more "degraded" to a machine than, to put it crudely, I myself become a machine by the fact that I can explain a partial function of my body, such as that of my elbow joint, rather completely in a mechanical way.

The obligation to analyze, imposed on us by the recognition of the special nature of the sought-after hereditary coordination and its special laws, leads us to make a sharp separation between the instinctive action and a whole series of phenomena that have been judged to be constituents of it by virtually all authors, including even some whose notion of the concept of instinct closely approximates our own in many other points.

B) THE SIMULTANEOUS ORIENTATION REACTION

In order to characterize the instinctive action, and, above all, to distinguish it from the reflex, various authors have emphasized the circumstance [46] that instinctive actions involve activating the whole organism, while the reflex only activates a part of it. One can confidently assert this of the instinctive action in our sense only to the extent that one can ever say at all that just one part of the organism participates in a reaction. Coordination always controls only the rhythm of the "impulse melody" for quite specific antagonistic muscle groups, but quite certainly never does so for the entire body musculature. In principle, therefore, it is possible for one group of antagonistic muscles to obey a central hereditary coordination that is independent of the receptors of the animal, while another simultaneously obeys a taxis guided by the receptors. For example, a mallard drake (*Anas platyrhynchos* L.) performs a courtship movement that is executed exactly in the median sagittal plane. This movement is a perfectly pure instinctive action and is carried out by the bending and stretching muscles of the rump and neck. During its entire course of action, however, the drake continues to swim upright, guided tropotactically by its semicircular canals according to gravitational stimuli. Another example is the egg rolling reaction of a greylag goose (*Anser anser* L.) and of many other birds. While the egg is being rolled slowly towards the nest, about all an observer directly in front of the goose sees is a lateral back and forth movement of the beak, the tip of which holds the egg in balance. Each of these small lateral movements is guided directly by the tactile stimulus coming from the egg as it slips to the right or to the left; this action is therefore a pure taxis, though certainly a quite complicated one. In profile, one sees a movement proceeding in the sagittal plane which is a pure instinctive action and, as such, lacks all capacity to adjust to guiding stimuli. For example, the movement literally locks if one gives the goose a dummy egg too large to roll, or the movement runs through to completion uninfluenced when the beak completely loses contact with the egg by chance or experimental intervention.

Such orientation reactions, functioning simultaneously with the instinctive action, can, in a certain sense, be conceived of as appetitive behaviors since they maintain, but do not create, the stimulus situation necessary for the instinctive reaction to run its course. A special case that seems to be particularly frequent is one in which a hereditarily coordinated movement is fixed only in one plane and guided by tactic movements in a direction in space perpendicular to it so that the optimal stimulus situation is maintained. Besides those examples given, many more can be cited, such as the chewing movements of humans, the friction movement of copulating male mammals, and others. Of course, it is certain that there [is] a very large number of such simultaneous interweavings [47] that cannot be analyzed in such a simple way as the movements mentioned, which are clearly distributed in different directions in space. But this has nothing to do with the fundamental duty to make conceptual distinctions and to attempt analysis.

C) APPETITIVE BEHAVIOR PRECEDING AND INTERWOVEN WITH MULTIPLE INSTINCTIVE ACTIONS

Wallace Craig conceives of appetitive behavior as an integral constituent of the instinctive action. If we sharply separate appetitive behavior here from the actual course of action sought after by it, then we do that because it can encompass absolutely *all* higher psychic accomplishments and can muster them to achieve the stimulus situation releasing the action. Thus, any chain of action, however complicated, that is performed for the sake of an instinctively pleasurable goal would consequently have to be considered an instinctive action. The necessity of distinguishing purpose-directed and instinctively automatic links in functionally unitary sequences of animal actions arises from the possibility of making such a distinction. It is a peculiarity of very many types of behavior of higher animals that, in these behaviors, purposively variable movements and instinctively innate movements follow immediately one upon the other. The knowledge of these instinctive movements is therefore important for the understanding of the motivation of the total behavior because every variable behavior chronologically preceding an instinctive action must be understood as an appetitive action seeking the subsequent instinctive portion of the whole behavior.

Besides orientation reactions in the broadest sense of the term, learning processes also play an important role in appetitive behavior. Even ignoring the fact that the release of an instinctive action frequently represents a learned or conditioned reaction (p. 33), one very often finds entrained types of behavior inserted as links *between* any two instinctive courses of action in fairly long, functionally unitary chains of action. The unity of the action is in such cases a chain of appetences following one after the other, and in this very type of action sequence, the appetitive behavior that follows the one instinctive component, and that must create the stimulus situation for the next, is represented by a *trained action*. For example, the movement pattern with which a sparrow picks up nesting material from the ground and stuffs its beak back to the corners of its mouth is a genuine instinctive action. When, after [48] several repetitions, the appetence for this first component action is fully satisfied and consequently stilled, the appetence for the next instinctive action, that of using the collected materials in building, awakens. The behavior that brings about the appropriate stimulus situation for this next instinctive action link is in many bird species a pathway training, a "mnemotaxis" in Kühn's sense. No one who has ever watched the establishment of such a pathway training can ever doubt that we are dealing with the same sort of process as in the development of training produced experimentally and intentionally by humans. Let me point out parenthetically that this very origination of training is one of the strongest arguments for the correctness of Craig's approach to appetitive behavior. One can scarcely imagine better proof of it than the fact that the appetence for the next nest-building reaction can train the bird to use a certain direct path or circuitous path, just as appetite for food trains a rat to do so in a maze experiment.

Repeated objection has been voiced to my attempts to separate out the movements based on hereditary coordinations from all other animal behavior. It is said that such a procedure is "atomistic" and not consistent with the modern biological holistic view. One could just as well say that it is "atomistic" to distinguish dermis and epidermis in the functional whole of the skin. If within a functional unit of animal behavior, a distinction can be made between parts of the behavior that can be recognized unambiguously as true trained actions or as taxes and other parts bearing all the characteristics of the instinctive action in the sense advocated here, then these distinctions are not only *possible*, but we also have the *duty* to draw the analytical conclusion from these facts. We do not have the right to broaden the concept of instinct in such a way that demonstrably different types of behavior are included.

LITERATURE

F. Alverdes, 1925. *Tiersoziologie*. Leipzig.

W. Bechterew, 1936. *Reflexologie des Menschen*. Leipzig und Wien.

A. Bethe, 1931. Anpassungsfähigkeit (Plastizität) des Nervensystems. *Bethe's Handbuch der normalen und pathologischen Physiologie*. Vol. XV/2. pp. 1045–1130.

J. A. Bierens de Haan, 1933. Der Stieglitz als Schöpfer. *J. Ornith.*, 81/1.

———. 1935. Probleme des tierischen Instinktes. *Naturwiss.*, 23/42,43.

G. Brown, 1916. *Erg. Physiol.*, 15.*

C. Bühler, 1927. Das Problem des Instinktes. *Z. Psychol.*, 103.

L. Carmichael, 1926. The development of behaviour in vertebrates experimentally removed from the influence of external stimulation. *Psychol. Rev.* 33,34,35.

W. Craig, 1908. The voices of pigeons regarded as a means of social control. *Amer. J. Sociol.*, 14.

———. 1909. The expression of emotion in the pigeons. *J. comp. Neurol. a. Psychol.*, 19.

———. 1912. Observations on young doves learning to drink. *J. Animal Behavior,* 2/4 (July-August).

———. 1914. Male doves reared in isolation. *J. Animal Behavior,* 4/2 (March-April).

———. 1918. Appetites and Aversions as Constituents of Instincts. *Biol. Bull.*, 34.

———. 1921–1922. A Note on Darwin's work on the expressions of emotions etc. *J. abnorm. a. soc. Psychol.* 1921 (December); 1922 (March).

F. Doflein, 1916. *Der Ameisenlöwe, eine biologische, tierpsychologische und relexbiologische Untersuchung*. Jena.

H. Driesch, 1928. *Philosophie des Organischen*. 4th ed. Leipzig.

H. Friedmann, 1934. The instinctive emotional life in birds. *Psychoanal. Rev.*, 21/3–4.

J. Grohmann, 1937. Reifung oder Lernvorgang. Dissertation. To appear in *Ztschr. Tierpsych.*

K. Groos, 1907. *Die Spiele der Tiere*. 2nd ed. Jena.

M. Hartmann, 1927. *Allgemeine Biologie*. Jena.

H. Hediger, 1935. Zur Biologie und Psychologie der Zahmheit. *Arch. f. Psychol.*, 93.

O. Heinroth, 1910. Beiträge zur Biologie, insbesondere Psychologie und Ethologie der Anatiden [Contributions to the Biology, especially the Psychology and Ethology of the Anatidae]. *Verh. d. V. Intern. Ornith. Kongr. Berlin* 1910.

———. 1918. Reflektorische Bewegungen bei Vögeln. *J. Ornith.*, Vol. 66.

———. 1930. Über bestimmte Bewegungsweisen bei Wirbeltieren [On Certain Patterns of Movement in Vertebrates]. Sitzungsber. *Ges. Naturforsch. Freunde Berlin* 1930 (February).

O. Heinroth and M. Heinroth, 1924–1928. *Die Vögel Mitteleuropas*. Berlin-Lichterfelde.

F. H. Herrick, 1919. Instinct. *Western Reserve Univ. Bull.* 22/6.

———. 1935. *Wild birds at home*. New York & London.

E. Hilgard, 1936. The nature of conditioned response: II. Alternatives to stimulus-substitution. *Psychologic. Rev.*, 43, pp. 547–564.

E. von Holst, 1935. Alles oder Nichts, Block, Alternans, Bigemini und verwandte Phänomene als Eigenschaften des Rückenmarks. *Pflügers Arch.*, Vol. 236/4,5,6.
———. 1936. Versuche zur relativen Koordination. ibid. Vol. 237,1.
———. 1936. Von Dualismus der motorischen und der automatisch-rhythmischer Funktion im Rückenmark und vom Wesen des automatischen Rhythmus. ibid. Vol. 237,3.
———. 1936. Über den "Magnet-Effekt" als koordinierendes Prinzip im Rückenmark. ibid. Vol. 237,6.
E. Howard, 1928. *An introduction to the study of bird behaviour.* Cambridge.
———. 1935. *The nature of a bird's world.* Cambridge.
H. S. Jennings, 1915. *Behaviour of the lower organisms.* 2nd ed. New York.
D. Katz, 1931. *Hunger und Appetit.* Leipzig.
O. Koehler, 1933. Die Ganzheitsbetrachtung in der modernen Biologie. *Verh. der Königsberger gelehrten Gesellschaft.*
G. Kramer, 1930. Bewegungsstudien an Vögeln des Berliner Zoologischen Gartens. *J. Ornith.*, 78/3.
A. Kühn, 1919. *Die Orientierung der Tiere im Raum.* Jena.
H. Lissmann, 1933. Die Umwelt des Kampffisches *Betta splendens Regan. Z. vergl. Physiol.*, 18/1.
K. Lorenz, 1931. Beiträge zur Ethologie sozialer Corviden. *J. Ornith.*, 79/1.
———. 1932. Betrachtungen über das Erkennen der arteigenen Triebhandlungen der Vögel. ibid. Vol. 80/1.
———. 1935. Der Kumpan in der Umwelt des Vogels (Der Artgenosse als auslösendes Moment sozialer Verhaltungsweisen). ibid. Vol. 83/2,3.
———. 1937. Über die Bildung des Instinktbegriffes. *Naturwiss.* 25/19,20,21(May).
E. Mach, 1923. *Popularwissenschaftliche Vorlesungen.* 5th ed. Leipzig.
W. McDougall, 1922. The use and abuse of instinct in social psychology. *J. abnorm. a. soc. psychol.* 16/5,6.
———. 1923. *An outline of psychology.* London.
C. Lloyd Morgan, 1913. *Instinkt und Erfahrung.* (Translator R. Thesing.) Berlin.
H. Poll, 1910. Über Vogelmischlinge. *Verh. d. V. Intern. Ornith. Kongr. Berlin* 1910, pp. 399-468.
E. S. Russell, 1934. *The behaviour of animals.* London.
E. C. Tolman, 1932. *Purposive behaviour in animals and men.* New York.
J. von Uexküll, 1909. *Umwelt und Innenwelt der Tiere* [Environment and Inner World of Animals]. Berlin.
L. Verlaine, 1936. La connaissance chez le singe inferieur, II. Le syncretique, et III. L'abstrait. *Actualites scientifiques et industrielles* 320 et 360.
J. Verwey, 1930. Die Paarungsbiologie des Fischreihers. *Zool. Jahrb.* (Abt. f. allg. Zool.), 48.
H. Volkelt, 1912. Die Vorstellungen der Tiere. Diss. Leipzig, 1912. (also in Kruegers Arbeiten zur Entwicklungspsychologie, 1, 1914).
C. O. Whitman, 1898. Animal behaviour. 16th Lecture from "Biological Lectures from the Marine Biological Laboratory." Woods Hole, Mass.
W. Wundt, 1919. *Vorlesungen über Menschen- und Tierseele.* 6th ed. Leipzig.
H. Ziegler, 1920. *Begriff des Instinktes einst und jetzt.* Jena.

* The starred paper is known to the author only in quotes.

AUTHOR CITATION INDEX

SUBJECT INDEX

Editor's Note: Due to limited space, the large number of authorities cited are not included in this index. Main entries for such general topics as behavior, evolution, experience, innate behavior, instinct, and intelligence are also omitted.

About the Editor

GORDON MARTIN BURGHARDT is a Professor of Psychology at the University of Tennessee at Knoxville with joint appointments in Zoology and Ecology. He currently directs the Ethology (Life Sciences) Graduate Program. President-Elect of the Animal Behavior Society, he is also American editor of the *Zeitschrift für Tierpsychologie,* a Fellow of the American Psychological Association, and a U.S. member of the International Ethological Conference Committee. Dr. Burghardt attended the University of Chicago, receiving the S.B. degree in 1963 and the Ph.D. in 1966.

Although publishing research articles in several areas, including animal play, behavioral ontogeny, history and theory in ethology, and bear and human behavior, his main interests are in reptile behavior and chemoreception. In addition to laboratory work Dr. Burghardt has been involved in extensive field studies on green iguana behavior in Panama and Venezuela and is a Research Associate of the Smithsonian Tropical Research Institute. His books include *The Development of Behavior: Comparative and Evolutionary Aspects* (co-edited with M. Bekoff, Garland STPM, 1978) and *Iguanas of the World: Their Behavior, Ecology, and Conservation (co-edited with A. S. Rand, Noyes, 1982).*